COMPLEX FUNCTIONS

To Order

1-800-872-7423

COMPLEX FUNCTIONS
an algebraic and geometric viewpoint

GARETH A.JONES and **DAVID SINGERMAN**

Lecturers in Mathematics, University of Southampton

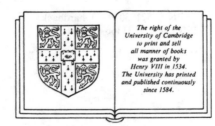

The right of the
University of Cambridge
to print and sell
all manner of books
was granted by
Henry VIII in 1534.
The University has printed
and published continuously
since 1584.

CAMBRIDGE UNIVERSITY PRESS

Cambridge

New York   Port Chester

Melbourne   Sydney

Published by the Press Syndicate of the University of Cambridge
The Pitt Building, Trumpington Street, Cambridge CB2 1RP
40 West 20th Street, New York, NY 10011—4211, USA
10 Stamford Road, Oakleigh, Melbourne 3166, Australia

First published 1987
Reprinted 1988, 1991

Printed in Great Britain by
Athenaeum Press Ltd, Newcastle upon Tyne

British Library cataloguing in publication data

Jones, Gareth A.
Complex functions: an algebraic and geometric viewpoint.

1. Functions of complex variables
I. Title   II. Singerman, David
515.9   QA331

Library of Congress cataloguing in publication data

Jones, Gareth A.
Complex functions.

Bibliography: p.
Includes index.
1. Functions of complex variables.   I. Singerman,
David.   II. Title.
QA331.J66   1986   515.9   85—19031

ISBN 0 521 30893 3 hardback
ISBN 0 521 31366 X paperback

TM

Contents

Introduction

1 The Riemann sphere   1
1.1 The sphere   1
1.2 Compactness   3
1.3 Behaviour of functions at infinity   4
1.4 Rational functions   8
1.5 Topological properties   11
     Exercises   15

2 Möbius transformations   17
2.1 Automorphisms of Σ 17
2.2 Linear and projective groups   19
2.3 Generators for $PGL(2, \mathbb{C})$ 20
2.4 Circles in Σ 21
2.5 Transitivity and cross-ratios   22
2.6 Cross-ratios and circles   26
2.7 Inversion   28
2.8 The stabilisers of a circle and a disc   30
2.9 Conjugacy classes in $PGL(2, \mathbb{C})$ 32
2.10 Geometric classification of Möbius transformations   34
2.11 Conformality   36
2.12 Rotations of Σ 40
2.13 Finite groups of Möbius transformations   42
2.14 The area of a spherical triangle   50
2.15 $PGL(2, \mathbb{C})$ as a Galois group   52
     Exercises   53

3 Elliptic functions   56
3.1 Periods   56
3.2 Topological groups   60
3.3 Simply periodic functions   62
3.4 Lattices and fundamental regions   65
3.5 The torus   70
3.6 General properties of elliptic functions   72
3.7 Uniform and normal convergence   78

3.8	Infinite products	83
3.9	Weierstrass functions	90
3.10	The differential equation for $\wp(z)$	95
3.11	The field of elliptic functions	98
3.12	Translation properties of $\zeta(z)$ and $\sigma(z)$	101
3.13	The construction of elliptic functions with given zeros and poles	103
3.14	The construction of elliptic functions with given principal parts	104
3.15	Topological properties of elliptic functions	106
3.16	Real elliptic curves	109
3.17	The addition theorem	115
	Exercises	120
4	**Meromorphic continuation and Riemann surfaces**	**123**
4.1	Meromorphic and analytic continuation	123
4.2	Analytic continuation using power series	130
4.3	Regular and singular points	133
4.4	Meromorphic continuation along a path	137
4.5	The monodromy theorem	141
4.6	The fundamental group	147
4.7	The Riemann surface of $\log(z)$	149
4.8	The Riemann surface of $z^{1/q}$	154
4.9	The Riemann surface of $\sqrt{p(z)}$, p a polynomial	157
4.10	Branch-points and the monodromy group	163
4.11	Abstract Riemann surfaces	167
4.12	Analytic, meromorphic and holomorphic functions on Riemann surfaces	172
4.13	The sheaf of germs of meromorphic functions	176
4.14	The Riemann surface of an algebraic function	184
4.15	Orientable and non-orientable surfaces	190
4.16	The genus of a compact Riemann surface	193
4.17	Conformal equivalence and automorphisms of Riemann surfaces	198
4.18	Conformal equivalence of tori	202
4.19	Covering surfaces of Riemann surfaces	206
	Exercises	214
5	**$PSL(2,\mathbb{R})$ and its discrete subgroups**	**217**
5.1	The transformations of $PSL(2,\mathbb{R})$	217
5.2	Transitivity, conjugacy and centralisers	218
5.3	The hyperbolic metric	221
5.4	Computation of $\rho(z,w)$	225
5.5	Hyperbolic area and the Gauss–Bonnet formula	228
5.6	Fuchsian groups	231
5.7	Elementary algebraic properties of Fuchsian groups	239
5.8	Fundamental regions	240
5.9	The quotient-space $\mathcal{U}/\Gamma$	248

5.10	The hyperbolic area of a fundamental region	255
5.11	Automorphisms of compact Riemann surfaces	263
5.12	Automorphic functions and uniformisation	266
	Exercises	267
6	**The modular group**	**271**
6.1	Lattices, tori and moduli	272
6.2	The discriminant of a cubic polynomial	274
6.3	The modular function J	275
6.4	Analytic properties of g_2, g_3, Δ and J	278
6.5	The Riemann surface of $\sqrt{p(z)}$, p a cubic polynomial	283
6.6	The mapping $J : \mathcal{U} \to \mathbb{C}$	289
6.7	The λ-function	293
6.8	A presentation for Γ	296
6.9	Homomorphic images of Γ	299
6.10	Quotient-surfaces for subgroups of Γ	302
	Exercises	315
	Appendix 1: A review of complex variable theory	318
	Appendix 2: Presentations of groups	325
	Appendix 3: Resultants	327
	Appendix 4: Modern developments	331
	References	334
	Index of symbols	337
	Index of names and definitions	339

Introduction

Throughout the nineteenth century, the attention of the mathematical world was, to a large extent, concentrated on complex function theory, that is, the study of meromorphic functions of a complex variable. Some of the greatest mathematicians of that period, including Gauss, Cauchy, Abel, Jacobi, Eisenstein, Riemann, Weierstrass, Klein and Poincaré, made substantial contributions to this theory, and their work (mainly on what we would now regard as specific, concrete problems) led to the subsequent development of more general and abstract theories throughout pure mathematics in the present century. Because of its central position, directly linked with analysis, algebra, number theory, potential theory, geometry and topology, complex function theory makes an interesting and important topic for study, especially at undergraduate level: it has a good balance between general theory and particular examples, it illustrates the development of mathematical thought, and it encourages the student to think of mathematics as a unified subject rather than (as it is often taught) as a collection of mutually disjoint topics.

Even though the subject matter of this book is classical, it has recently assumed great importance in several different areas of mathematics. For example, the recent work on W. Thurston on 3-manifolds shows the vital importance of hyperbolic geometry and Möbius transformations to this rapidly developing subject; a totally different example is given by the work of J.G. Thompson, J.H. Conway and others on the 'monster' simple group, where the J-function, studied in Chapter 6, seems to play an important (and, at the time of writing, rather mysterious) role. Thus many active mathematicians, whose work may not involve classical complex function theory directly, will nevertheless need to become familiar with certain aspects of the theory, and we hope that they find our elementary approach of use, at least initially.

This book is based on a final-year undergraduate course at the University of Southampton, taught first by D.S. and then by G.A.J., though we have also included some additional material, generally at the end of a chapter, suitable for graduates or for more advanced undergraduates. Our aim, both in the lecture-course and in this book, is to teach some of the main ideas about complex functions and Riemann surfaces, assuming only the basic algebraic, analytic and topological theories covered by students in their first and second years at university, and to show how these three subjects can be combined to throw light on a single, specific topic. (Of course, this involves reversing the historical development of the subject: to the modern mind, general theories often appear more elementary and accessible than the particular examples from which they grew.) Shortages of space and time forced us to ignore the connections with, for example, number theory and potential theory, interesting though they are; in any case, there are excellent books on these topics.

In Chapter 1 we use stereographic projection to show how the addition of a single point ∞ to $\mathbb{C}$ transforms the plane into a sphere, the Riemann sphere $\Sigma = \mathbb{C} \cup \{\infty\}$, and we describe the meromorphic functions $f : \Sigma \to \Sigma$ from both an algebraic and a topological point of view. The main result, which is a typical connection between analytic and algebraic concepts, is that $f : \Sigma \to \Sigma$ is meromorphic if and only if it is a rational function.

Chapter 2 concerns the automorphisms of Σ, that is, the meromorphic bijections $f : \Sigma \to \Sigma$, or equivalently the Möbius transformations

$$f(z) = \frac{az + b}{cz + d}, \tag{*}$$

with $a, b, c, d \in \mathbb{C}$ and $ad - bc \neq 0$. These transformations form a group Aut Σ under composition, and the emphasis of this chapter is mainly group-theoretic; for example, the finite subgroups of Aut Σ are determined, and the cross-ratio λ is introduced in order to study the transitivity properties of Aut Σ. We also consider some of the geometric properties of Möbius transformations (especially their relationship with circles in Σ), and the way in which Aut Σ acts as the Galois group of the field of all meromorphic functions on Σ.

In Chapter 3 we study periodic meromorphic functions on $\mathbb{C}$; these fall into two classes, the simply and doubly periodic, according to whether the group of periods has one or two generators. After briefly considering simply periodic functions (such as the exponential and trigonometric functions), and their Fourier series expansions, we devote the rest of the chapter to doubly periodic functions, called elliptic functions because they first arose from attempts to evaluate certain integrals associated with the formula for the circumference of an ellipse. The periods of such a function form a lattice, that is, a subgroup of $\mathbb{C}$ (under addition) generated by two complex numbers which are linearly independent over $\mathbb{R}$. Just as the rational functions are the meromorphic functions on the sphere Σ, the elliptic functions can be regarded as the meromorphic functions on the torus $\mathbb{C}/\Omega$ whose elements are the cosets in $\mathbb{C}$ of a lattice Ω. There are many close analogies between rational and elliptic functions, mainly based on the fact that both Σ and $\mathbb{C}/\Omega$ are compact surfaces: for example, an important consequence of Liouville's theorem is that an analytic function on either of these surfaces must be constant. However, in the case of the torus (as opposed to the sphere) the construction of non-constant meromorphic functions represents a substantial problem: by imitating the infinite product expansion of the simply periodic function $\sin(z)$, we introduce the Weierstrass function $\sigma(z)$, and then by successive differentiation we obtain the Weierstrass functions $\zeta(z)$ and $\wp(z)$, the last of these being elliptic and not constant. This approach is an alternative to the now-traditional direct construction of $\wp$ (outlined in the exercises) by infinite series, and it involves some elementary properties of uniform and normal convergence of infinite series and products; these properties, important in their own right, are outlined in §3.7 and §3.8. The rest of this chapter is concerned with deriving properties of the functions $\wp$, ζ and σ, and hence of all elliptic functions. For example, the elliptic functions are precisely the rational functions of $\wp$ and its derivative $\wp'$, these two functions being related by an ordinary differential equation $\wp' = \sqrt{(p(\wp))}$, where p is a cubic polynomial; the functions ζ and σ, though not themselves elliptic, are important for the construction of elliptic functions with certain properties such as specific zeros, poles or principal parts. The chapter closes with the addition theorem, expressing $\wp(z_1 + z_2)$ in terms of $\wp(z_1)$ and $\wp(z_2)$; historically this should come first, since it was the work of Fagnano and Euler on addition theorems for elliptic integrals which eventually led to the discovery of elliptic functions.

Whereas Chapters 1–3 can be regarded as concerned with meromorphic functions on two specific surfaces Σ and $\mathbb{C}/\Omega$, the theme of Chapter 4 is to take a function f (possibly many-valued, such as $\log(z)$) and to find the most natural surface to regard as its domain of definition. More precisely, we replace f by a single-valued function ϕ which represents the different branches of f; the domain of ϕ, chosen to be as large as possible subject to ϕ representing f locally, is called the Riemann surface S of f. The construction of ϕ and S involves the concepts of analytic and meromorphic continuation, together with the monodromy theorem which allow us to construct single-valued functions on simply connected regions; several examples, such as $\log(z)$ and $\sqrt{p(z)}$ (p a polynomial) are studied in detail. In the second half of the chapter we consider Riemann surfaces as abstract topological objects in their own right, not necessarily obtained from functions. By introducing the concept of the germ of a meromorphic function we show that every algebraic function determines a compact Riemann surface, and we prove the Riemann–Hurwitz formula for the genus of such a surface. Every Riemann surface is conformally equivalent (that is, isomorphic) to a quotient surface $\hat{S}/G$, where $\hat{S}$ (the universal covering space of S) is a simply connected Riemann surface and G is a discrete group of automorphisms of $\hat{S}$; for example, a torus S has the form $\mathbb{C}/\Omega$ for some lattice Ω which acts as a discrete group of translations of $\hat{S} = \mathbb{C}$. By the uniformisation theorem of Poincaré and Koebe (the proof of which is beyond the scope of this book), $\hat{S}$ is conformally equivalent to $\mathbb{C}$, Σ or $\mathcal{U} = \{z \in \mathbb{C} \mid \text{Im}(z) > 0\}$, so we conclude the chapter by determining the automorphism groups of these three important surfaces.

With just a few exceptions, most Riemann surfaces S have as their universal covering space $\hat{S}$ the upper half-plane $\mathcal{U}$, and Chapter 5 is devoted to the study of this particular surface and its discrete groups of automorphisms. These are the Fuchsian groups, consisting of Möbius transformations (*) with $a, b, c, d \in \mathbb{R}$ and $ad - bc = 1$; by defining an appropriate metric on $\mathcal{U}$ (the hyperbolic metric) we can regard $\mathcal{U}$ as a model of the hyperbolic plane, with these transformations acting as isometries. This situation is similar to, but considerably more complicated than earlier cases where we considered automorphisms of Σ and of $\mathbb{C}$. Using hyperbolic geometry we study Fuchsian groups G, the associated quotient surfaces $S = \mathcal{U}/G$, and their automorphism groups Aut S. For example, if S is compact and has genus $g > 1$, then $|\text{Aut } S| \leqslant 84(g - 1)$, and we shall give an algebraic description of the Fuchsian groups G and the groups Aut S (the Hurwitz groups) for which this bound is attained.

Chapter 6 concerns perhaps the most important of all Fuchsian groups, the modular group Γ consisting of the Möbius transformations (*) with $a, b, c, d \in \mathbb{Z}$ and $ad - bc = 1$. This group and its action on $\mathcal{U}$ arise from the problem of determining all Riemann surfaces of genus 1, or equivalently, all similarity classes of lattices $\Omega \subset \mathbb{C}$; there is one conformal equivalence class of such surfaces for each orbit of Γ on $\mathcal{U}$. For example, if $p(z)$ is a cubic polynomial with distinct roots then the Riemann surface S of $\sqrt{p(z)}$ has genus 1, and we shall show that S is conformally equivalent to a torus $\mathbb{C}/\Omega$ by finding a lattice Ω for which the associated Weierstrass elliptic function $\wp$ satisfies the differential equation $\wp' = \sqrt{p(\wp)}$; this is done by constructing an analytic function $J:\mathcal{U} \to \mathbb{C}$, invariant under the action of Γ on $\mathcal{U}$, and using J to select the orbit of Γ on $\mathcal{U}$ corresponding to the appropriate lattice

Ω. (This function J is closely associated with the cross-ratio function λ introduced in Chapter 2.) From its action on $\mathcal{U}$ we obtain generators and relations for Γ, and hence we are able to consider its homomorphic images, many of which (such as the Hurwitz groups) have already appeared in earlier chapters. Finally we consider the quotient surfaces of $\mathcal{U}$ corresponding to normal subgroups of Γ, including the congruence subgroups obtained by mapping the coefficients a, b, c, d in (*) into the ring of integers mod (n), for positive integers n.

The Appendix contains statements of the main elementary results we have assumed about complex functions, and also some of the basic facts (less well known than they should be) about polynomials and their discriminants.

Clearly, this book contains considerably more material than could possibly be taught in the 36-lecture course on which it is based: a typical course would cover Chapter 1 and about half each of Chapters 2, 3 and 4. In fact, since the chapters are fairly self-contained, this book could be used as the basis for more specialised courses on several different subjects, such as the Riemann sphere and its Möbius transformations (Chapters 1 and 2), elliptic functions (Chapters 1 and 3), analytic continuation and Riemann surfaces (Chapters 1 and 4), and hyperbolic geometry (Chapter 5 and parts of Chapter 4), while for more advanced students Chapter 6 would serve as an introduction to the modular group, leading on to the more detailed treatments in the books by Rankin and Schoeneberg.

In writing a book of this nature, one acquires many debts of gratitude. Our first is to the great men, named above, who founded this subject; the ideas in this book are all theirs, and our only contribution has been to become sufficiently enthusiastic to wish to teach, and then to write down, what they did. One learns mathematics and how to communicate it from many sources and people, far too numerous to mention here; let us simply say that without Murray Macbeath and Peter Neumann we could never have written this book. Alan Beardon, who read the early drafts of the manuscript, saved us from a number of embarrassing solecisms and ambiguities with his detailed criticisms and generous advice, while Robin Bryant, John Thornton and Mary Tyrer-Jones also gave us invaluable help by checking some of the later drafts and the exercises; any remaining blemishes are entirely of our own making. Beryl Betts, June Kerry and Marie Turner deserve our heartfelt thanks for transforming our hand-written scrawls into presentable typescript, and similarly Rose Cassell for her careful drawing of the diagrams; we are also grateful to the staff of the

Cambridge University Press, especially David Tranah, for their infinite patience and cooperation during the writing of this book. Finally, our eternal gratitude is due to our wives, who, during our several years of writing, have had to display toleration and understanding well beyond that specified in the marriage service.

Numbering of theorems

Theorems are numbered according to their chapter and section. For example, Theorem 5.7.2 is in Chapter 5, Section 7. Equations are numbered in the same way. The only exceptions are the theorems in the appendix, which are numbered Theorem A.1, Theorem A.2, etc.

1
The Riemann sphere

1.1 The sphere

There are several advantages in using the set $\mathbb{C}$ of complex numbers as the domain of definition of functions. The complex numbers form a field which is algebraically closed, that is, polynomials of degree n have n roots in $\mathbb{C}$, counting multiplicities. Geometrically, $\mathbb{C}$ can be regarded as the Euclidean plane $\mathbb{R}^2$, probably the most familiar geometric structure of all (hence we sometimes call $\mathbb{C}$ the *complex plane*). As a domain of definition of functions, $\mathbb{C}$ has the following remarkable property: if f is a function of a complex variable and is differentiable on some region $R \subseteq \mathbb{C}$ (recall that a *region* is a non-empty, path-connected, open set), then f is infinitely differentiable on R, and for each $a \in R$ we can expand f as a convergent power series in some sufficiently small disc containing a. (In contrast, there are functions of a *real* variable which are once but not twice differentiable, or which are infinitely differentiable but cannot be represented by power series.) When f is differentiable on a region R, we will say that f is *analytic* on R; in some books the words 'holomorphic' or 'regular' are used instead of 'analytic'. A function whose only singularities in R are poles is called *meromorphic* in R.

There are, however, some disadvantages in using $\mathbb{C}$. Division by 0 is impossible, and so some standard functions are not defined everywhere; for example, z^{-1} is undefined at $z = 0$. There is a rather less obvious disadvantage in that $\mathbb{C}$ is not compact, so that certain sequences $(1, 2, 3, \ldots$ for example) have no convergent subsequences. When we study the connections between functions and surfaces, we will see that compactness of the surface is an important property for proving theorems about functions defined on that surface. There is a classification of compact surfaces, closely related to the classification of certain types of complex functions which we will consider in this book; there is, however, no straightforward classification for non-compact surfaces.

We can avoid these disadvantages, and retain some of the advantages, by using the *extended complex plane* $\Sigma = \mathbb{C} \cup \{\infty\}$, where ∞ is an extra point called the point at infinity. Geometrically, Σ is still very well behaved, for as we shall now show, Σ may be regarded as being a sphere.

Consider the 2-sphere

$$S^2 = \{(x_1, x_2, x_3) \in \mathbb{R}^3 \mid x_1^2 + x_2^2 + x_3^2 = 1\}$$

in $\mathbb{R}^3$, and identify the complex plane $\mathbb{C}$ with the plane $x_3 = 0$ by identifying $z = x + iy$ $(x, y \in \mathbb{R})$ with $(x, y, 0)$ for all $z \in \mathbb{C}$. If $N = (0, 0, 1)$ is the 'north pole' of S^2, then stereographic projection from N gives a bijective map $\pi: S^2 \setminus \{N\} \to \mathbb{C}$, $Q \mapsto P$, where $P \in \mathbb{C}$, $Q \in S^2 \setminus \{N\}$, and P, Q and N are collinear (see Fig. 1.1).

Fig. 1.1

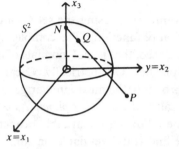

The following argument shows that π is a homeomorphism between $S^2 \setminus \{N\}$ and $\mathbb{C}$. Let $P = (x, y, 0)$ where $z = x + iy \in \mathbb{C}$, and let $Q = (x_1, x_2, x_3) \in S^2 \setminus \{N\}$. Since P, Q and N are collinear we have

$$\frac{x}{x_1} = \frac{y}{x_2} = \frac{1}{1 - x_3}$$

(all equal to NP/NQ, by projection onto each of the coordinate axes). Thus $x = x_1/(1 - x_3)$ and $y = x_2/(1 - x_3)$, and so $\pi: Q \mapsto P$ is given by

$$z = x + iy = \frac{x_1 + ix_2}{1 - x_3}. \tag{1.1.1}$$

Using $x_1^2 + x_2^2 + x_3^2 = 1$, we have

$$x^2 + y^2 + 1 = \frac{2 - 2x_3}{(1 - x_3)^2} = \frac{2}{1 - x_3},$$

so that $\pi^{-1}: P \mapsto Q$ is given by

$$x_1 = \frac{2x}{x^2 + y^2 + 1}, \quad x_2 = \frac{2y}{x^2 + y^2 + 1}, \quad x_3 = \frac{x^2 + y^2 - 1}{x^2 + y^2 + 1}. \tag{1.1.2}$$

These expressions show that both π and π^{-1} are continuous, so π is a homeomorphism.

Now let Σ denote the *extended complex plane* $\mathbb{C} \cup \{\infty\}$, where ∞ (called the *point at infinity*) is a symbol which does not represent an element of $\mathbb{C}$; we extend $\pi: S^2 \setminus \{N\} \to \mathbb{C}$ to a bijection $\pi: S^2 \to \Sigma$ by defining $\pi(N)$ to be ∞. Thus points Q on S^2 close to N correspond under π to complex numbers $P = z$ with $|z|$ large, so that z is in some sense 'close to ∞' (we will express this idea more precisely in the next section when we consider the topological properties of Σ); similarly points Q' close to the south pole $S = (0, 0, -1)$ correspond to complex numbers z' with $|z'|$ small, while the equator $x_3 = 0$ of S^2 corresponds to the unit circle $|z| = 1$ of Σ. This is illustrated in Fig. 1.2.

Fig. 1.2

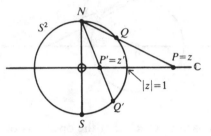

We can use the bijection $\pi : S^2 \to \Sigma$ to transfer algebraic and topological properties from Σ to S^2 and vice-versa. For example, since Σ has the same topological properties as the sphere S^2, Σ is often referred to as the *Riemann sphere* (after B. Riemann, 1826–66). Indeed, it is often convenient (if imprecise) to regard S^2 and Σ as identical, by identifying each point $Q \in S^2$ with $P = \pi(Q) \in \Sigma$.

1.2 Compactness

A topological space X is *compact* if every open cover has a finite subcover, that is, whenever $\mathscr{A}$ is a family of open sets whose union is X, then there is a finite subfamily $\mathscr{B} \subseteq \mathscr{A}$ whose union is also X. In a metric space (for example, if $X \subseteq \mathbb{R}^n$), this is equivalent to the property that every infinite sequence $x_1, x_2, x_3, \ldots$ of points in X has a subsequence which converges in X, that is, there is a subsequence $x_{n_1}, x_{n_2}, \ldots$ with a limit $\lim_{k \to \infty} x_{n_k} \in X$. The Heine–Borel theorem states that a subspace X of $\mathbb{R}^n$ is compact if and only if it is closed and bounded; for example, S^2 is compact.

We use the bijection $\pi : S^2 \to \Sigma = \mathbb{C} \cup \{\infty\}$ to define a topology on Σ by defining the open sets to be the images under π of the open sets of S^2 (in its usual topology as a subspace of $\mathbb{R}^3$). Then Σ is a topological space and π is a homeomorphism, so we have:

Theorem 1.2.1. Σ *is compact.* $\square$

Theorem 1.2.2.   *Every infinite sequence in* Σ *has a convergent subsequence.*

$\square$

For example, the sequence $1, 2, 3, \ldots$ converges to ∞ in Σ (since $\pi^{-1}(n) \to N$ in S^2), although this sequence has no convergent subsequence in the non-compact space $\mathbb{C}$ (see Fig. 1.3).

Fig. 1.3

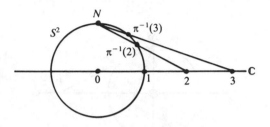

It is straightforward to check that the open sets of Σ are of two types: open sets in $\mathbb{C}$ (in its usual topology as $\mathbb{R}^2$), and sets of the form $(\mathbb{C} \backslash K) \cup \{\infty\}$ where K is any compact subset of $\mathbb{C}$. Thus the subspace topology of $\mathbb{C}$ (induced by its inclusion in Σ) agrees with the usual topology. This shows that Σ is the one-point compactification of $\mathbb{C}$: we can embed any topological space X in a compact space $X \cup \{\infty\}$, called its *one-point compactification*, by adding a single point ∞ and defining the open sets of $X \cup \{\infty\}$ to be the open sets of X together with those subsets which contain ∞ and have a closed, compact complement in X.

1.3 Behaviour of functions at infinity

If a subset D of Σ does not contain ∞, then since $D \subseteq \mathbb{C}$ we can refer to functions on D as being analytic, meromorphic, having poles, Taylor expansions, etc. Our aim is to define similar concepts at ∞, so that all points of Σ have equal status. We do this by using the transformation $J(z) = z^{-1}$ of Σ; this is well defined on $\mathbb{C} \backslash \{0\}$, and we use the convention that $J(0) = \infty$ and $J(\infty) = 0$. Thus $J : \Sigma \to \Sigma$ is a bijection and J^2 is the identity.

Now let P be the point $z = x + iy \in \mathbb{C} \backslash \{0\}$, with $x, y \in \mathbb{R}$, and let P^* be the point $J(z) = z^{-1} = (x - iy)/z\bar{z}$. Then the point $Q = \pi^{-1}(P)$ of S^2 corresponding to P has coordinates

$$x_1 = \frac{2x}{z\bar{z} + 1}, \quad x_2 = \frac{2y}{z\bar{z} + 1}, \quad x_3 = \frac{z\bar{z} - 1}{z\bar{z} + 1}$$

in $\mathbb{R}^3$, and the coordinates of $Q^* = \pi^{-1}(P^*)$ are

$$x_1^* = \frac{2x(z\bar{z})^{-1}}{(z\bar{z})^{-1} + 1} = \frac{2x}{1 + z\bar{z}} = x_1,$$

$$x_2^* = \frac{-2y(z\bar{z})^{-1}}{(z\bar{z})^{-1} + 1} = \frac{-2y}{1 + z\bar{z}} = -x_2,$$

$$x_3^* = \frac{(z\bar{z})^{-1} - 1}{(z\bar{z})^{-1} + 1} = \frac{1 - z\bar{z}}{1 + z\bar{z}} = -x_3.$$

Thus J induces the transformation $\pi^{-1}J\pi : Q = (x_1, x_2, x_3) \mapsto Q^* = (x_1, -x_2, -x_3)$ of S^2 (a separate but simple argument is needed for $Q = N$ or S, corresponding to $z = \infty$ or 0), and this is the rotation of S^2 by the angle π about the x_1-axis. From now on, we will abuse the notation and refer to the rotation $J : S^2 \to S^2$, rather than $\pi^{-1}J\pi$; equivalently, we are identifying S^2 and Σ by means of π, and regarding J as a transformation of each of these two spaces (Fig. 1.4).

Fig. 1.4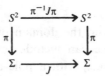

Suppose that a function $f(z)$ is defined on $D \backslash \{\infty\}$, where D is a neighbourhood of ∞ in Σ; equivalently, $f(z)$ is defined provided $|z|$ is sufficiently large. We can extend the domain of f to include ∞ by defining $f(\infty)$ to be $\lim_{z \to \infty} f(z)$, provided this limit exists. Then f is continuous at ∞, and $f(\infty) = \lim_{z \to 0} f(z^{-1}) = \lim_{z \to 0} (f \circ J)(z)$. We say that f is *analytic, meromorphic, etc., at* ∞ provided $f \circ J$ has the corresponding property at 0. For example:

(i) $f(z) = (z^2 + 1)^{-1}$ is analytic at ∞ with a zero of order 2 there, since $(f \circ J)(z) = z^2(z^2 + 1)^{-1}$ is analytic with a zero of order 2 at 0;
(ii) $f(z) = z^3$ is meromorphic at ∞, with a pole of order 3 there;
(iii) $\sin(z)$ has an isolated essential singularity at ∞, and is therefore not analytic at ∞.

We define a *region* of Σ to be a non-empty, path-connected, open subset. Using these definitions, we can extend certain theorems about functions defined on $\mathbb{C}$ to functions defined on Σ. For example:

Theorem 1.3.1. *Let f be an analytic function on a region R of Σ. If* *f has zeros at an infinite sequence of points z_n in R with a limit $z^* = \lim_{n \to \infty} z_n$ in R, then f is identically zero on R.*

Proof. If $z^* \neq \infty$, then $z_n \neq \infty$ for all sufficiently large n, so by omitting finitely many terms we may assume that $z_n \in \mathbb{C}$ for all n. Now $R' = R \setminus \{\infty\}$ is a region in $\mathbb{C}$ (why?), and f is analytic on R' with zeros at an infinite sequence of points $z_n \in R'$ with a limit $z^* \in R'$, so by Theorem A.8 f is identically zero on R'. If $\infty \notin R$ then $R = R'$ and the result is proved. If $\infty \in R$, then since f is analytic at ∞ and vanishes identically in a neighbourhood of ∞, we have $f(\infty) = 0$ as required, by continuity.

Now suppose that $z^* = \infty$. Omitting finitely many terms, we may assume that $z_n \neq 0$ for all n. Since f is analytic on the region $\tilde{R} = R \setminus \{0\}$, $f \circ J$ is analytic on the region $R^* = \{z^{-1} | z \in \tilde{R}\}$. Now $f \circ J$ has zeros at the points z_n^{-1} of R^*, and these have a limit $J(z^*) = 0$ in R^*, so $f \circ J$ is identically zero on R^* and hence f is identically zero on $\tilde{R}$. If $0 \notin R$ then $R = \tilde{R}$ and the result is proved. If $0 \in R$ then $f(0) = 0$ by continuity, so f is identically zero on R. $\square$

We have seen how to extend the domain of definition of a function f to include ∞. Similarly, we can include ∞ in the image of f: if f is meromorphic at a point $a \in \Sigma$, with a pole at a (that is, $(z - a)^k f(z)$ is analytic at a for some $k \in \mathbb{N}$, but not for $k = 0$), then we write $f(a) = \infty$. Thus the poles of f correspond to the zeros of $J \circ f$. We define $f : \Sigma \to \Sigma$ to be *meromorphic on* Σ if it is meromorphic at each $a \in \Sigma$. We immediately have:

(i) if f is meromorphic on Σ then f is continuous on Σ;
(ii) each constant function $f(z) \equiv c \in \mathbb{C}$ is meromorphic on Σ (however, the constant function $f(z) \equiv \infty$ is *not* meromorphic on Σ);
(iii) the meromorphic functions on Σ form a field, that is, if f and g are meromorphic on Σ then so are $f \pm g$, fg and f/g provided $g \not\equiv 0$.

Suppose that f is analytic at $a \in \mathbb{C}$, with $f(a) = c \in \mathbb{C}$; if f is not constant then $f^{(k)}(a) \neq 0$ for some $k \geq 1$, and we call the least such k the *multiplicity* of the solution of $f(z) = c$ at $z = a$. Thus

$$f(z) = c + \sum_{j=k}^{\infty} \frac{f^{(j)}(a)}{j!} (z - a)^j$$

near $z = a$, with $f^{(k)}(a) \neq 0$. If f is meromorphic at $a \in \mathbb{C}$, with a pole of order k at a, we say that $f(a) = \infty$ with multiplicity k; then $f(z) = \sum_{j=-k}^{\infty} a_j (z - a)^j$ near $z = a$, with constants a_j such that $a_{-k} \neq 0$, and we call $\sum_{j=-k}^{-1} a_j(z-a)^j$ the *principal part* of f at a. Similarly, if $a = \infty$ then we say that $f(\infty) = c$ with multiplicity k if $(f \circ J)(0) = c$ with multiplicity k; for example if f has a pole of order k at ∞ then $f(z) = \sum_{j=-\infty}^{k} a_j z^j$ near $z = \infty$, with $a_k \neq 0$; we call $\sum_{j=1}^{k} a_j z^j$ the principal part of f at ∞. We say that a is a *simple* point for f if it has multiplicity $k = 1$, and a *multiple point* if $k > 1$.

Corollary 1.3.2. *A non-constant meromorphic function $f : \Sigma \to \Sigma$ takes any given value $c \in \Sigma$ only finitely many times, counting multiplicities (that is, the sum of the multiplicities of the solutions of $f(z) = c$ is finite).*

Proof. If $z \in \Sigma$ and $f(z) = c$, then there exists a neighbourhood N_z of z such that f does not take the value c on $N_z \setminus \{z\}$: for if $c = \infty$ then the poles of f are the zeros of $J \circ f$, and since $J \circ f$ is meromorphic and not constant these are isolated, by Theorem 1.3.1; if $c \neq \infty$ we use the fact that the zeros of $f - c$ are isolated. Being compact, Σ is covered by finitely many such neighbourhoods $N_{z_1}, \dots, N_{z_k}$, so $f^{-1}(c) = \{z_1, \dots, z_k\}$, a finite set. Since f is meromorphic, each solution of $f(z) = c$ has finite multiplicity, so f takes the value c only finitely many times. $\square$

(We shall shortly see that f takes each value $c \in \Sigma$ the *same* number of times, counting multiplicities.)

Theorem 1.3.3. *Let f and g be meromorphic functions on Σ with poles at the same points in Σ, and with the same principal parts at these points. Then $f(z) = g(z) + c$ for some constant c. (Thus meromorphic functions on Σ are determined, up to additive constants, by their principal parts.)*

Proof. The function $h = f - g$ is meromorphic and therefore continuous on Σ; since Σ is compact, the image $h(\Sigma)$ is compact. Since the principal parts of f and g cancel, h has no poles, so $h(\Sigma)$ is a subset of $\mathbb{C}$, and being compact it is bounded. Liouville's theorem (Theorem A.4) shows that h, being analytic and bounded, must be constant on $\mathbb{C}$, and hence (by continuity) on Σ, so that $f = g + c$ for some constant c. $\square$

The following is a multiplicative version of this result:

Theorem 1.3.4. *Let f and g be meromorphic functions on Σ with zeros and poles of the same orders at the same points of $\mathbb{C}$. Then $f(z) = cg(z)$ for some constant $c \neq 0$.*

Proof. We may assume that neither f nor g is identically zero. Then both f/g and g/f are meromorphic on Σ, and neither of them has any poles in $\mathbb{C}$, so both are analytic on $\mathbb{C}$. At least one of them (call it h) is finite at ∞, so h is analytic on Σ. As in Theorem 1.3.3, Liouville's theorem implies that h is constant, so that $f = cg$ for some constant c. We have $c \neq 0$ since f is not identically zero. $\square$

(Notice that in the hypotheses of Theorem 1.3.4 we do not need to assume that f and g behave similarly at ∞: this follows from the fact that $f = cg$.)

1.4 Rational functions

A *rational function* is a function of the form $f(z) = p(z)/q(z)$, where $p(z)$ and $q(z)$ are polynomials with complex coefficients and $q(z)$ is not identically zero. When $z \in \mathbb{C}$ and $q(z) \neq 0$, $f(z)$ is a well-defined element of $\mathbb{C}$; when $q(z) = 0$ or $z = \infty$, we define $f(z) = \lim_{z' \to z} f(z')$ as in §1.3. Thus f is a function $\Sigma \to \Sigma$.

The rational functions form a field which we denote by $\mathbb{C}(z)$. For each fixed $a \in \mathbb{C}$ the constant function $f_a : z \mapsto a$ is a rational function, and these functions form a field isomorphic to $\mathbb{C}$ under the isomorphism $f_a \mapsto a$. Thus $\mathbb{C}(z)$ contains a subfield isomorphic to $\mathbb{C}$, so $\mathbb{C}(z)$ may be regarded as an extension field of $\mathbb{C}$.

Two polynomials p and q are *co-prime* if there is no non-constant polynomial r dividing both p and q. If $f = p/q$ is a rational function then we can cancel any common factors and hence assume that p and q are co-prime. By the fundamental theorem of algebra we can express every polynomial as a product of linear factors. Hence we can write

$$f(z) = c(z - \alpha_1)^{m_1} \ldots (z - \alpha_r)^{m_r}(z - \beta_1)^{-n_1} \ldots (z - \beta_s)^{-n_s},$$

where $c \in \mathbb{C}$, $\alpha_1, \ldots, \alpha_r$ are the zeros of p of orders $m_1, \ldots, m_r$, and $\beta_1, \ldots, \beta_s$ are the zeros of q of orders $n_1, \ldots, n_s$. As p and q are co-prime, the zeros of p are distinct from those of q. Thus $\alpha_1, \ldots, \alpha_r$ are zeros of f of orders $m_1, \ldots, m_r$, and $\beta_1, \ldots, \beta_s$ are poles of f of orders $n_1, \ldots, n_s$. These are the only zeros and poles of f on $\mathbb{C}$, and ∞ is a zero or a pole as $(m_1 + \ldots + m_r) - (n_1 + \ldots + n_s)$ is negative or positive. For example, $f(z) = (z - 1)/(z^2 + 4)$ has zeros of order 1 at 1 and ∞, and poles of order 1 at $\pm 2i$.

The next result shows that the *algebraic* definition of a rational function is equivalent to an *analytic* condition:

Theorem 1.4.1. *A function $f:\Sigma \to \Sigma$ is rational if and only if it is mero-morphic on Σ.*

Proof. If we decompose a rational function f as above, then f is differenti-able at each $z \neq \infty$, $\beta_j (1 \leqslant j \leqslant s)$, so f is analytic on $\mathbb{C}\backslash\{\beta_1, \ldots, \beta_s\}$. At each β_j, f has a pole of order n_j, while at ∞, f is analytic if $\deg(p) \leqslant \deg(q)$ and f has a pole of order $\deg(p) - \deg(q)$ if $\deg(p) > \deg(q)$. Thus f is meromorphic on Σ.

Conversely, suppose that f is meromorphic on Σ. By Corollary 1.3.2, f has finitely many poles in $\mathbb{C}$, say $\beta_1, \ldots, \beta_s$ of orders $n_1, \ldots, n_s$. Then the function

$$g(z) = (z - \beta_1)^{n_1}(z - \beta_2)^{n_2} \ldots (z - \beta_s)^{n_s} f(z)$$

is analytic on $\mathbb{C}$, so g has a Taylor expansion

$$g(z) = a_0 + a_1 z + a_2 z^2 + \ldots,$$

valid for all $z \in \mathbb{C}$. Now g is meromorphic at ∞ (since f is), so

$$(g \circ J)(z) = a_0 + a_1 z^{-1} + a_2 z^{-2} + \ldots$$

is meromorphic at 0, and hence $a_j = 0$ for all sufficiently large j. Thus g is a polynomial, so

$$f(z) = g(z)(z - \beta_1)^{-n_1} \ldots (z - \beta_s)^{-n_s}$$

is a rational function. $\square$

If $f = p/q$ is a rational function, with p and q co-prime polynomials, then the *degree* (or *order*) $\deg(f)$ of f is the maximum of the degrees of p and q. Thus f is constant if and only if $\deg(f) = 0$.

Theorem 1.4.2. *If $f:\Sigma \to \Sigma$ is a rational function of degree $d > 0$, then f takes each value $c \in \Sigma$ exactly d times, counting multiplicities.*

Proof. Let $f = p/q$, where p and q are co-prime polynomials. First suppose that $c = \infty$. For $z \in \mathbb{C}$ we have $f(z) = \infty$ if and only if $q(z) = 0$, and by the fundamental theorem of algebra this equation has $\deg(q)$ solutions, counting multiplicities. If $\deg(p) \leqslant \deg(q)$ then these are the only poles of f; if $\deg(p) > \deg(q)$ then f has an additional pole of order $\deg(p) - \deg(q)$ at ∞. In either case, the number of solutions (counting multiplicities) of $f(z) = \infty$ is $\max(\deg(p), \deg(q))$, which is the degree of f.

Now suppose that $c \neq \infty$. Since $\deg(f) > 0, f$ is not identically equal to c, so there is a rational function

$$g = \frac{1}{f - c} = \frac{q}{p - cq};$$

the solutions of $f(z) = c$ are the poles of g, and by the previous argument there are $\deg(g)$ of these counting multiplicities. Now q and $p - cq$ are co-prime (since p and q are), so $\deg(g) = \max(\deg(q), \deg(p - cq)) = \max(\deg(q), \deg(p)) = \deg(f)$, as required. $\square$

Let f be meromorphic at $a \in \Sigma$ and let $f(a) = c$. Recall that a is a multiple point for f if the equation $f(z) = c$ has a multiple solution at $z = a$; if $c \neq \infty$ this is equivalent to $f'(a) = 0$, while if $c = \infty$ then this is equivalent to f having a pole of order at least two at a. All other points are called simple points for f.

Corollary 1.4.3. *Let $f : \Sigma \to \Sigma$ be a rational function of degree $d > 0$. Then*

(i) *f has only finitely many multiple points in Σ;*
(ii) *$|f^{-1}(c)| = d$ for all but finitely many points $c \in \Sigma$, and $1 \leqslant |f^{-1}(c)| < d$ for the remaining points c.*

Proof

(i) Since the derivative f' is rational and not identically zero, f' has only finitely many zeros in Σ; since f has only finitely many poles, (i) follows.
(ii) By Theorem, 1.4.2, if $c \in \Sigma$ then there are solutions $z = a_1, \ldots, a_r$ of $f(z) = c$ with multiplicities $k_1, \ldots, k_r$ satisfying $k_1 + \ldots + k_r = d$. Thus $|f^{-1}(c)| = r$ so that $1 \leqslant |f^{-1}(c)| \leqslant d$, and we have $|f^{-1}(c)| = d$ unless some $k_j \geqslant 2$. Since f has only finitely many multiple points, by (i), the result follows. $\square$

(Notice that this characterises $d = \deg(f)$ purely set-theoretically as the maximum value attained by $|f^{-1}(c)|$ as c ranges over Σ, provided $d > 0$.)

Suppose that a function f is meromorphic at a point $a \in \Sigma$. We define the *order* of f at a to be

$$v_a(f) = \begin{cases} k & \text{if } f \text{ has a zero of multiplicity } k \text{ at } a, \\ 0 & \text{if } f(a) \neq 0, \infty, \\ -k & \text{if } f \text{ has a pole of order } k \text{ at } a. \end{cases}$$

Thus for $a \in \mathbb{C}$, $v_a(f)$ is the lowest exponent of $(z - a)$ in the Laurent expansion of f near $z = a$, while for $a = \infty$, $v_\infty(f)$ is the lowest exponent of z in the Laurent expansion of $(f \circ J)(z) = f(z^{-1})$ near $z = 0$. In the case of a non-zero rational function f, it is clear that for $a \neq \infty$, $v_a(f)$ is the exponent of $(z - a)$ in the factorisation of f into powers of linear polynomials, so that

$$f(z) = c \prod_{a \in \mathbb{C}} (z - a)^{v_a(f)} \quad (c \in \mathbb{C} \setminus \{0\}),$$

this product being finite since $v_a(f) = 0$ for all but finitely many $a \in \mathbb{C}$; for $a = \infty$, we have $v_\infty(f) = \deg(q) - \deg(p)$ where $f = p/q$, p and q co-prime polynomials. Thus, if the zeros and poles of f have multiplicities m_j, n_j respectively, then

$$v_\infty(f) = \sum n_j - \sum m_j = - \sum_{a \in \mathbb{C}} v_a(f),$$

so that

$$\sum_{a \in \Sigma} v_a(f) = 0.$$

This shows that the number of zeros of f on Σ is equal to the number of poles (counting multiplicities). We can regard this as a special case of Theorem 1.4.2, and the full statement of this theorem follows easily. We shall see that a similar equation holds for functions f which are meromorphic on other compact surfaces, such as the torus.

1.5 Topological properties

We now examine the topological properties of meromorphic (or equivalently rational) functions as maps from the sphere to itself; the main technical difficulty arises in dealing with multiple points.

Suppose that $f: \Sigma \to \Sigma$ is meromorphic and not constant. If $f(a) = c$ with multiplicity k, where $a, c \in \mathbb{C}$, then Theorem A.10 shows that f is a k-to-one function near a; more precisely, if U is any sufficiently small open set containing a then there exists an open set V containing c such that for each $c' \in V \setminus \{c\}$ the equation $f(z) = c'$ has exactly k solutions in U, all simple (see Fig. 1.5). Thus we may think of the solution of multiplicity k at a as splitting into k simple solutions near a. By considering the functions $f \circ J$,

Fig. 1.5

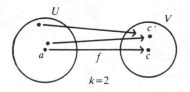

$k = 2$

$J \circ f$ and $J \circ f \circ J$, it is easily seen that we have the same behaviour near a when either a, c or both are equal to ∞. (For functions of a *real* variable, the situation is less satisfactory: for example, as shown in Fig. 1.6, the function $f : \mathbb{R} \to \mathbb{R}$ given by $f(x) = x^3$ has a triple root at $x = 0$, but is locally one-to-one; the function $f(x) = x^2$ has a double root at $x = 0$, but the equations $f(x) = c'$ have two solutions for $c' > 0$, and none for $c' < 0$.)

A mapping between topological spaces is called *open* if the image of each open set is open; we shall show that non-constant meromorphic functions are open. Let X be an open subset of Σ and let $c \in f(X)$, say $c = f(a)$ with multiplicity k for some $a \in X$. Then by Theorem A.10 (extended to Σ as above), c is contained in an open set $V \subseteq f(U) \subseteq f(X)$, where U is a neighbourhood of a in X. Thus $f(X)$ is open, as required.

If $f(a) = c$ with multiplicity $k = 1$, then f maps a neighbourhood $W = U \cap f^{-1}(V)$ of a bijectively onto a neighbourhood V of c, so we can define an inverse function $f^{-1} : V \to W$; since f is open, f^{-1} is continuous, so f, being

Fig. 1.6

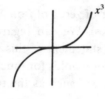

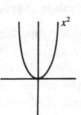

Fig. 1.7

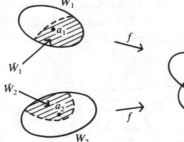

itself continuous, induces a homeomorphism of W onto V. Thus, at simple points, f is locally a homeomorphism.

A point of multiplicity $k > 1$ is often called a *branch-point* of *order* $k - 1$. By Corollary 1.4.3, the set B of branch-points of f is finite, and for each $c \in \Sigma \setminus f(B)$ the equation $f(z) = c$ has only simple solutions, so that if f has degree d then there are exactly d points $a_1, \ldots, a_d \in f^{-1}(\Sigma \setminus f(B)) = \Sigma \setminus f^{-1}(f(B))$ such that $f(a_j) = c$. Each a_j has a neighbourhood W_j mapped homeomorphically onto a neighbourhood V_j of c, and we can choose these neighbourhoods W_j to be mutually disjoint, so that c has a neighbourhood $V = V_1 \cap \ldots \cap V_d$ such that $f^{-1}(V)$ consists of d disjoint open sets $\tilde{W}_j = W_j \cap f^{-1}(V)$, each mapped homeomorphically onto V by f (see Fig. 1.7 for the case $d = 2$). This shows that $f: \Sigma \setminus f^{-1}(f(B)) \to \Sigma \setminus f(B)$ is an example of a *covering map*: a continuous map $f: X \to Y$ is a covering map if every $y \in Y$ has a neighbourhood V such that $f^{-1}(V)$ consists of disjoint open sets mapped homeomorphically onto V by f, in which case X is a *covering space* of Y. If we include the branch-points, we have a *branched covering* $f: \Sigma \to \Sigma$. (For more details about covering spaces see [Massey, 1967].)

It is useful to imagine the neighbourhoods $\tilde{W}_1, \ldots, \tilde{W}_d$ forming parallel layers lying above V, so that f projects each $\tilde{W}_j$ homeomorphically down-

Fig. 1.8

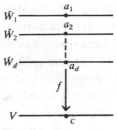

Fig. 1.9

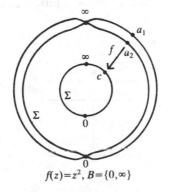

$f(z) = z^2$, $B = \{0, \infty\}$

wards onto V, as in Fig. 1.8. As c moves around $\Sigma \setminus f(B)$, these neighbourhoods $\tilde{W}_j$ form the d sheets of the surface $\Sigma \setminus f^{-1}(f(B))$ which is 'wrapped around' the image sphere Σ, with d points $a_1, \ldots, a_d$ lying above each $c \in \Sigma \setminus f(B)$; at branch-points of order $k-1$, we have k sheets coming together to form a single point on the domain sphere Σ (see Fig. 1.9 for an example with $d = 2$). This construction, in which the domain sphere is wrapped d times around the image sphere, is difficult to visualise, and is in fact impossible to perform in $\mathbb{R}^3$ without the domain sphere intersecting itself, though it can be done in spaces of higher dimension. We will return later to these ideas on covering surfaces; for example, in Chapter 3 we will see that elliptic functions lead to branched coverings of the sphere by the torus. For the time being, we will try to make these abstract ideas a little more concrete by considering specific examples.

Examples (1) Let $f(z) = z^n$. We have $f(0) = 0$ with multiplicity n, and as $f'(z) \neq 0$ for $z \neq 0$, there are no other branch-points in $\mathbb{C}$. To see if ∞ is a branch-point we consider $(J \circ f \circ J)(z) = z^n$ which has a zero of order n at $z = 0$, so $f(\infty) = \infty$ with multiplicity n. Hence the covering surface has n sheets which come together at branch-points of order $n-1$ at 0 and ∞.

(2) Let $f(z) = z/(z^3 + 2)$, a rational function of degree 3. Since $f'(z) = (2 - 2z^3)/(z^3 + 2)^2$, there are branch-points at the cube roots of unity $z = 1$, ω, ω^2 ($\omega = e^{2\pi i/3}$). These branch-points lie over the points $\frac{1}{3}, \omega/3, \omega^2/3$ in the image sphere. We can see that these branch-points have order 1 either by showing that $f''(z) \neq 0$, or alternatively by solving the equations $f(z) = \frac{1}{3}$, $f(z) = \omega/3$ and $f(z) = \omega^2/3$. For example, the first of these equations is equivalent to $(z - 1)^2(z + 2) = 0$, showing that over the point $\frac{1}{3}$ two of the three sheets come together to give a double solution at $z = 1$ (that is, a branch-point of order $2 - 1 = 1$), while there is a simple solution at $z = -2$ on the third sheet (see Fig. 1.10). Similarly, over the point $\omega/3$ there is a double solution at $z = \omega$ and a simple solution at $z = -2\omega$, while over

Fig. 1.10

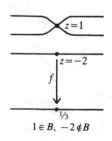

$1 \in B, \ -2 \notin B$

$\omega^2/3$ there is a double solution at $z = \omega^2$ and a simple solution at $z = -2\omega^2$.

We have to examine the cases $z = \infty$ and the poles of f separately. The function $(f \circ J)(z) = z^2/(1 + 2z^3)$ has a double zero at $z = 0$, so f has a branch-point of order $2 - 1 = 1$ at $z = \infty$, the other solution of $f(z) = 0$ being a simple one at $z = 0$. Finally, the poles of f are all simple (at the cube roots of -2), so these are not branch-points of f.

EXERCISES

1A. Show that the eight points whose coordinates are $(\pm 1/\sqrt{3}, \pm 1/\sqrt{3}, \pm 1/\sqrt{3})$ form the vertices of a cube lying on the Riemann sphere. Find the images of these points under stereographic projection and plot them in the complex plane.

1B. Use 1A to find four points on the Riemann sphere which are vertices of a regular tetrahedron and find their images under stereographic projection.

1C. Find the points of the Riemann sphere which project under stereographic projection to

    (i) the circle $C_r = \{z \in \mathbb{C} \,|\, |z| = r\}$;

    (ii) the real axis;

    (iii) the imaginary axis;

    (iv) the points of the unit square whose vertices are $\pm 1 \pm i$.

1D. Let $z_1, z_2 \in \mathbb{C}$ correspond (under stereographic projection) to $Q_1, Q_2 \in S^2$. Find the distance $d(Q_1, Q_2)$ between Q_1 and Q_2 in $\mathbb{R}^3$ in terms of z_1 and z_2, and also find the angle subtended by Q_1 and Q_2 at the origin. Use this to show that stereographic projection from $S^2 \setminus \{N\}$ onto $\mathbb{C}$ is a homeomorphism.

1E. Show that complex conjugation of Σ (that is complex conjugation of $\mathbb{C}$ extended to Σ by defining $\bar{\infty} = \infty$) corresponds to reflection of S^2 in the plane $x_2 = 0$. What transformations of Σ correspond to reflections in the planes $x_1 = 0$ and $x_3 = 0$? Show that the antipodal map $Q \mapsto -Q$ of S^2 is the composition of the above three reflections in any order and hence express it as a transformation of Σ.

1F. Show that the one-point compactification of the real line is homeomorphic to the circle.

1G. Investigate the nature of the following functions at ∞:

    (i) $f_1(z) = 3z + 4$,

    (ii) $f_2(z) = \exp(z/(z + 1))$,

    (iii) $f_3(z) = \sin^2(1/(z + 2))$,

    (iv) $f_4(z) = z \sin(1/z)$.

1H. Give an example of a meromorphic function $f : \mathbb{C} \to \Sigma$ with a simple pole at each $z \in \mathbb{Z}$; does it extend to a meromorphic function $\Sigma \to \Sigma$?

1I. Find the zeros and poles of

$$f(z) = \frac{z^2 + 3}{z^4 + 3z^2 + 2}$$

and check that $\sum_{a \in \Sigma} v_a(f) = 0$.

1J. Suppose that f is a rational function with a pole of order 3 at $z = i$, a zero of order 2 at $z = 1$, a zero of order 3 at ∞, and a pole at $z = 2$ (and no other zeros or poles). Find the order of the pole at 2 and determine f to within a constant factor.

1K. Let $a_1, \ldots, a_r, b_1, \ldots, b_s \in \Sigma$ and let $k_1, \ldots, k_r, l_1, \ldots, l_s$ be positive integers. Find necessary and sufficient conditions for the existence of a rational function with a zero of order k_j at each a_j, a pole of order l_j at each b_j, and no other zeros or poles.

1L. Let f be a rational function whose poles in $\mathbb{C}$ are at $\beta_1, \ldots, \beta_q$. Prove that there exist unique polynomials $\phi_0, \phi_1, \ldots, \phi_q$ with zero constant term such that

$$f(z) = \phi_0(z) + \sum_{i=1}^{q} \phi_i\left(\frac{1}{z - \beta_i}\right) + \text{constant}$$

(use Theorem 1.3.3). Illustrate this result with reference to the function

$$f(z) = \frac{z^2}{(z-1)^2(z-2)}.$$

(This question shows that every rational function can be decomposed into partial fractions.)

1M. Let $f(z)$ be a rational function such that $|z| = 1$ implies $|f(z)| = 1$. Show that α is a zero of $f(z)$ if and only if $1/\bar{\alpha}$ is a pole of $f(z)$, and hence find the most general form of $f(z)$.

1N. Investigate the covering of the sphere by the sphere associated with the rational function

$$f(z) = \frac{z^3}{z^4 + 27}.$$

(Find the number of sheets, the branch points and the nature of branching.)

2

Möbius transformations

2.1 Automorphisms of Σ

An *automorphism* of the Riemann sphere Σ is a meromorphic bijection $T:\Sigma \to \Sigma$; we denote the set of all automorphisms of Σ by Aut(Σ).

Theorem 2.1.1. Aut(Σ) *consists of the functions*

$$T(z) = \frac{az + b}{cz + d} \quad (a, b, c, d \in \mathbb{C}, ad - bc \neq 0).$$

Proof. By Theorem 1.4.1 a function $T:\Sigma \to \Sigma$ is meromorphic if and only if it is rational, and by Corollary 1.4.3 such a function T is a bijection if and only if it has degree 1. The automorphisms of Σ are therefore the functions $T(z) = (az + b)/(cz + d)$ where $az + b$ and $cz + d$ are co-prime polynomials, that is, $ad - bc \neq 0$. $\square$

The transformations $w = T(z)$ of the above type are known as *linear fractional* or *Möbius transformations* (A.F. Möbius, 1790–1868). It is important to notice that T does not determine the coefficients a, b, c, d uniquely: if $\lambda \in \mathbb{C} \setminus \{0\}$ then the coefficients λa, λb, λc, λd correspond to the same transformation T. The Möbius transformations form a group under composition: if

$$U(z) = \frac{a'z + b'}{c'z + d'}$$

is also a Möbius transformation, then $U \circ T$ is the Möbius transformation

$$(U \circ T)(z) = \frac{(a'a + b'c)z + (a'b + b'd)}{(c'a + d'c)z + (c'b + d'd)},$$

with $(a'a + b'c)(c'b + d'd) - (a'b + b'd)(c'a + d'c) = (a'd' - b'c')(ad - bc) \neq 0$. The identity transformation has $a = d \neq 0$ and $b = c = 0$, and the inverse of T is the transformation

$$T^{-1}(z) = \frac{dz - b}{-cz + a}.$$

Corollary 2.1.2. Aut (Σ) *is a group of homeomorphisms from Σ to itself.*

Proof. We have just shown that Aut (Σ) is a group. If $T \in \text{Aut}(\Sigma)$ then, being meromorphic, T is continuous; since $T^{-1} \in \text{Aut}(\Sigma)$, T^{-1} is also continuous, so T is a homeomorphism from Σ to itself. $\square$

There is a strong connection between Möbius transformations and matrices: if T and U are expressed in the above form, and if M and N are the corresponding matrices

$$M = \begin{pmatrix} a & b \\ c & d \end{pmatrix}, \quad N = \begin{pmatrix} a' & b' \\ c' & d' \end{pmatrix},$$

then $U \circ T$ corresponds to the matrix product

$$NM = \begin{pmatrix} a'a + b'c & a'b + b'd \\ c'a + d'c & c'b + d'd \end{pmatrix}.$$

More precisely, let $GL(2, \mathbb{C})$ denote the *general linear group* consisting of all 2×2 complex matrices

$$M = \begin{pmatrix} a & b \\ c & d \end{pmatrix},$$

with $\det(M) = ad - bc \neq 0$, and for each such M let $\theta(M)$ be the Möbius transformation $T(z) = (az + b)/(cz + d)$. Then $\theta(NM) = U \circ T = \theta(N) \circ \theta(M)$, so that $\theta : GL(2, \mathbb{C}) \to \text{Aut}(\Sigma)$ is a group-homomorphism, and by Theorem 2.1.1 θ is an epimorphism. The kernel $K = \ker(\theta)$ consists of those $M \in GL(2, \mathbb{C})$ such that $T(z) = z$ for all $z \in \Sigma$, or equivalently $a = d \neq 0$ and $b = c = 0$, so K consists of the matrices

$$M = \begin{pmatrix} \lambda & 0 \\ 0 & \lambda \end{pmatrix} = \lambda I \quad (\lambda \in \mathbb{C} \setminus \{0\});$$

thus two matrices $M, N \in GL(2, \mathbb{C})$ determine the same automorphism of Σ if and only if $M^{-1}N \in K$, that is, $N = \lambda M$ for some $\lambda \neq 0$. Applying the first isomorphism theorem to θ, we get

$$\text{Aut}(\Sigma) \cong GL(2, \mathbb{C})/K = GL(2, \mathbb{C})/\{\lambda I \,|\, \lambda \neq 0\}.$$

The quotient group $GL(2, \mathbb{C})/K$ is the *projective general linear group*, denoted by $PGL(2, \mathbb{C})$.

Since $\det(NM) = \det(N)\det(M)$ for all $M, N \in GL(2, \mathbb{C})$, the function

$$\det : GL(2, \mathbb{C}) \to \mathbb{C}^* = \mathbb{C} \setminus \{0\}$$

is a group-homomorphism; its kernel is a normal subgroup of $GL(2, \mathbb{C})$, the *special linear group* $SL(2, \mathbb{C})$ consisting of all $M \in GL(2, \mathbb{C})$ such that

$\det(M) = 1$. Since det is onto, we have

$$GL(2, \mathbb{C})/SL(2, \mathbb{C}) \cong \mathbb{C}^*.$$

If $N \in GL(2, \mathbb{C})$ then we can write $N = \lambda M$ where $\lambda^2 = \det(N)$ and $M \in SL(2, \mathbb{C})$; since $\theta(N) = \theta(M)$, this shows that every automorphism of Σ has the form

$$T(z) = \frac{az + b}{cz + d} \quad \text{with} \quad ad - bc = 1;$$

equivalently, θ maps $SL(2, \mathbb{C})$ onto $\text{Aut}(\Sigma)$. Thus $PGL(2, \mathbb{C})$ coincides with the *projective special linear group* $PSL(2, \mathbb{C})$, the image of $SL(2, \mathbb{C})$ in the quotient-group $PGL(2, \mathbb{C})/K$, and we have proved:

Theorem 2.1.3. $\text{Aut}(\Sigma) \cong PGL(2, \mathbb{C}) = PSL(2, \mathbb{C})$. $\square$

From now on, we will use this isomorphism to identify $\text{Aut}(\Sigma)$ with $PGL(2, \mathbb{C})$.

The transformations

$$T(z) = \frac{a\bar{z} + b}{c\bar{z} + d},$$

with $a, b, c, d \in \mathbb{C}$ and $ad - bc \neq 0$ (or equivalently $ad - bc = 1$) are known as *anti-automorphisms* of Σ. Each anti-automorphism T is the composition of complex conjugation with an automorphism of Σ; since both of these are homeomorphisms of Σ onto itself (complex conjugation being given by reflection in the plane through $\mathbb{R} \cup \{\infty\}$), so is T. The composition of two anti-automorphisms is an automorphism, and the composition of an anti-automorphism with an automorphism is an anti-automorphism, so the automorphisms and anti-automorphisms of Σ form a group denoted by $\overline{\text{Aut}}(\Sigma) = \overline{PGL}(2, \mathbb{C})$, in which $\text{Aut}(\Sigma)$ is a normal subgroup of index 2. There is a topological distinction between automorphisms and anti-automorphisms in that the former preserve the orientation of Σ while the latter reverse it.

2.2 Linear and projective groups (This section may be omitted on first reading)

Some of the ideas on matrix groups in §2.1 may be extended to higher dimensions and to arbitrary fields. If n is any positive integer and F is any field, then $GL(n, F)$ is the *general linear group* of $n \times n$ invertible matrices with coefficients in F, and $SL(n, F)$ is the *special linear group* consisting of the *unimodular* matrices, those of determinant 1; we have $GL(n, F)/SL(n, F) \cong F^* = F \setminus \{0\}$. If $n \geqslant 2$, then since $GL(n, F)$ acts on the n-dimensional vector space F^n as a group of linear transformations, $GL(n, F)$ permutes the set $PG(n-1, F)$ of 1-dimensional subspaces of F^n. This set $PG(n-1, F)$ has the structure of an $(n-1)$-dimensional projective geometry, on which $GL(n, F)$ induces a group of projective transformations. The kernel K of this action of $GL(n, F)$ is the set of $n \times n$ scalar matrices $\lambda I_n (\lambda \in F^*)$, and the group induced by $GL(n, F)$ is the *projective general linear group* $PGL(n, F) = GL(n, F)/K$. The subgroup induced by $SL(n, F)$ is the *projective special linear group* $PSL(n, F) = SL(n, F)/(K \cap SL(n, F))$; we have $PSL(n, F) = PGL(n, F)$ if and only if every element of F has an nth root in F (for example, $PSL(2, \mathbb{C}) = PGL(2, \mathbb{C})$, but $PSL(2, \mathbb{R}) < PGL(2, \mathbb{R})$).

If $n = 2$ then each 1-dimensional subspace V of F^2 is spanned by a non-zero vector $v = (\alpha, \beta)$ with $\alpha, \beta \in F$. The ratio $z = \alpha/\beta \in F \cup \{\infty\}$ is independent of the choice of $v \in V$, so we may identify $PG(1, F)$ with $F \cup \{\infty\}$, identifying V with z. If $M = \begin{pmatrix} a & b \\ c & d \end{pmatrix} \in GL(2, F)$, then $M(v) = (a\alpha + b\beta, c\alpha + d\beta)$, so that the subspace $M(V)$ has ratio $(a\alpha + b\beta)/(c\alpha + d\beta) = (az + b)/(cz + d)$. Thus the actions of $GL(2, F)$ on $PG(1, F)$ (by projective transformations) and on $F \cup \{\infty\}$ (by linear fractional transformations) are identical. Taking $F = \mathbb{C}$, we see that the Riemann sphere $\Sigma = \mathbb{C} \cup \{\infty\}$ is identified with the *complex projective line* $PG(1, \mathbb{C})$, and that $GL(2, \mathbb{C})$ acts in the same way on Σ and on $PG(1, \mathbb{C})$, inducing the group $PGL(2, \mathbb{C})$ in each case.

2.3 Generators for $PGL(2, \mathbb{C})$

It is useful to consider the following special types of Möbius transformation:

(i) the transformation $R_\theta(z) = e^{i\theta}z$ $(\theta \in \mathbb{R})$ represents a rotation of the sphere Σ by an angle θ about the vertical axis through 0 and ∞ (we shall see later that *every* rotation of Σ corresponds to some Möbius transformation);

(ii) the transformation $J(z) = 1/z$ represents a rotation of Σ by an angle π about the axis through 1 and -1, as shown in §1.3;

(iii) the transformation $S_r(z) = rz$ $(r \in \mathbb{R}, r > 0)$ fixes 0 and ∞, and acts on the plane $\mathbb{C}$ as a similarity transformation, expanding or contracting distances by a factor r;

(iv) the transformation $T_t(z) = z + t$ $(t \in \mathbb{C})$ fixes ∞ and acts on the plane
$\mathbb{C}$ as a translation.

The next result shows that the Möbius transformations of types (i) to
(iv) generate $PGL(2, \mathbb{C})$.

Theorem 2.3.1. *Every Möbius transformation is a composition of finitely
many Möbius transformations of types* (i), (ii), (iii) *or* (iv).

Proof. Each Möbius transformation has the form $T(z) = (az + b)/(cz + d)$
with $ad - bc = 1$. If $c = 0$ then $T(z) = (az + b)/d$ with $a, d \neq 0$, so let $a/d = re^{i\theta}$
and $b/d = t$. Then $T(z) = re^{i\theta}z + t$, so $T = T_t \circ S_r \circ R_\theta$. If $c \neq 0$ then

$$T(z) = \frac{a}{c} - \frac{1}{c(cz + d)}$$
$$= (T_t \circ J)(-c^2z - cd),$$

with $t = a/c$; by the method used in the case $c = 0$, the transformation
$z \mapsto -c^2z - cd$ can be expressed in terms of the given generators, and hence
so can T. □

2.4 Circles in Σ

A *circle* in the sphere $S^2 \subset \mathbb{R}^3$ is defined to be any intersection $S^2 \cap \Pi$
where Π is a plane in $\mathbb{R}^3$ and $|S^2 \cap \Pi| > 1$ (that is, Π meets S^2 and is not
a tangent plane to S^2); this is illustrated in Fig. 2.1. Using the bijection
$\pi : S^2 \to \Sigma$, we define a *circle in* Σ to be the image under π of any circle in
S^2. If C is a circle in Σ corresponding to a plane Π with equation

$$\alpha x_1 + \beta x_2 + \gamma x_3 = \delta \quad (\alpha, \beta, \gamma, \delta \in \mathbb{R}),$$

then using equations (1.1.2) we find that C is given by

$$2\alpha x + 2\beta y + \gamma(|z|^2 - 1) = \delta(|z|^2 + 1),$$

where $z = x + iy$. Writing $a = \gamma - \delta \in \mathbb{R}$, $b = \alpha - i\beta \in \mathbb{C}$, and $c = -(\gamma + \delta) \in \mathbb{R}$,

Fig. 2.1

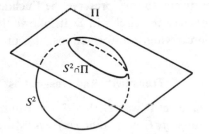

this becomes

$$azz̄ + bz + b̄z̄ + c = 0,$$

or in (x, y) coordinates,

$$ax^2 + ay^2 + 2\alpha x + 2\beta y + c = 0.$$

The condition $|S^2 \cap \Pi| > 1$ is equivalent to the existence of a point $(x_1, x_2, x_3) \in \Pi$ with $x_1^2 + x_2^2 + x_3^2 < 1$, and this is equivalent to $\alpha^2 + \beta^2 + \gamma^2 > \delta^2$; in terms of a, b and c this is equivalent to $b\bar{b} > ac$. Conversely, if $b\bar{b} > ac$ then the above equations always define a circle in Σ.

If $a \neq 0$ we have

$$\left(x + \frac{\alpha}{a}\right)^2 + \left(y + \frac{\beta}{a}\right)^2 = \frac{\alpha^2 + \beta^2 - ac}{a^2},$$

and this represents a circle in $\mathbb{R}^2$ with centre $(-\alpha/a, -\beta/a)$ and radius $(\alpha^2 + \beta^2 - ac)^{1/2}/a$. In complex coordinates, the centre is $-\bar{b}/a$ and the radius is $(b\bar{b} - ac)^{1/2}/a$.

If $a = 0$ then the equations represent a straight line in $\mathbb{R}^2$; this case corresponds to circles in Σ containing ∞ or equivalently circles in S^2 containing the north pole $N = (0, 0, 1)$.

Thus circles in Σ are of two types: circles in $\mathbb{C}$ (in the usual Euclidean sense), and sets of the form $\Lambda \cup \{\infty\}$, where Λ is a straight line in $\mathbb{C}$. The connection between $PGL(2, \mathbb{C})$ and circles in Σ is given by the following result.

Theorem 2.4.1. *If C is a circle in Σ, and if $T \in PGL(2, \mathbb{C})$, then $T(C)$ is a circle of Σ.*

Proof. By Theorem 2.3.1, T is a composition of transformations R_θ, J, S_r and T_t defined in §2.3, and it is easy to see, either geometrically or by a simple calculation, that these map circles to circles. For example, R_θ and J induce rotations of S^2, so they preserve the circles in S^2 and hence in Σ; S_r and T_t fix ∞ and act on $\mathbb{C}$ as a similarity transformation and as a translation respectively, so they preserve the Euclidean lines and circles in $\mathbb{C}$ and hence preserve the circles in Σ. It follows that T, being a composition of circle-preserving transformations, also preserves circles. $\square$

2.5 Transitivity and cross-ratios

If a group G acts on a set Ω we say that G acts *transitively* if, for each α, $\beta \in \Omega$ there exists some $g \in G$ such that $g(\alpha) = \beta$. More generally, we say that G acts k-*transitively* on Ω if, whenever $(\alpha_1, \ldots, \alpha_k)$ and $(\beta_1, \ldots, \beta_k)$ are k-tuples of distinct elements of Ω (that is, $\alpha_i \neq \alpha_j$ and $\beta_i \neq \beta_j$ whenever $i \neq j$), there exists some $g \in G$ such that $g(\alpha_j) = \beta_j$ for $j = 1, 2, \ldots, k$. Clearly, 1-transitivity means the same as transitivity, and k-transitivity implies $(k-1)$-transitivity for $k \geq 2$. We shall show that $PGL(2, \mathbb{C})$ acts 3- but not 4-transitively on Σ.

Theorem 2.5.1. *If z_1, z_2, z_3 are three distinct elements of Σ, then there is a unique $T \in PGL(2, \mathbb{C})$ such that $T(z_1) = 0$, $T(z_2) = 1$, and $T(z_3) = \infty$.*

Proof. If $z_1, z_2, z_3 \neq \infty$, let

$$T(z) = \frac{(z - z_1)(z_2 - z_3)}{(z_1 - z_2)(z_3 - z)}.$$

Then T sends z_1, z_2, z_3 to $0, 1, \infty$ respectively, and $T \in PGL(2, \mathbb{C})$ since T has the form $T(z) = (az + b)/(cz + d)$ with $ad - bc = (z_1 - z_2)(z_2 - z_3) \times (z_3 - z_1) \neq 0$.

If some $z_j = \infty$, we take the limit of the above transformation T as $z_j \to \infty$: if $z_1 = \infty$ we take $T(z) = -(z_2 - z_3)/(z_3 - z)$; if $z_2 = \infty$ we take $T(z) = -(z - z_1)/(z_3 - z)$; if $z_3 = \infty$ we take $T(z) = -(z - z_1)/(z_1 - z_2)$. In each case, $T \in PGL(2, \mathbb{C})$ and T sends z_1, z_2, z_3 to $0, 1, \infty$.

Finally, T is unique: for if $U \in PGL(2, \mathbb{C})$ also sends z_1, z_2, z_3 to $0, 1, \infty$, then UT^{-1} fixes $0, 1, \infty$; putting $UT^{-1}(z) = (az + b)/(cz + d)$ and solving the equations $UT^{-1}(z) = z$ ($z = 0, 1, \infty$) for a, b, c and d, we see that UT^{-1} is the identity, so that $U = T$. $\square$

Corollary 2.5.2. *If (z_1, z_2, z_3) and (w_1, w_2, w_3) are triples of distinct points in Σ, then there is a unique $T \in PGL(2, \mathbb{C})$ such that $T(z_j) = w_j$ for $j = 1, 2, 3$.*

Proof. Let $T_1, T_2 \in PGL(2, \mathbb{C})$ satisfy $T_1(z_j) = T_2(w_j) = 0, 1, \infty$ for $j = 1, 2, 3$. Then the element $T = T_2^{-1} T_1$ of $PGL(2, \mathbb{C})$ sends z_j to w_j for $j = 1, 2, 3$. If $U \in PGL(2, \mathbb{C})$ also satisfies $U(z_j) = w_j$ for $j = 1, 2, 3$, then both T_1 and $T_2 U$ send z_j to $0, 1, \infty$, so that $T_1 = T_2 U$ and hence $U = T_2^{-1} T_1 = T$. $\square$

Corollary 2.5.3. *If $T \in PGL(2, \mathbb{C})$ and T fixes three distinct points of Σ, then T is the identity.*

Proof. If T fixes z_1, z_2 and z_3, then both T and the identity $I \in PGL(2, \mathbb{C})$ send z_j to itself ($j = 1, 2, 3$), so $T = I$ by the uniqueness result in Corollary 2.5.2. $\square$

By Corollary 2.5.2, $PGL(2,\mathbb{C})$ acts 3-transitively on Σ. We shall now determine the elements of $PGL(2,\mathbb{C})$ which leave invariant a subset Δ of Σ containing three elements. They form a subgroup

$$G(\Delta) = \{T \in PGL(2,\mathbb{C}) \mid T(\Delta) = \Delta\} \leqslant PGL(2,\mathbb{C}),$$

and if $\Theta = \{0, 1, \infty\}$ then $G(\Delta) = U^{-1}G(\Theta)U$ where $U \in PGL(2,\mathbb{C})$ is chosen so that $U(\Delta) = \Theta$. Thus $G(\Delta)$ is isomorphic to $G(\Theta)$, so it is sufficient to determine the subgroup $G(\Theta)$ leaving Θ invariant.

It follows from Corollary 2.5.2 that each permutation π of Θ, there is a unique $T \in PGL(2,\mathbb{C})$ inducing the permutation π on Θ; we shall write $T = T_\pi$. Then $T_\pi \mapsto \pi$ gives an isomorphism between $G(\Theta)$ and the group of all permutations of Θ, that is, $G(\Theta) \cong S_3$. We can easily check that the following elements of $PGL(2,\mathbb{C})$ leave Θ invariant:

$$T(z) = z, 1-z, \frac{1}{z}, \frac{z}{z-1}, \frac{1}{1-z}, \frac{z-1}{z},$$

so these must be the elements of $G(\Theta)$, corresponding to the permutations

$$\pi = (0)(1)(\infty), \quad (01)(\infty), \quad (0\infty)(1), \quad (0)(1\infty), \quad (01\infty), \quad (0\infty1)$$

respectively.

By Corollary 2.5.3, $PGL(2,\mathbb{C})$ cannot be 4-transitive: for example, there is no Möbius transformation mapping $0, 1, \infty, 2$ to $0, 1, \infty, -1$ respectively. In order to determine which 4-tuples are equivalent under the action of $PGL(2,\mathbb{C})$, we make the following

Definition. If (z_0, z_1, z_2, z_3) is a 4-tuple of distinct points in Σ, the *cross-ratio*

$$\lambda = (z_0, z_1; z_2, z_3)$$

is defined to be $T(z_0)$, where T is the unique element of $PGL(2,\mathbb{C})$ satisfying $T(z_1) = 0$, $T(z_2) = 1$, $T(z_3) = \infty$.

Since $z_0 \neq z_1, z_2, z_3$, we have $\lambda \neq 0, 1, \infty$. By the proof of Theorem 2.5.1, we have $T(z) = (z-z_1)(z_2-z_3)/(z_1-z_2)(z_3-z)$, so putting $z = z_0$ we get

$$\lambda = (z_0, z_1; z_2, z_3) = \frac{(z_0-z_1)(z_2-z_3)}{(z_1-z_2)(z_3-z_0)}, \qquad (2.5.4)$$

with the usual convention of taking limits if some $z_j = \infty$.

Theorem 2.5.5. *Let* (z_0, z_1, z_2, z_3) *and* (w_0, w_1, w_2, w_3) *be 4-tuples of distinct elements of* Σ. *Then there exists some* $T \in PGL(2,\mathbb{C})$ *with* $T(z_j) = w_j$ $(j = 0, 1, 2, 3)$ *if and only if* $(z_0, z_1; z_2, z_3) = (w_0, w_1; w_2, w_3)$.

Proof. Suppose that $T(z_j) = w_j$ ($j = 0, 1, 2, 3$). Let U be the unique element of $PGL(2, \mathbb{C})$ sending w_1, w_2, w_3 to $0, 1, \infty$ respectively, so that $(w_0, w_1; w_2, w_3) = U(w_0)$. Now UT is an element of $PGL(2, \mathbb{C})$ sending z_1, z_2, z_3 to $0, 1, \infty$ respectively, and is therefore unique in doing this, so

$$(z_0, z_1; z_2, z_3) = UT(z_0) = U(w_0) = (w_0, w_1; w_2, w_3).$$

Conversely, if $(z_0, z_1; z_2, z_3) = \lambda = (w_0, w_1; w_2, w_3)$, then there exist U, $V \in PGL(2, \mathbb{C})$ with $U(w_j) = V(z_j) = \lambda, 0, 1, \infty$ for $j = 0, 1, 2, 3$. Hence $U^{-1}V(z_j) = w_j$ so take $T = U^{-1}V$. □

If z_0, z_1, z_2, z_3 are four distinct elements of Σ, then by permuting them we must expect to obtain different values for their cross-ratio. Without loss of generality, we may apply a Möbius transformation replacing these points by $\lambda, 0, 1, \infty$, where $\lambda = (z_0, z_1; z_2, z_3)$. The permutations of $\{\lambda, 0, 1, \infty\}$ form a group $G \cong S_4$, and for each $\pi \in G$ we define

$$\lambda_\pi = (\pi(\lambda), \pi(0); \pi(1), \pi(\infty)).$$

For example, if π is the identity then $\lambda_\pi = \lambda$, while if $\pi = (\lambda 0 1 \infty)$ then

$$\lambda_\pi = (0, 1; \infty, \lambda) = \lim_{z \to \infty} \frac{(0 - 1)(z - \lambda)}{(1 - z)(\lambda - 0)} = \frac{1}{\lambda}.$$

We might expect to obtain $|G| = 4! = 24$ different values of λ_π for $\pi \in G$, but some of these values coincide: for example, if $\pi' = (\lambda)(0\infty)(1)$ then

$$\lambda_{\pi'} = (\lambda, \infty; 1, 0) = \frac{1}{\lambda}$$

also, even though $(\lambda)(0\infty)(1) \neq (\lambda 0 1 \infty)$. The permutations $v = (01)(\lambda\infty)$, $(\lambda 0)(1\infty)$, $(\lambda 1)(0\infty)$, together with the identity, form a subgroup N of G, and satisfy $\lambda_\pi = \lambda_{\pi v}$ for all $\pi \in G$, $v \in N$. For example, if $v = (01)(\lambda\infty)$ then

$$\lambda_v = (\infty, 1; 0, \lambda),$$

and so

$$\lambda_{\pi v} = (\pi(\infty), \pi(1); \pi(0), \pi(\lambda))$$
$$= \frac{(\pi(\infty) - \pi(1))(\pi(0) - \pi(\lambda))}{(\pi(1) - \pi(0))(\pi(\lambda) - \pi(\infty))}$$
$$= \frac{(\pi(\lambda) - \pi(0))(\pi(1) - \pi(\infty))}{(\pi(0) - \pi(1))(\pi(\infty) - \pi(\lambda))}$$
$$= \lambda_\pi.$$

Thus $\lambda_\pi = \lambda_{\pi'}$ whenever $\pi N = \pi' N$ (as illustrated in the above example, where $\pi' = (\lambda)(0\infty)(1) = (\lambda 0 1 \infty).(01)(\lambda\infty) = \pi v$), so we obtain at most $\frac{24}{4} = 6$ different values of λ_π, one for each coset πN of N in G.

In general, these six cosets give different values for λ_π. For example, the permutations $\pi \in G$ which fix λ and permute $\Theta = \{0, 1, \infty\}$ form a subgroup $H = G(\Theta) \cong S_3$ of G, and for each $\pi \in G$ we can evaluate $\lambda_\pi = (\lambda, \pi(0); \pi(1), \pi(\infty))$ by applying a Möbius transformation T sending $\pi(0)$, $\pi(1)$, $\pi(\infty)$ to 0, 1, ∞ respectively; then $\lambda_\pi = T(\lambda)$ by definition of the cross-ratio. We have seen that there is a unique such transformation, namely $T = T_{(\pi^{-1})} = T_\pi^{-1}$, so corresponding to

$$\pi = (0)(1)(\infty), (01)(\infty), (0\infty)(1), (0)(1\infty), (01\infty), (0\infty1),$$

we have

$$\lambda_\pi = T_{(\pi^{-1})}(\lambda) = \lambda, 1 - \lambda, \frac{1}{\lambda}, \frac{\lambda}{\lambda - 1}, \frac{\lambda - 1}{\lambda}, \frac{1}{1 - \lambda}.$$

For all but finitely many λ, these values of λ_π are all distinct: for example, $\lambda \neq 1 - \lambda$ provided $\lambda \neq \frac{1}{2}$. Thus the six elements $\pi \in H$ lie in distinct cosets of N, so each coset of N contains a unique element of H.

Suppose that λ is chosen so that the six cosets πN determine distinct values of λ_π. For each $\pi \in G$ the coset πN contains a unique element $\pi' \in H$, and we have $\lambda_\pi = \lambda_{\pi'}$ since $\pi N = \pi' N$. Being a union of conjugacy classes (see §2.9 for conjugacy), N is a normal subgroup of G, that is, $\pi N = N\pi$ for all $\pi \in G$. Now if $\pi_1, \pi_2 \in G$ correspond in the above way to $\pi'_1, \pi'_2 \in H$, so that $\pi_j N = \pi'_j N$ for $j = 1, 2$, then we have

$$\pi_1 \pi_2 N = \pi_1 N \pi_2 = \pi'_1 N \pi_2 = \pi'_2 \pi_2 N = \pi'_1 \pi'_2 N.$$

Since H is a subgroup, we have $\pi'_1 \pi'_2 \in H$, so $(\pi_1 \pi_2)' = \pi'_1 \pi'_2$. Thus the function $\pi \mapsto \pi'$ is a homomorphism from G to H; the kernel is N, and the image is H since $\pi' = \pi$ for all $\pi \in H$. This shows that S_4 has a normal subgroup N with $S_4/N \cong H \cong S_3$. This behaviour of S_4 is exceptional: for each n, the symmetric group S_n has a proper normal subgroup A_n, with $S_n/A_n \cong C_2$, and there are no other proper normal subgroups, with the single exception of N in the case $n = 4$.

2.6 Cross-ratios and circles

Any three distinct points in S^2 lie on a unique plane $\Pi \in \mathbb{R}^3$, and hence any three distinct points in Σ lie on a unique circle $C = \pi(S^2 \cap \Pi)$ in Σ. Using the 3-transitivity of the action of $PGL(2, \mathbb{C})$ on Σ, we deduce

Theorem 2.6.1. *$PGL(2, \mathbb{C})$ permutes the circles in Σ transitively, that is, if C and C' are circles in Σ then there exists some $T \in PGL(2, \mathbb{C})$ such that $T(C) = C'$.*

Proof. Let z_1, z_2, z_3 be any three points on C, and w_1, w_2, w_3 any three points on C'. By Corollary 2.5.2 there exists some $T \in PGL(2, \mathbb{C})$ with $T(z_j) = w_j$ for $j = 1, 2, 3$. By Theorem 2.4.1, $T(C)$ is a circle, so both $T(C)$ and C' are circles containing w_1, w_2, w_3 and hence $T(C) = C'$. $\square$

If z_1, z_2 and z_3 are distinct, then the cross-ratio $\lambda = (z, z_1; z_2, z_3)$ is defined for all $z \neq z_j (j = 1, 2, 3)$. We can extend this definition to the cases $z = z_j$ by defining

$$\lambda = (z_j, z_1; z_2, z_3) = \lim_{z \to z_j} (z, z_1; z_2, z_3),$$

so that $\lambda = 0, 1, \infty$ for $j = 1, 2, 3$. As before, λ is invariant under Möbius transformations. We can now use cross-ratios to characterise the circles in Σ:

Theorem 2.6.2. *Let C be the circle through three distinct points z_1, z_2, z_3 in Σ. Then $C = \{z \in \Sigma \,|\, (z, z_1; z_2, z_3) \in \mathbb{R} \cup \{\infty\}\}$.*

Proof. Choose $T \in PGL(2, \mathbb{C})$ so that $T(z_j) = 0, 1, \infty$ for $j = 1, 2, 3$; then $T(z) = (z, z_1; z_2, z_3)$, and so $z \in C$ if and only if $(z, z_1; z_2, z_3) \in T(C)$. But $T(C)$ is the circle through $0, 1, \infty$, so $T(C) = \mathbb{R} \cup \{\infty\}$, as required. $\square$

We can interpret this result geometrically. Assume that $\infty \notin C$, so that C is a Euclidean circle in $\mathbb{C}$. Let θ be the angle between the vectors $z - z_1$ and $z - z_3$ in $\mathbb{C}$, and ϕ the angle between $z_2 - z_1$ and $z_2 - z_3$. Since

$$\lambda = \frac{(z - z_1)(z_2 - z_3)}{(z_1 - z_2)(z_3 - z)} = \frac{(z - z_1)}{(z - z_3)} \bigg/ \frac{(z_2 - z_1)}{(z_2 - z_3)},$$

we have

$$\arg(\lambda) = (\arg(z - z_1) - \arg(z - z_3)) - (\arg(z_2 - z_1) - \arg(z_2 - z_3))$$
$$= \theta - \phi,$$

Fig. 2.2

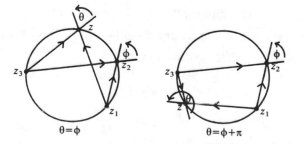

$\theta = \phi$ $\theta = \phi + \pi$

so the theorem expresses the result (familiar in Euclidean geometry) that the points z, z_1, z_2, z_3 are concyclic (lie on a circle) if and only if $\theta - \phi$ is a multiple of π, or equivalently, that a quadrilateral is cyclic if and only if opposite internal angles add up to π (Fig. 2.2 illustrates two cases).

If $\infty \in C$, then $C\backslash\{\infty\}$ is a Euclidean line in $\mathbb{C}$, and the theorem expresses a condition for the points z, z_1, z_2, z_3 to be collinear.

2.7. Inversion

Let C be a circle in Σ with equation

$$az\bar{z} + bz + \bar{b}\bar{z} + c = 0 \quad (a, c \in \mathbb{R}, b \in \mathbb{C}).$$

If $a \neq 0$, then C is a Euclidean circle in $\mathbb{C}$, and we define a transformation I_C, called *inversion* in C, as follows. Let p be the centre and r the radius of C. For each $z \in \mathbb{C}\backslash\{p\}$ there is a unique point w on the line through p and z such that

$$|z - p| \cdot |w - p| = r^2,$$

and such that z and w lie on the same side of p (see Fig. 2.3). We define $I_C(z) = w$, and we call w the *conjugate point* of z with respect to C. As $z \to p$, $w \to \infty$, and as $z \to \infty$, $w \to p$, so we extend I_C to a transformation of Σ by defining $I_C(p) = \infty$ and $I_C(\infty) = p$. Then I_C^2 is the identity, and $I_C(z) = z$ if and only if $z \in C$.

Fig. 2.3

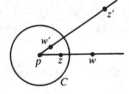

We now derive an equation for I_C. If $z \neq p, \infty$ we have

$$|(\bar{z} - \bar{p})(w - p)| = |z - p| \cdot |w - p| = r^2;$$

now $\arg(z - p) = \arg(w - p)$, so $\arg(\bar{z} - \bar{p})(w - p) = 0$, and hence

$$(\bar{z} - \bar{p})(w - p) = r^2,$$

so that

$$w = I_C(z)$$

$$= p + \frac{r^2}{(\bar{z} - \bar{p})}.$$

As shown in §2.4 we have $p = -\bar{b}/a$ and $r^2 = (b\bar{b} - ac)/a^2$, so

$$w = I_C(z) = -\frac{\bar{b}\bar{z} + c}{a\bar{z} + b}. \tag{2.7.1}$$

This equation, also valid for $z = p$ and $z = \infty$, and hence for all $z \in \Sigma$, shows that I_C is an anti-automorphism of Σ, that is, $I_C \in \overline{PGL}(2, \mathbb{C})$ (see §2.1).

If $a = 0$, then $C \backslash \{\infty\}$ is the Euclidean line

$$bz + \bar{b}\bar{z} + c = 0;$$

we can use equation (2.7.1) to define a transformation

$$w = I_C(z) = -\frac{\bar{b}\bar{z} + c}{b}$$

of Σ, and we shall show that in this case, I_C represents reflection of $\mathbb{C}$ in $C \backslash \{\infty\}$. Firstly, a point $z \in \mathbb{C}$ is fixed by I_C if and only if $z = (-\bar{b}\bar{z} - c)/b$, that is, if and only if $z \in C$. Secondly, if $z \in \mathbb{C} \backslash C$ and $q \in C \backslash \{\infty\}$, then

$$|I_C(z) - q| = |I_C(z) - I_C(q)|$$
$$= \left| -\frac{\bar{b}\bar{z} + c}{b} + \frac{\bar{b}\bar{q} + c}{b} \right|$$
$$= \left| \frac{\bar{b}}{b}(\bar{q} - \bar{z}) \right|$$
$$= |z - q|.$$

Thus z and $I_C(z)$ are equidistant from any point $q \in C \backslash \{\infty\}$, so $I_C(z)$ must be z or its mirror-image in $C \backslash \{\infty\}$. Since $z \notin C$, $I_C(z) \neq z$, so I_C reflects $\mathbb{C}$ in $C \backslash \{\infty\}$, as shown in Fig. 2.4. In particular, if $C = \mathbb{R} \cup \{\infty\}$ then I_C represents complex conjugation.

Fig. 2.4

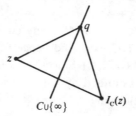

To sum up, the transformation, I_C given by (2.7.1) represents inversion in C if C is a circle in $\mathbb{C}$, and reflection in $C \backslash \{\infty\}$ if $C \backslash \{\infty\}$ is a line in $\mathbb{C}$.

By Theorem 2.4.1, if C is a circle in Σ and if $T \in PGL(2, \mathbb{C})$, then $T(C)$ is a circle C' in Σ. Under these circumstances, we have

Lemma 2.7.2. $I_{C'} = T I_C T^{-1}$.

Proof. If $z \in C'$, then $I_{C'}(z) = z$; moreover, since $T^{-1}(z) \in C$, we have $I_C T^{-1}(z) = T^{-1}(z)$. It follows that if we define $S = T I_C T^{-1} I_{C'}$, then

$$S(z) = T I_C T^{-1} I_{C'}(z) = T I_C T^{-1}(z) = T T^{-1}(z) = z$$

for all $z \in C'$, so S fixes C'. Being a composition of two automorphisms and two anti-automorphisms, S is an automorphism of Σ (by the comments at the end of §2.1), so $S \in PGL(2, \mathbb{C})$. Since S fixes C', it fixes at least three points of Σ, so S is the identity automorphism by Corollary 2.5.3, and therefore

$$T I_C T^{-1} = I_{C'}^{-1} = I_{C'}. \qquad \square$$

It follows that elements $T \in PGL(2, \mathbb{C})$ map conjugate pairs with respect to C to conjugate pairs with respect to $C' = T(C)$:

Theorem 2.7.3. *Let* $T \in PGL(2, \mathbb{C})$ *and let* C *and* $C' = T(C)$ *be circles in* Σ*. If* $w = I_C(z)$ *then* $T(w) = I_{C'}(T(z))$.

Proof. If $w = I_C(z)$ then $I_{C'}(T(z)) = T I_C T^{-1} T(z) = T I_C(z) = T(w)$. $\quad \square$

2.8 The stabilisers of a circle and a disc

If X is any subset of Σ then the Möbius transformations mapping X onto itself form a subgroup of $PGL(2, \mathbb{C})$, the *stabiliser* of X, which we denote by $G(X)$. In §2.5 we considered the cases where $|X| = 3$ or 4, and here we determine $G(X)$ where X is a circle or a disc. (A *disc* D in Σ is an open region bounded by a circle; equivalently, D is an open disc or an open half-plane in $\mathbb{C}$, or $D = E \cup \{\infty\}$ where E is the complement of a closed disc in $\mathbb{C}$.)

Let $\hat{\mathbb{R}}$ be the circle $\mathbb{R} \cup \{\infty\}$. If C is any circle in Σ, then there exists $T \in PGL(2, \mathbb{C})$ such that $T(C) = \hat{\mathbb{R}}$, and hence $G(C) = T^{-1}G(\hat{\mathbb{R}})T$, so $G(C)$ is conjugate to $G(\hat{\mathbb{R}})$ in $PGL(2, \mathbb{C})$. Since T is a homeomorphism, it maps each of the two discs D bounded by C to a disc bounded by $\hat{\mathbb{R}}$, so $T(D)$ is the upper or lower half-plane

$$\mathcal{U} = \{z \in \mathbb{C} \mid \text{Im}(z) > 0\} \quad \text{or} \quad \mathcal{L} = \{z \in \mathbb{C} \mid \text{Im}(z) < 0\}.$$

Since $J: z \mapsto 1/z$ interchanges $\mathcal{U}$ and $\mathcal{L}$, there exists $S(= T$ or $JT)$ in $PGL(2, \mathbb{C})$ mapping D onto $\mathcal{U}$, so that $G(D)$ is conjugate to $G(\mathcal{U})$ in $PGL(2, \mathbb{C})$. It is therefore sufficient to determine the stabilisers $G(\hat{\mathbb{R}})$ and $G(\mathcal{U})$.

We recall from §2.2 that $PGL(2, \mathbb{R})$ and $PSL(2, \mathbb{R})$ are the groups of transformations $z \mapsto (az + b)/(cz + d)$ $(a, b, c, d \in \mathbb{R})$ satisfying $\Delta \neq 0$ and $\Delta = 1$ respectively, where $\Delta = ad - bc$. By dividing all four coefficients by $\sqrt{|\Delta|}$, we may assume that the elements of $PGL(2, \mathbb{R})$ satisfy $\Delta = \pm 1$; those with $\Delta = 1$ form the subgroup $PSL(2, \mathbb{R})$, while those with $\Delta = -1$ (such as $J : z \mapsto 1/z$) form a coset of $PSL(2, \mathbb{R})$. Thus $PGL(2, \mathbb{R})$ is a union of two cosets of $PSL(2, \mathbb{R})$, so $PSL(2, \mathbb{R})$ is a normal subgroup of index two in $PGL(2, \mathbb{R})$.

Theorem 2.8.1.

(i) $G(\hat{\mathbb{R}}) = PGL(2, \mathbb{R})$;

(ii) $G(\mathcal{U}) = PSL(2, \mathbb{R})$.

Proof. (i) Let $T \in G(\hat{\mathbb{R}})$. Choose distinct elements $z_1, z_2, z_3 \in \mathbb{R} \backslash \{T^{-1}(\infty)\}$, and put $w_j = T(z_j)$, so $w_j \in \mathbb{R}$. Then the transformations

$$U(z) = \frac{(z - z_1)(z_2 - z_3)}{(z_1 - z_2)(z_3 - z)} \quad \text{and} \quad V(z) = \frac{(z - w_1)(w_2 - w_3)}{(w_1 - w_2)(w_3 - z)}$$

are in $PGL(2, \mathbb{R})$; since $V^{-1}U(z_j) = w_j = T(z_j)$ for $j = 1, 2, 3$, we have $T = V^{-1}U$ by Corollary 2.5.2, so $T \in PGL(2, \mathbb{R})$. Thus $G(\hat{\mathbb{R}}) \leqslant PGL(2, \mathbb{R})$, and the reverse inclusion is obvious.

(ii) Since Möbius transformations are homeomorphisms of Σ, any element of $G(\mathcal{U})$ must leave invariant the frontier $\hat{\mathbb{R}}$ of $\mathcal{U}$, so $G(\mathcal{U}) \leqslant G(\hat{\mathbb{R}}) = PGL(2, \mathbb{R})$. Let $T \in PGL(2, \mathbb{R})$, say

$$w = T(z) = \frac{az + b}{cz + d},$$

with $a, b, c, d \in \mathbb{R}$ and $ad - bc = \pm 1$. Then

$$w = \frac{(az + b)(c\bar{z} + d)}{|cz + d|^2} = \frac{ac|z|^2 + adz + bc\bar{z} + bd}{|cz + d|^2},$$

so that

$$\text{Im}(w) = \frac{(ad - bc)}{|cz + d|^2} \text{Im}(z).$$

Hence T leaves $\mathcal{U}$ invariant if and only if $ad - bc > 0$, and this is equivalent to $ad - bc = 1$, so $G(\mathcal{U}) = PSL(2, \mathbb{R})$. $\square$

There is an alternative proof of Theorem 2.8.1(i) using inversion: one uses Theorem 2.7.3 (with $C = \hat{\mathbb{R}}$) to show that if $T \in G(\hat{\mathbb{R}})$ then $\overline{T(z)} = T(\bar{z})$ for all z, and this implies that the coefficients of T are real.

2.9 Conjugacy classes in $PGL(2, \mathbb{C})$

If G is any group, then elements $g, h \in G$ are *conjugate* in G if there exists $a \in G$ such that $g = aha^{-1}$. Conjugacy is an equivalence relation on G, and the equivalence classes are called *conjugacy classes*. In many cases it is important to determine these classes: for example, if G occurs geometrically (as does $PGL(2, \mathbb{C})$) then conjugate elements have similar geometric properties, such as numbers of fixed-points. We say that $z \in \Sigma$ is a *fixed-point* of some $T \in PGL(2, \mathbb{C})$ if $T(z) = z$; then $U(z)$ is a fixed-point of the conjugate transformation $UTU^{-1} \in PGL(2, \mathbb{C})$.

To consider fixed-points and conjugacy classes of elements of $PGL(2, \mathbb{C})$, it is more convenient to work with $PSL(2, \mathbb{C})$: the two groups are identical, but we now write Möbius transformations in the form $T(z) = (az + b)/(cz + d)$ with $ad - bc = 1$, not simply $ad - bc \neq 0$.

We have $T(\infty) = a/c$, so T fixes ∞ if and only if $c = 0$. If $c \neq 0$, then $z \in \mathbb{C}$ is a fixed-point of T if and only if

$$cz^2 + (d - a)z - b = 0;$$

this equation has two roots z, and so T has two fixed-points in Σ, unless

$$(d - a)^2 + 4bc = 0,$$

in which case the equation has a double root and T has a unique fixed-point. Using $ad - bc = 1$, this condition becomes

$$(a + d)^2 - 4 = 0,$$

so T has a single fixed-point if and only if $(a + d)^2 = 4$.

If $c = 0$ then T fixes ∞; we have $ad = 1$ and $T(z) = a^2 z + ab$, so there is a second fixed-point $z = ab/(1 - a^2) \neq \infty$ if and only if $a^2 \neq 1$, or equivalently $(a + d)^2 \neq 4$. When $a^2 = 1$ we have $T(z) = z \pm b$, so that T is the identity for $b = 0$, and T has a unique fixed-point (at ∞) for $b \neq 0$. Thus we have

Theorem 2.9.1.   *Let* $T(z) = (az + b)/(cz + d)$,   *with* $ad - bc = 1$.   *If* $(a + d)^2 \neq 4$, *then T has two fixed-points in Σ. If $(a + d)^2 = 4$ and $T \neq I$, then T has one fixed-point in Σ.* $\square$

(Note that this gives an alternative proof of Corollary 2.5.3, that if T fixes more than two points then $T = I$.)

We can use the function $(a + d)^2$, considered above, to determine the conjugacy classes in $PSL(2, \mathbb{C})$. If

$$A = \begin{pmatrix} a & b \\ c & d \end{pmatrix} \in GL(2, \mathbb{C}),$$

we define the *trace* of A by

$$\mathrm{tr}\,(A) = a + d$$

By direct calculation, we observe that $\mathrm{tr}\,(AB) = \mathrm{tr}\,(BA)$, so that if B is invertible then

$$\mathrm{tr}\,(BAB^{-1}) = \mathrm{tr}\,(B^{-1}.BA) = \mathrm{tr}\,(A).$$

Thus $\mathrm{tr}\,(A)$ depends only on the conjugacy class of an element $A \in GL(2, \mathbb{C})$. Now each Möbius transformation T is represented by a pair $\pm A$ of matrices in $SL(2, \mathbb{C})$, and we have $\mathrm{tr}\,(-A) = -\mathrm{tr}\,(A)$, so that

$$\mathrm{tr}^2\,(T) = (\mathrm{tr}\,(A))^2 = (a + d)^2$$

is a well-defined function of T, depending only on the conjugacy class of T in $PSL(2, \mathbb{C})$.

Example.   Let $U(z) = \lambda z$, $\lambda \in \mathbb{C} \backslash \{0\}$. In $SL(2, \mathbb{C})$, U is represented by the matrices

$$\pm \begin{pmatrix} \sqrt{\lambda} & 0 \\ 0 & 1/\sqrt{\lambda} \end{pmatrix},$$

and so $\mathrm{tr}^2\,(U) = (\sqrt{\lambda} + 1/\sqrt{\lambda})^2 = \lambda + \lambda^{-1} + 2$.

We now describe the conjugacy classes in $PSL(2, \mathbb{C})$. In any group, the identity element forms a conjugacy class; the remaining conjugacy classes of $PSL(2, \mathbb{C})$ are described by selecting a representative from each class. If $\lambda \in \mathbb{C} \backslash \{0\}$ we define

$$U_\lambda(z) = \begin{cases} \lambda z & \text{if } \lambda \neq 1, \\ z + 1 & \text{if } \lambda = 1. \end{cases}$$

Theorem 2.9.2.   *If T is a non-identity element of $PSL(2, \mathbb{C})$, then there exists some $\lambda \in \mathbb{C} \backslash \{0\}$ such that T is conjugate to U_λ.*

Proof.   Suppose that T has just one fixed-point z_0. By Theorem 2.5.1, there exists some $S \in PSL(2, \mathbb{C})$ such that $S(z_0) = \infty$. Then STS^{-1} fixes only ∞, and so $STS^{-1}(z) = z + t$ for some $t \in \mathbb{C} \backslash \{0\}$, that is, $STS^{-1} = T_t$ in the notation of §2.3. Let $V(z) = z/t$. Then $VT_t V^{-1}(z) = z + 1$, and hence

$$(VS)T(VS)^{-1} = U_1,$$

so that T is conjugate to U_1.

Now suppose that T has two fixed-points z_1, z_2. By Theorem 2.5.1, there exists some $W \in PSL(2, \mathbb{C})$ such that $W(z_1) = 0$ and $W(z_2) = \infty$. Then WTW^{-1} fixes 0 and ∞, and hence it is easily seen that

$$WTW^{-1} = U_\lambda$$

for some $\lambda \in \mathbb{C} \backslash \{0, 1\}$.   □

To describe the conjugacy classes completely, we have to determine when U_κ and U_λ are conjugate.

Theorem 2.9.3. *U_κ is conjugate to U_λ if and only if $\kappa = \lambda$ or $\kappa = \lambda^{-1}$.*

Proof. We deal with U_1 first. Since U_1 fixes only ∞, SU_1S^{-1} fixes only $S(\infty)$ for each $S \in PSL(2, \mathbb{C})$. Thus U_1 cannot be conjugate to any $U_\lambda(\lambda \neq 1)$, since these elements fix 0 and ∞.

Now suppose that U_κ and U_λ are conjugate, where κ, $\lambda \neq 1$. Then $\text{tr}^2(U_\kappa) = \text{tr}^2(U_\lambda)$ and hence

$$\kappa + \frac{1}{\kappa} + 2 = \lambda + \frac{1}{\lambda} + 2,$$

giving $\kappa = \lambda$ or $\kappa = 1/\lambda$. Conversely, U_κ is conjugate to $U_{1/\kappa}$, for if $J(z) = 1/z$ then

$$JU_\kappa J^{-1} = U_{1/\kappa}. \qquad \square$$

Corollary 2.9.4. *Two non-identity elements T_1, T_2 of $PSL(2, \mathbb{C})$ are conjugate if and only if $\text{tr}^2(T_1) = \text{tr}^2(T_2)$.*

Proof. We need show only that $\text{tr}^2(T_1) = \text{tr}^2(T_2)$ implies that T_1 and T_2 are conjugate. Let T_1 and T_2 be conjugate to U_κ and U_λ respectively; then $\text{tr}^2(T_1) = \text{tr}^2(T_2)$ implies that $\text{tr}^2(U_\kappa) = \text{tr}^2(U_\lambda)$, so that $\kappa = \lambda$ or $\kappa\lambda = 1$, and hence U_κ and U_λ are conjugate, so T_1 and T_2 are conjugate. $\square$

2.10 Geometric classification of Möbius transformations

We now consider the geometric behaviour of a non-identity Möbius transformation T, and in particular the limiting behaviour of T^n as $n \to \infty$. We saw in §2.9 that T is conjugate in $PSL(2, \mathbb{C})$ to some $U_\lambda(\lambda \neq 0)$. Now T does not determine λ uniquely since T is also conjugate to $U_{1/\lambda}$; we call the pair $\{\lambda, 1/\lambda\}$, which *is* uniquely determined by T, the *multiplier* of T. It follows from §2.9 that two non-identity Möbius transformations are conjugate if and only if they have the same multiplier, so the multiplier is just as effective as the function tr^2 for determining conjugacy. The connection between these two invariants is that

$$\text{tr}^2(T) = \text{tr}^2(U_\lambda) = \lambda + \frac{1}{\lambda} + 2,$$

so that λ and $1/\lambda$ are the roots of the quadratic equation

$$z^2 + (2 - \text{tr}^2(T))z + 1 = 0.$$

We have seen that T has a unique fixed-point $z_0 \in \Sigma$ if and only if $\mathrm{tr}^2(T) = 4$, or equivalently, $\lambda = 1$; we call such transformations T *parabolic*. In this case, we have $T = V^{-1} U_1 V$ for some $V \in PSL(2, \mathbb{C})$ satisfying $V(z_0) = \infty$. Now

$$\lim_{n \to \infty} U_1^n(z) = \lim_{n \to \infty} (z + n) = \infty$$

for all $z \in \Sigma$, and hence

$$\lim_{n \to \infty} T^n(z) = \lim_{n \to \infty} V^{-1} U_1^n V(z) = V^{-1}(\infty) = z_0$$

for all $z \in \Sigma$. Thus each $z \in \Sigma$ is eventually moved by T^n towards the fixed-point z_0, as n increases.

If T is not parabolic, then T has two fixed-points z_1 and z_2, so let $V(z) = (z - z_1)/(z - z_2)$, mapping these fixed-points to 0 and ∞. Then $VTV^{-1} = U_\lambda$, fixing 0 and ∞, for some $\lambda \neq 0, 1$. We have $U_\lambda^n(z) = \lambda^n z$, so if $|\lambda| < 1$ then $\lim_{n \to \infty} U_\lambda^n(z) = 0$ for all $z \neq \infty$, and hence $\lim_{n \to \infty} T^n(z) = z_1$ for all $z \neq z_2$; similarly, if $|\lambda| > 1$ then $\lim_{n \to \infty} T^n(z) = z_2$ for all $z \neq z_1$. (Of course, these two cases are essentially the same, since transposing z_1 and z_2 corresponds to replacing λ by $1/\lambda$.) Thus if $|\lambda| \neq 1$ then T progressively moves all points $z \neq z_1, z_2$ away from one of these fixed-points and towards the other; we call such a transformation T *hyperbolic* if λ is real and positive, and *loxodromic* otherwise. If $|\lambda| = 1$ then U_λ is a rotation R_θ of Σ, using the notation of §2.3, with $\lambda = e^{i\theta}$, so $U_\lambda^n(z)$ has no limit for $z \neq 0$, ∞, and hence neither has $T^n(z)$ for $z \neq z_1, z_2$; we call T *elliptic* in this case.

Now T is conjugate to U_λ if and only if $\mathrm{tr}^2(T) = \mathrm{tr}^2(U_\lambda)$, and we have $\mathrm{tr}^2(U_\lambda) = \lambda + \lambda^{-1} + 2$, so:

T is elliptic if and only if $0 \leqslant \mathrm{tr}^2(T) < 4$;

T is parabolic if and only if $\mathrm{tr}^2(T) = 4$;

T is hyperbolic if and only if $\mathrm{tr}^2(T) > 4$;

T is loxodromic if and only if $\mathrm{tr}^2(T) < 0$ or $\mathrm{tr}^2(T) \notin \mathbb{R}$.

The distinction between hyperbolic and loxodromic transformations becomes clear when we consider discs in Σ: a non-identity transformation T leaves a disc invariant if and only if it is conjugate to an element of $PSL(2, \mathbb{R})$, by Theorem 2.8.1(ii). Elements of $PSL(2, \mathbb{R})$ (and hence their conjugates) must have real traces, and therefore cannot be loxodromic, whereas every hyperbolic element is conjugate to some $U_\lambda \in PSL(2, \mathbb{R})$, $\lambda > 0$. Thus hyperbolic elements leave a disc invariant, loxodromic elements do not. Similarly, all elliptic and parabolic elements leave a disc invariant.

The *period* (or *order*) of a transformation T is the least positive integer

m such that $T^m = I$, provided such an integer exists; if there is no such integer m, we say that T has *infinite period*.

Theorem 2.10.1. *If T is a non-identity element of $PSL(2, \mathbb{C})$ with finite period, then T is elliptic.*

Proof. T is conjugate to some U_λ, so T^n is conjugate to U_λ^n for each integer n, and hence U_λ has finite period. Now $U_1^n(z) = z + n$, so U_1 has infinite period, and hence $\lambda \neq 1$. Thus $U_\lambda^n(z) = \lambda^n z$, so putting $U_\lambda^m = I$ (where m is the period of T and hence of U_λ), we get $\lambda^m = 1$ and therefore $|\lambda| = 1$. Thus T is elliptic. $\square$

However, not every elliptic transformation has finite period: putting $\lambda = e^{i\theta}$, with $\theta \in \mathbb{R}$, we see that U_λ is elliptic if and only if θ is not an integer multiple of 2π, and that U_λ has finite period if and only if θ is a rational multiple of 2π, so if θ is an irrational multiple of 2π then U_λ is an elliptic element of infinite period.

2.11 Conformality

If S_1 and S_2 are surfaces, then we say that a map $f : S_1 \to S_2$ is *conformal* if it preserves angles, that is, whenever curves c_1 and c_2 on S_1 meet at a point Q with an angle θ, then $f(c_1)$ and $f(c_2)$ meet at $f(Q)$ with the same

Fig. 2.5

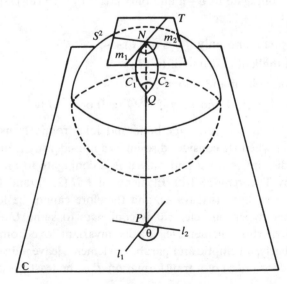

angle θ (the angle between two curves is defined to be the angle between their tangents). A map f is an *isometry* if it preserves the distance between each pair of points; for example, rotations and reflections of Euclidean space $\mathbb{R}^n$ are isometries. Since angles in $\mathbb{R}^n$ can be expressed in terms of distances, isometries of $\mathbb{R}^n$ induce conformal maps between surfaces in $\mathbb{R}^n$.

Theorem 2.11.1.   *Stereographic projection $\pi^{-1}:\mathbb{C} \to S^2\backslash\{N\}$ is conformal.*

Proof.   Let $P \in \mathbb{C}$, and let l_1 and l_2 be straight lines in $\mathbb{C}$, meeting with an angle θ at P (see Fig. 2.5). If Π_j is the plane through N and $l_j (j = 1, 2)$, then $S^2 \cap \dot{\Pi}_j$ is a circle C_j in S^2. For each point R on l_j, the line NR meets S^2 at $\pi^{-1}(R) \in S^2 \cap \Pi_j$, so C_j is the projection $\pi^{-1}(l_j \cup \{\infty\})$ of the circle $l_j \cup \{\infty\}$ in Σ; in particular, C_1 and C_2 meet at N and at $Q = \pi^{-1}(P)$. The tangent plane T to S^2 at N is parallel to the equatorial plane $\mathbb{C}$, so the lines $m_j = T \cap \Pi_j$, which meet at N, are parallel to the lines $l_j = \mathbb{C} \cap \Pi_j$, and hence they make an angle θ at N. Now the circles C_j meet at the same angle at their two points of intersection N and Q (since C_1 and C_2 are mapped to themselves by reflection in the plane perpendicularly bisecting the segment NQ); since m_j is the tangent to C_j at N, this angle must be θ, so $\pi^{-1}(l_1)$ and $\pi^{-1}(l_2)$ meet with an angle θ at $Q = \pi^{-1}(P)$, as required. $\square$

Corollary 2.11.2.   *Stereographic projection $\pi:S^2\backslash\{N\} \to \mathbb{C}$ is conformal.*

Proof.   If curves c_1, c_2 on S^2 meet at $Q \neq N$ with an angle θ, then (by definition) so do their tangents t_j at $Q(j = 1, 2)$. Let Π_j be the plane through N and t_j, so that $S^2 \cap \Pi_j$ is a circle C_j in S^2 passing through N and Q, and $\mathbb{C} \cap \Pi_j$ is a line l_j in $\mathbb{C}$ passing through $P = \pi(Q)$. Now C_1 and C_2 make an angle θ at Q (since their tangents t_j do), so the argument used in the proof of Theorem 2.11.1 shows that l_1 and l_2 make an angle θ at P. Since l_j is the tangent to $\pi(C_j)$ at P, this shows that π is conformal. $\square$

We define the angle at ∞ between two curves c_1, c_2 in $\mathbb{C}$ to be the angle at N between the curves $\pi^{-1}(c_1)$ and $\pi^{-1}(c_2)$, provided this angle exists; it immediately follows that the stereographic projections $\pi:S^2 \to \Sigma$ and $\pi^{-1}:\Sigma \to S^2$ are conformal. Since $J:z \mapsto 1/z$ induces a rotation $\pi^{-1}J\pi$ of S^2, sending an angle θ at N to an angle θ at S, the angle between c_1 and c_2 at ∞ is equal to the angle between $J(c_1)$ and $J(c_2)$ at 0, so a function f on Σ is conformal at ∞ if and only if $f \circ J$ is conformal at 0.

A conformal map from an oriented surface to itself is *directly* or *indirectly* conformal if it preserves or reverses the orientation; for example, rotations and reflections of S^2 are directly and indirectly conformal, respectively. Since $\pi:S^2 \to \Sigma$ and $\pi^{-1}:\Sigma \to S^2$ are conformal, it follows that a function $f:\Sigma \to \Sigma$ is directly (indirectly) conformal if and only if the induced map $\pi^{-1}f\pi:S^2 \to S^2$, shown in Fig. 2.6, is directly (indirectly) conformal. The next two results show that these functions f are the automorphisms (anti-automorphisms) of Σ.

Fig. 2.6

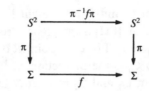

Theorem 2.11.3.   *Each automorphism (anti-automorphism) T of Σ is a directly (indirectly) conformal homeomorphism of Σ onto itself.*

Proof.   By §2.1, T is a homeomorphism of Σ onto itself.

If $T \in \text{Aut}(\Sigma)$, then $T \in PSL(2, \mathbb{C})$, so putting $T(z) = (az + b)/(cz + d)$ with $ad - bc = 1$, we have the derivative

$$T'(z) = \frac{1}{(cz + d)^2}.$$

Thus $T'(z) \neq 0, \infty$ for each $z \in \mathbb{C} \setminus \{ -d/c \}$, so by Theorem A.12, T is directly conformal at z. The remaining cases are:

(i) $z = \infty$ and $T(z) \neq \infty$;
(ii) $z = \infty$ and $T(z) = \infty$;
(iii) $z = -d/c \neq \infty$ and $T(z) = \infty$.

In case (i) we have $c \neq 0$. The transformation

$$U = T \circ J : z \mapsto \frac{a + bz}{c + dz}$$

satisfies $U'(0) = -1/c^2 \neq 0, \infty$, so U is directly conformal at 0, and hence T is directly conformal at ∞.

In case (ii) we have $c = 0$ and hence $a \neq 0$. The transformation

$$V = J \circ T \circ J : z \mapsto \frac{c + dz}{a + bz}$$

satisfies $V'(0) = 1/a^2 \neq 0, \infty$, so V is directly conformal at 0, and hence so is T at ∞.

In case (iii) we have $c \neq 0$. The transformation

$$W = J \circ T : z \mapsto \frac{cz + d}{az + b}$$

satisfies $W'(-d/c) = -c^2 \neq 0, \infty$, so W is directly conformal at $-d/c$, and hence so is T.

Thus T is directly conformal on Σ. Each *anti*-automorphism T of Σ has the form $z \mapsto T_+(\bar{z})$, where $T_+ \in \text{Aut}(\Sigma)$. Since complex conjugation is an indirectly conformal homeomorphism of Σ onto itself (corresponding to reflection of S^2 in the plane $x_2 = 0$), and since T_+ is directly conformal, it follows that T is an indirectly conformal homeomorphism of Σ onto itself. $\square$

Theorem 2.11.4 is a little stronger than the converse of Theorem 2.11.3, since we do not require f to be a bijection:

Theorem 2.11.4. *Each directly (indirectly) conformal map f from Σ to itself is an automorphism (anti-automorphism) of Σ.*

Proof. Let $f : \Sigma \to \Sigma$ be directly conformal. We may assume that $f(\infty) = \infty$, by composing f with a suitable automorphism. We now show that f is meromorphic at each point $a \in \Sigma$.

(i) If $a \in \mathbb{C}$ and $f(a) \in \mathbb{C}$, then since f is directly conformal at a there exists $f'(a) \neq 0, \infty$ by Theorem A.12, so f is analytic at a, and a is a simple point for f.

(ii) If $a \in \mathbb{C}$ and $f(a) = \infty$, then since the rotation J is directly conformal, $J \circ f$ is directly conformal; since $(J \circ f)(a) = 0$, the previous argument shows that $J \circ f$ is analytic with a simple zero at a, so f is meromorphic with a simple pole at a.

(iii) Since $f(\infty) = \infty$ we have $(J \circ f \circ J)(0) = 0$, and a similar argument shows that f is meromorphic with a simple pole at ∞.

Thus f is meromorphic on Σ, so f' is meromorphic and hence rational, by Theorem 1.4.1. In case (i) we have $f'(a) \neq 0$, and in (ii) f' has a (double) pole at a, so $f'(a) = \infty \neq 0$. In (iii),

$$f(z) = a_1 z + a_0 + a_{-1} z^{-1} + \dots$$

for large z, with $a_1 \neq 0$, so

$$f'(z) = a_1 - a_{-1} z^{-2} - \dots$$

for large z, and hence $f'(\infty) = a_1 \neq 0$. Thus f' is a rational function which does not take the value 0, so f' is a non-zero constant (either by Theorem 1.4.2, or because $1/f'$ is analytic and bounded, and hence constant). Hence f is a polynomial of degree 1, so $f \in \text{Aut}(\Sigma)$ by Theorem 2.1.1.

If f is indirectly conformal, then the map $z \mapsto f(\bar{z})$ is directly conformal and therefore an automorphism, so f is an anti-automorphism. $\square$

2.12 Rotations of Σ

We define a transformation $T:\Sigma \to \Sigma$ to be a *rotation* of Σ if the induced transformation $\pi^{-1}T\pi:S^2 \to S^2$ is a rotation of the sphere S^2. Under composition, the rotations of Σ form a group $\text{Rot}(\Sigma)$, isomorphic to the group of rotations of S^2. We shall show that every rotation of Σ is a Möbius transformation, and then determine which Möbius transformations arise in this way.

The rotations of S^2 are directly conformal, and hence so are the rotations of Σ, so Theorem 2.11.4 implies that $\text{Rot}(\Sigma)$ is a subgroup of $\text{Aut}(\Sigma)$.

If $Q = (x_1, x_2, x_3) \in S^2$, then the antipodal point of Q is the point $\tilde{Q} = (-x_1, -x_2, -x_3) \in S^2$; if $z \in \Sigma$ then we define the *antipodal point* of z to be $\tilde{z} = \pi(\tilde{Q})$, where $Q = \pi^{-1}(z)$. It follows easily from equation 1.1.1 that $\tilde{z} = -1/\bar{z}$, so the *antipodal map* $z \mapsto \tilde{z}$ is an anti-automorphism of Σ.

If $T \in \text{Rot}(\Sigma)$, then T maps each antipodal pair $z, \tilde{z}$ to an antipodal pair $T(z)$, $\widetilde{T(z)}$, so $T(-1/\bar{z}) = -1/\overline{T(z)}$ for all $z \in \Sigma$. Putting $T(z) = (az + b)/(cz + d)$, with $ad - bc = 1$, this implies that

$$\frac{b\bar{z} - a}{d\bar{z} - c} = \frac{-\bar{c}\bar{z} - \bar{d}}{\bar{a}\bar{z} + \bar{b}}$$

for all $z \in \Sigma$, and hence

$$\begin{pmatrix} b & -a \\ d & -c \end{pmatrix} = \lambda \begin{pmatrix} -\bar{c} & -\bar{d} \\ \bar{a} & \bar{b} \end{pmatrix}$$

for some $\lambda \in \mathbb{C} \setminus \{0\}$. Taking determinants we have $\lambda^2 = 1$, so either $b = -\bar{c}$ and $a = \bar{d}$, or else $b = \bar{c}$ and $a = -\bar{d}$. However, in this second case we have

$$1 = ad - bc = -a\bar{a} - b\bar{b} = -|a|^2 - |b|^2,$$

which is impossible. Hence

$$T(z) = \frac{az + b}{-\bar{b}z + \bar{a}}, \quad a\bar{a} + b\bar{b} = 1. \tag{2.12.1}$$

It is easily verified that the transformations of type 2.12.1 form a subgroup of $PSL(2, \mathbb{C})$. For reasons we shall explain shortly, this subgroup is denoted by $PSU(2, \mathbb{C})$. We have shown that $\text{Rot}(\Sigma) \leqslant PSU(2, \mathbb{C})$ and now we shall show that these two groups are equal.

First observe that if a transformation T of type 2.12.1 fixes 0, then $b = 0$ and so $a\bar{a} = 1$. Thus $T(z) = a^2 z$ with $|a^2| = 1$, so T is a rotation R_θ of Σ (in the notation of §2.3), with $a^2 = e^{i\theta}$. Now let T be *any* transformation of type (2.12.1), and let $T(0) = z_0$. Since the rotation group of S^2 acts transitively on S^2, there is a rotation R of Σ satisfying $R(z_0) = 0$, so RT is an element of $PSU(2, \mathbb{C})$ fixing 0. By the above argument, $RT = R_\theta \in \mathrm{Rot}(\Sigma)$ for some θ, so $T = R^{-1}R_\theta \in \mathrm{Rot}(\Sigma)$. Thus we have proved

Theorem 2.12.2. $\mathrm{Rot}(\Sigma) = \{T : z \mapsto (az + b)/(-\bar{b}z + \bar{a}) \mid a\bar{a} + b\bar{b} = 1\} = PSU(2, \mathbb{C})$. □

Each rotation of S^2 extends to a unique rotation of $\mathbb{R}^3$ about the origin, and conversely each such rotation of $\mathbb{R}^3$ induces a unique rotation of S^2. These rotations of $\mathbb{R}^3$ are the linear transformations represented by 3×3 real orthogonal matrices of determinant 1 (using the standard basis of $\mathbb{R}^3$), and such matrices form a group $SO(3, \mathbb{R})$, the *special orthogonal group*, isomorphic to the rotation group of S^2. We therefore have

Corollary 2.12.3. $PSU(2, C) \cong SO(3, \mathbb{R})$. □

$PSU(2, \mathbb{C})$ arises algebraically in the following way. A matrix $A \in GL(n, \mathbb{C})$ is called *unitary* if

$$A\bar{A}^t = I,$$

where $\bar{A}^t$ denotes the transpose of the complex conjugate of A. Now $\det(\bar{A}^t) = \overline{\det(A)}$, so if A is unitary then $|\det(A)| = 1$; thus A is invertible, and one easily sees that the unitary matrices in $GL(n, \mathbb{C})$ form a group under multiplication. We call this the *unitary group* $U(n, \mathbb{C})$; the *special unitary group* $SU(n, \mathbb{C})$ consists of those unitary matrices of determinant 1. If $A = \begin{pmatrix} a & b \\ c & d \end{pmatrix} \in SU(2, \mathbb{C})$, then

$$\begin{pmatrix} d & -b \\ -c & a \end{pmatrix} = A^{-1} = \bar{A}^t = \begin{pmatrix} \bar{a} & \bar{c} \\ \bar{b} & \bar{d} \end{pmatrix},$$

so $A = \begin{pmatrix} a & b \\ -\bar{b} & \bar{a} \end{pmatrix}$ with $a\bar{a} + b\bar{b} = \det(A) = 1$; conversely, every such matrix A is in $SU(2, \mathbb{C})$. The scalar matrices $\lambda I \in SU(2, \mathbb{C})$ are just $\pm I$, so the group of Möbius transformations of Σ induced by $SU(2, \mathbb{C})$ is $SU(2, \mathbb{C})/\{\pm I\}$, the *projective special unitary group* $PSU(2, \mathbb{C})$.

The matrices $\begin{pmatrix} a & b \\ -\bar{b} & \bar{a} \end{pmatrix}$, with $a, b \in \mathbb{C}$, form a 4-dimensional vector space Q over $\mathbb{R}$, with basis

$$\mathbf{e} = \begin{pmatrix} 1 & 0 \\ 0 & 1 \end{pmatrix}, \mathbf{i} = \begin{pmatrix} i & 0 \\ 0 & -i \end{pmatrix}, \mathbf{j} = \begin{pmatrix} 0 & 1 \\ -1 & 0 \end{pmatrix}, \mathbf{k} = \begin{pmatrix} 0 & i \\ i & 0 \end{pmatrix}.$$

These four matrices, together with their negatives, form a non-abelian group of order 8, the *quaternion group* Q_8, satisfying

$$\mathbf{i}^2 = \mathbf{j}^2 = \mathbf{k}^2 = -\mathbf{e},$$

$$\mathbf{ij} = \mathbf{k}, \mathbf{ji} = -\mathbf{k}, \text{ etc.},$$

where $\mathbf{i}, \mathbf{j}$ and $\mathbf{k}$ may be permuted cyclically in the last two equations. This shows that Q is closed under multiplication, so it is an algebra, isomorphic to the *quaternion algebra* of W.R. Hamilton (1805–65). An element $A = \alpha\mathbf{e} + \beta\mathbf{i} + \gamma\mathbf{j} + \delta\mathbf{k}$ of Q lies in $SU(2, \mathbb{C})$ if and only if $\alpha^2 + \beta^2 + \gamma^2 + \delta^2 = 1$, so $SU(2, \mathbb{C})$ is homeomorphic to the 3-sphere $S^3 \subset \mathbb{R}^4$. We obtain $PSU(2, \mathbb{C})$ from $SU(2, \mathbb{C})$ by identifying each $A \in SU(2, \mathbb{C})$ with $-A$, so $PSU(2, \mathbb{C})$ is homeomorphic to the real projective space $P_3(\mathbb{R})$ obtained by identifying antipodal points of S^3.

For further details, see P. du Val [1964], also relevant to the next section.

2.13 Finite groups of Möbius transformations

The aim of this section is to determine the finite groups of Möbius transformations of Σ. By Theorem 2.10.1, every such group consists of elliptic transformations (together with the identity), and by §2.10, each elliptic transformation is conjugate to some rotation R_θ of Σ (conversely, it is easily seen that each non-identity rotation is elliptic). The following classical result shows that if Γ is a group (finite or infinite) of elliptic transformations, then its elements are *simultaneously* conjugate to rotations, that is $V^{-1}\Gamma V \leqslant PSU(2, \mathbb{C})$ for some $V \in PSL(2, \mathbb{C})$; the proof we give is due to R.C. Lyndon and J.L. Ullman [1967]. For notational convenience, we will identify Σ with S^2 by means of stereographic projection π.

Theorem 2.13.1. *Let Γ be a subgroup of $PSL(2, \mathbb{C})$ consisting of elliptic transformations together with the identity. Then Γ is conjugate in $PSL(2, \mathbb{C})$ to a subgroup of $PSU(2, \mathbb{C})$.*

Proof. Let $\hat{\Gamma}$ be the inverse image of Γ in $SL(2, \mathbb{C})$. Then apart from $T = \pm I$, the matrices $T \in \hat{\Gamma}$ satisfy $\operatorname{tr}(T) \in \mathbb{R}$ and $|\operatorname{tr}(T)| < 2$; since conjugate matrices have the same trace, any conjugate of $\hat{\Gamma}$ in $SL(2, \mathbb{C})$ shares this property.

We may assume that $|\Gamma| > 1$, so that $\hat{\Gamma}$ contains some $T \neq \pm I$. Then

T induces an elliptic transformation of Σ. If $U \in SL(2, \mathbb{C})$ is chosen to send $\{0, \infty\}$ to the two fixed-points of T in Σ, then $S = U^{-1}TU$ fixes 0 and ∞, so $S = \begin{pmatrix} \lambda & 0 \\ 0 & \lambda^{-1} \end{pmatrix}$ for some $\lambda \in \mathbb{C} \setminus \{0\}$. Since $T \neq \pm I$, $\lambda \neq \pm 1$.

Since S is elliptic, $|\lambda| = 1$, and hence $S = \begin{pmatrix} \lambda & 0 \\ 0 & \bar{\lambda} \end{pmatrix}$. Replacing $\hat{\Gamma}$ by $U^{-1}\hat{\Gamma}U$, that is, replacing $\hat{\Gamma}$ by a conjugate subgroup of $PSL(2, \mathbb{C})$, we may assume that $S \in \hat{\Gamma}$.

From now on, let $T = \begin{pmatrix} a & b \\ c & d \end{pmatrix}$ be an arbitrary element of $\hat{\Gamma}$. First we shall show that $d = \bar{a}$. Since $S, T \in \hat{\Gamma}$ we have $\begin{pmatrix} \lambda a & \lambda b \\ \bar{\lambda} c & \bar{\lambda} d \end{pmatrix} = STE\hat{\Gamma}$. Now $a + d = \text{tr}(T) \in \mathbb{R}$, so $d - \bar{a} \in \mathbb{R}$ since $\text{Im}(\bar{a}) = -\text{Im}(a) = \text{Im}(d)$; similarly, $\bar{\lambda}(d - \bar{a}) \in \mathbb{R}$ using ST. Since $\bar{\lambda} \notin \mathbb{R}$, this implies that $d = \bar{a}$.

Next we show that $b = 0$ if and only if $c = 0$. If $c = 0$ then $T = \begin{pmatrix} a & b \\ 0 & \bar{a} \end{pmatrix}$ and so $\hat{\Gamma}$ contains

$$STS^{-1}T^{-1} = \begin{pmatrix} 1 & (\lambda^2 - 1)ab \\ 0 & 1 \end{pmatrix};$$

this has trace 2, so $(\lambda^2 - 1)ab = 0$ since Γ contains no parabolic transformations. Now $\lambda^2 \neq 1$, and $a \neq 0$ since T is invertible, so $b = 0$. The converse is similar.

Now we show that there exists $r \in \mathbb{R} \setminus \{0\}$, depending on $\hat{\Gamma}$ but not on T, such that $c = r\bar{b}$ for all $T \in \hat{\Gamma}$. If T is not a diagonal matrix, then $b \neq 0 \neq c$, so we can take $r = c/\bar{b} \neq 0$; since $a\bar{a} - rb\bar{b} = \det(T) = 1$, we have $r \in \mathbb{R}$. If

$$T_j = \begin{pmatrix} a_j & b_j \\ r_j\bar{b}_j & \bar{a}_j \end{pmatrix}$$

are non-diagonal elements of $\hat{\Gamma}(j = 1, 2)$, then $\hat{\Gamma}$ contains

$$T_1T_2 = \begin{pmatrix} a_1a_2 + r_2b_1\bar{b}_2 & * \\ * & \bar{a}_1\bar{a}_2 + r_1\bar{b}_1b_2 \end{pmatrix};$$

we have shown that the diagonal elements of this matrix must be complex conjugates, so $r_2 = \bar{r}_1 = r_1$. Thus the non-diagonal matrices in $\hat{\Gamma}$ all determine the same value of r, which we may also use for the diagonal matrices since they satisfy $b = c = 0$. (If $\hat{\Gamma}$ consists entirely of diagonal matrices, we can make an arbitrary choice of $r \in \mathbb{R} \setminus \{0\}$.)

Let $v = |r|^{1/4}$ and $V = \begin{pmatrix} v & 0 \\ 0 & v^{-1} \end{pmatrix} \in SL(2, \mathbb{C})$. Putting $b_1 = v^2b$, we have

$$VTV^{-1} = \begin{pmatrix} a & b_1 \\ v^{-4}r\overline{b}_1 & \overline{a} \end{pmatrix} = \begin{cases} \begin{pmatrix} a & b_1 \\ \overline{b}_1 & \overline{a} \end{pmatrix} & \text{if } r > 0, \\[10pt] \begin{pmatrix} a & b_1 \\ -\overline{b}_1 & \overline{a} \end{pmatrix} & \text{if } r < 0. \end{cases}$$

Call these matrices types (1) and (2) respectively. Then, replacing $\hat{\Gamma}$ by $V\hat{\Gamma}V^{-1}$, we may assume that the elements of $\hat{\Gamma}$ are all of type (1) or all of type (2), that is, $r = \pm 1$. Since

$$SU(2,\mathbb{C}) = \left\{ \begin{pmatrix} a & b \\ -\overline{b} & \overline{a} \end{pmatrix} \middle| a\overline{a} + b\overline{b} = 1 \right\},$$

it is sufficient to prove that all matrices $T \in \hat{\Gamma}$ are of type (2), so let $T = \begin{pmatrix} a & b \\ \overline{b} & \overline{a} \end{pmatrix} \in \hat{\Gamma}$, with $b \neq 0$. Since $VSV^{-1} = S$, we still have $S \in \hat{\Gamma}$, so $\hat{\Gamma}$ contains

$$W = STS^{-1}T^{-1} = \begin{pmatrix} a\overline{a} - \lambda^2 b\overline{b} & * \\ * & a\overline{a} - \overline{\lambda}^2 b\overline{b} \end{pmatrix}.$$

Now

$$\tfrac{1}{2}\mathrm{tr}(W) = \mathrm{Re}(a\overline{a} - \lambda^2 b\overline{b}) = |a|^2 - |b|^2 \mathrm{Re}(\lambda^2) = 1 + |b|^2(1 - \mathrm{Re}(\lambda^2)),$$

using the fact that $|a|^2 - |b|^2 = \det(T) = 1$. Since $|\lambda| = 1$ and $\lambda^2 \neq 1$, we have $\mathrm{Re}(\lambda^2) < 1$ and hence $\mathrm{tr}(W) > 2$. Thus W induces a Möbius transformation which is neither the identity nor elliptic. This contradicts our hypothesis on Γ, so the theorem is proved. $\square$

Corollary 2.13.2. *Every finite group of Möbius transformations is conjugate to a group of rotations of Σ.*

Proof. By Theorem 2.10.1, every such group consists of elliptic transformations, together with the identity, so the result follows from Theorem 2.13.1. $\square$

We have shown that every finite subgroup of $PSL(2,\mathbb{C})$ is conjugate to a subgroup of $PSU(2,\mathbb{C})$, the rotation group, and our aim is now to determine the finite groups of rotations. First we need to determine the finite groups of rotations which have an orbit of length 1 or 2 (the *orbit* of a point z under a group Γ is the set of all images $g(z)$, $g \in \Gamma$; the *length* is the number of points in the orbit).

Lemma 2.13.3. *Let Γ be a finite group of rotations of Σ with a fixed-point $z_0 \in \Sigma$; then Γ is conjugate within $PSU(2,\mathbb{C})$ to the subgroup generated by some rotation $R_\phi : z \mapsto e^{i\phi}z$ $(0 \leqslant \phi < 2\pi)$, and in particular, Γ is cyclic.*

Proof. $PSU(2, \mathbb{C})$ acts transitively on Σ, so there is some $U \in PSU(2, \mathbb{C})$ with $U(z_0) = 0$, as illustrated in Fig. 2.7. Then $\tilde{\Gamma} = U\Gamma U^{-1}$ is a finite group of rotations fixing 0, so $\tilde{\Gamma}$ also fixes the antipodal point ∞, and $\tilde{\Gamma}$ consists of transformations

$$R_\theta : z \mapsto e^{i\theta} z$$

for various θ, $0 \leqslant \theta < 2\pi$.

Fig. 2.7

Let $\phi = \min \{\theta > 0 \,|\, R_\theta \in \tilde{\Gamma}\}$. Then $\tilde{\Gamma}$ contains R_ϕ, and therefore contains all powers $R_\phi^n = R_{n\phi}, n \in \mathbb{Z}$. Suppose that $\tilde{\Gamma}$ contains some R_θ where θ is not a multiple of ϕ, as in Fig. 2.8. Then $m\phi < \theta < (m+1)\phi$ for some $m \in \mathbb{Z}$, and since $\tilde{\Gamma}$ contains both R_θ and $R_{m\phi}$, $\tilde{\Gamma}$ contains $R_\theta R_{m\phi}^{-1} = R_{\theta - m\phi}$, with $0 < \theta - m\phi < \phi$, contradicting the minimality of ϕ. Hence $\tilde{\Gamma}$ consists of the powers of R_ϕ, so $\tilde{\Gamma}$ is cyclic. Since $\Gamma \cong \tilde{\Gamma}$, Γ is also cyclic. $\quad\square$

Fig. 2.8

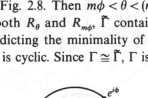

(In fact, if $|\Gamma| = n$, then $\phi = 2\pi/n$.)

For any integer $n \geqslant 3$, we define the dihedral group D_n to be the symmetry group of a regular polygon with n sides; D_n consists of n rotations (including the identity) and n reflections, so $|D_n| = 2n$ (hence this group is sometimes denoted by D_{2n}, especially by group-theorists). D_n is generated by a rotation a (though $2\pi/n$) and a reflection b, satisfying

$$a^n = b^2 = 1, \quad ba = a^{-1}b.$$

For $n = 1$ and 2, these generators and relations also define a group D_n:

for $n = 1$, D_1 is a cyclic group of order 2, while for $n = 2$, D_2 is a non-cyclic group of order 4. By a *dihedral group*, we shall mean any group isomorphic to D_n for some $n \geqslant 1$.

Lemma 2.13.4. *Let Δ be a finite group of rotations of Σ with an orbit of length 2 in Σ; then Δ is a dihedral group.*

Proof. Let $\Omega = \{z_0, z_1\}$ be the orbit of length 2. Since Δ permutes Ω transitively, Δ has a subgroup Γ of index 2 fixing z_0 and z_1; by Lemma 2.13.3, Γ is cyclic.

If $|\Gamma| = 1$ then $|\Delta| = 2$, so Δ is a dihedral group, being isomorphic to D_1. We therefore assume that $|\Gamma| > 1$, so that z_0 and z_1, being the only fixed-points of Γ, must be antipodal. We now choose $U \in PSU(2, \mathbb{C})$ so that $U(z_0) = 0$ (and hence $U(z_1) = \infty$), and we define $\tilde{\Delta} = U\Delta U^{-1}$, $\tilde{\Gamma} = U\Gamma U^{-1}$. By Lemma 2.13.3, $\tilde{\Gamma}$ is generated by some R_ϕ. Now $\tilde{\Delta}$ permutes $\{0, \infty\}$ transitively, so there is some $T \in \tilde{\Delta} \setminus \tilde{\Gamma}$ transposing 0 and ∞, as illustrated in Fig. 2.9. Such a rotation T must be through an angle π, about an axis passing through the equatorial circle $|z| = 1$, so

$$T^2 = I, \, T R_\phi = R_{-\phi} T = R_\phi^{-1} T.$$

Now R_ϕ generates $\tilde{\Gamma}$, and $\tilde{\Delta} \setminus \tilde{\Gamma} = T\tilde{\Gamma}$, so R_ϕ and T generate $\tilde{\Delta}$. Thus $\tilde{\Delta}$ is a dihedral group, and hence so is the conjugate group Δ. $\square$

Fig. 2.9

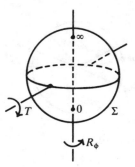

Theorem 2.13.5. *Let Γ be a finite group of rotations of Σ. Then one of the following holds:*

(i) *Γ is cyclic,*

(ii) *Γ is dihedral,*

(iii) *Γ is the rotation group of a regular tetrahedron, octahedron, or icosahedron inscribed in Σ.*

Proof. Let $|\Gamma| = N$, and let

$$\Phi = \{(z, T) \mid z \in \Sigma, T \in \Gamma, T \neq I, \text{ and } T(z) = z\}.$$

There are $N - 1$ elements $T \in \Gamma \setminus \{I\}$, each having two fixed-points, so

$$|\Phi| = 2(N - 1).$$

For each $z \in \Sigma$, let $N_z = |\Gamma_z|$, where $\Gamma_z = \{T \in \Gamma \mid T(z) = z\}$ is the subgroup of Γ fixing z. Then

$$|\Phi| = \sum_z (N_z - 1),$$

this sum being finite since $N_z = 1$ for all but finitely many $z \in \Sigma$. Each $z \in \Sigma$ lies in an orbit Ω of length

$$|\Omega| = |\Gamma : \Gamma_z| = N/N_z,$$

by the orbit–stabiliser theorem; the value of N_z depends only on the orbit Ω, and not on the particular element $z \in \Omega$. Let $\Omega_1, \ldots, \Omega_k$ be those orbits of Γ on Σ with lengths $|\Omega_j| < N$ (equivalently, containing points z with $N_z > 1$), and let n_j be the common value of N_z for $z \in \Omega_j$, so $n_j > 1$. Each Ω_j contributes $|\Omega_j| = N/n_j$ summands equal to $n_j - 1$ in $\sum_z (N_z - 1)$ so we have

$$|\Phi| = \sum_{j=1}^{k} \frac{N}{n_j}(n_j - 1) = N \sum_{j=1}^{k} \left(1 - \frac{1}{n_j}\right).$$

Combining this with $|\Phi| = 2(N - 1)$, we have

$$2\left(1 - \frac{1}{N}\right) = \sum_{j=1}^{k} \left(1 - \frac{1}{n_j}\right). \tag{2.13.6}$$

Now each $n_j \geqslant 2$, so $1 - n_j^{-1} \geqslant \frac{1}{2}$, and hence

$$\sum_{j=1}^{k} \left(1 - \frac{1}{n_j}\right) \geqslant \frac{k}{2}.$$

On the other hand, $2(1 - N^{-1}) < 2$, and so $k/2 < 2$, giving $k = 1, 2$ or 3.

If $k = 1$ then equation 2.13.6 gives $2(1 - N^{-1}) = 1 - n_1^{-1} < 1$, so $1 - N^{-1} < \frac{1}{2}$ and hence $N = 1$. Thus Γ is a cyclic group of order 1.

If $k = 2$ then equation 2.13.6 gives $2(1 - N^{-1}) = 2 - n_1^{-1} - n_2^{-1}$, so that $2N^{-1} = n_1^{-1} + n_2^{-1}$ and hence $(N/n_1) + (N/n_2) = 2$. Now $N/n_j = |\Omega_j|$ is a positive integer, so in this case each $|\Omega_j| = 1$. Thus Γ has a fixed-point (two, in fact), so Lemma 2.13.3 implies that Γ is cyclic.

If $k = 3$ then $2(1 - N^{-1}) = 3 - n_1^{-1} - n_2^{-1} - n_3^{-1}$, giving

$$\frac{1}{n_1} + \frac{1}{n_2} + \frac{1}{n_3} = 1 + \frac{2}{N} > 1.$$

We may choose the numbering so that $n_1 \leqslant n_2 \leqslant n_3$. Then $n_1 = 2$, for otherwise each $n_j \geqslant 3$ and hence $\sum n_j^{-1} \leqslant 1$. Putting $n_1 = 2$, we get

$$\frac{1}{n_2} + \frac{1}{n_3} = \frac{1}{2} + \frac{2}{N} > \frac{1}{2},$$

and simple arithmetic gives the following possibilities:

	n_1	n_2	n_3	N
(a)	2	2	n	$2n$
(b)	2	3	3	12
(c)	2	3	4	24
(d)	2	3	5	60

Fig. 2.10

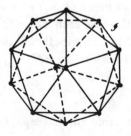

Fig. 2.11

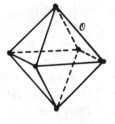

Fig. 2.12

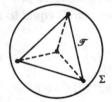

In case (a), n is any integer $n \geqslant 2$. Then $|\Omega_3| = N/n_3 = 2$, so Lemma 2.13.4 implies that Γ is a dihedral group.

In case (b), Γ has an orbit Ω_3 of length $|\Omega_3| = N/n_3 = 4$. For each $z \in \Omega_3$, the stabiliser Γ_z is a cyclic group of order $N_z = n_3 = 3$, fixing z and its antipodal point, and permuting all other points of Σ in cycles of length 3. Since Γ_z leaves $\Omega_3 \backslash \{z\}$ invariant, it permutes these three points transitively, so they are all equidistant from z. Applying this argument to each $z \in \Omega_3$, we see that the four points in Ω_3 are equidistant from each other, so they are the vertices of a regular tetrahedron $\mathcal{T}$ inscribed in Σ (see Fig. 2.10). Since Γ consists of rotations permuting these vertices, Γ is a group of rotations of $\mathcal{T}$. Now any regular tetrahedron has just 12 rotations, and $|\Gamma| = N = 12$, so Γ is the group of all rotations of $\mathcal{T}$.

Similar arguments may be used in cases (c) and (d). In case (c) one shows that Ω_3 consists of $N/n_3 = 6$ points of Σ, each of which is antipodal to one point of Ω_3 and is equidistant from the remaining four. Thus Ω_3 is the set of vertices of a regular octahedron $\mathcal{O}$, as in Fig. 2.11, and Γ consists of the 24 rotations of $\mathcal{O}$. In case (d), Ω_3 is the set of vertices of a regular icosahedron $\mathcal{I}$ as in Fig. 2.12, and Γ consists of the 60 rotations of $\mathcal{I}$. (In cases (b), (c) and (d), the orbits Ω_1 and Ω_2 consist of the mid-points of the edges and of the faces of the regular solid.) □

One can show that the groups Γ in case (iii) are isomorphic to A_4, S_4 and A_5 respectively: the rotation group of the tetrahedron $\mathcal{T}$ induces all the even permutations of the four vertices, so $\Gamma \cong A_4$; the rotation group of the octahedron $\mathcal{O}$ induces all the permutations of the four pairs of opposite faces, so $\Gamma \cong S_4$; the twenty faces of the icosahedron may be painted with five colours, no pair of adjacent faces having the same colour, in such a way that Γ induces all the even permutations of these five colours, giving $\Gamma \cong A_5$. Combining this with Corollary 2.13.2 and Theorem 2.13.5:

Corollary 2.13.7. *Every finite subgroup of $PSL(2, \mathbb{C})$ is cyclic, dihedral, or isomorphic to A_4, S_4 or A_5.* □

For each of the regular solids $\mathcal{T}$, $\mathcal{O}$ and $\mathcal{I}$, we have a single conjugacy class of finite subgroups $\Gamma \leqslant PSL(2, \mathbb{C})$. For example, suppose that Γ is conjugate to the rotation group $\text{Rot}(\mathcal{O})$ of a regular octahedron $\mathcal{O}$ inscribed in Σ; there exists a rotation $U \in PSL(2, \mathbb{C})$ sending $\mathcal{O}$ to the octahedron $\mathcal{O}^*$ with vertices $\{0, \infty, \pm 1, \pm i\}$, so that $\text{Rot}(\mathcal{O}) = U^{-1}\text{Rot}(\mathcal{O}^*)U$ and hence Γ is conjugate to $\text{Rot}(\mathcal{O}^*)$. Similarly, the tetrahedral and icosahedral subgroups Γ each form a single conjugacy class, as do the cyclic and the dihedral groups of any given order. Thus two finite subgroups of $PSL(2, \mathbb{C})$ are conjugate if and only if they are isomorphic.

We close Chapter 2 with two sections which are only loosely related to the rest of the chapter, but which will give us some useful analogies when we consider later generalisations.

2.14 The area of a spherical triangle

The geodesics on the sphere S^2 are the *great circles*, that is, the circles $C = S^2 \cap \Pi$ where Π is a plane containing the origin of $\mathbb{R}^3$. The antipodal map $K: Q \mapsto -Q$ of S^2 leaves each great circle C invariant, and interchanges the two hemispheres bounded by C.

Any two great circles C_1 and C_2 intersect in a pair of antipodal points Q and $\tilde{Q}$ (since $C_1 \cap C_2$ is a set of two points invariant under K), and C_1 and C_2 divide S^2 into two pairs of antipodal regions; we call any such region a *lune*. In Fig. 2.13 the lunes Λ_1 and Λ_2 are antipodal, as are Λ_3 and Λ_4.

Fig. 2.13

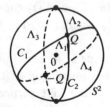

We denote the area of any measurable region $R \subseteq S^2$ by $\mu(R)$; clearly, $\mu(R) = \mu(K(R))$, since K is an isometry of S^2. If great circles C_1 and C_2 enclose a lune Λ, and meet at an angle θ within Λ, then $\mu(\Lambda)$ is proportional to θ; now S^2, having radius 1, has area $\mu(S^2) = 4\pi$, so $\mu(\Lambda) \to \frac{1}{2}\mu(S^2) = 2\pi$ as $\theta \to \pi$, and hence $\mu(\Lambda) = 2\theta$.

Three great circles C_1, C_2 and C_3 divide S^2 into four pairs of antipodal regions, called *spherical triangles*. Let Δ be a spherical triangle, illustrated

Fig. 2.14

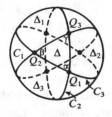

in Fig. 2.14, with internal angles α, β, γ at its vertices $Q_1 \in C_2 \cap C_3$, $Q_2 \in C_3 \cap C_1$, $Q_3 \in C_1 \cap C_2$. Then we have

Theorem 2.14.1. $\mu(\Delta) = \alpha + \beta + \gamma - \pi$.

Proof. Of the eight triangles bounded by C_1, C_2 and C_3, let Δ_1, Δ_2 and Δ_3 be the three which meet Δ across its sides Q_2Q_3, Q_3Q_1 and Q_1Q_2 respectively. Then Δ and Δ_j form a lune Λ_j with internal angle α, β, γ for $j = 1, 2, 3$, so we have

$$\mu(\Delta) + \mu(\Delta_1) = \mu(\Lambda_1) = 2\alpha,$$
$$\mu(\Delta) + \mu(\Delta_2) = \mu(\Lambda_2) = 2\beta,$$
$$\mu(\Delta) + \mu(\Delta_3) = \mu(\Lambda_3) = 2\gamma,$$

and hence

$$3\mu(\Delta) + \sum_{j=1}^{3} \mu(\Delta_j) = 2(\alpha + \beta + \gamma).$$

Now the triangles Δ, Δ_1, Δ_2, Δ_3, together with their antipodal triangles, form a partitition of $S^2 \backslash (C_1 \cup C_2 \cup C_3)$, and this has area 4π. Since antipodal triangles have the same area, this gives

$$2\mu(\Delta) + 2 \sum_{j=1}^{3} \mu(\Delta_j) = 4\pi.$$

Eliminating $\sum \mu(\Delta_j)$ from these two equations, we have

$$\mu(\Delta) = \alpha + \beta + \gamma - \pi. \qquad \square$$

For example, suppose that S^2 is tessellated by F congruent equilateral spherical triangles, with v triangles meeting at each vertex. Each triangle Δ has area $\mu(\Delta) = \mu(S^2)/F = 4\pi/F$, and has interior angles $\alpha = \beta = \gamma = 2\pi/v$, so

$$\frac{4\pi}{F} = \frac{6\pi}{v} - \pi,$$

giving

$$\frac{4}{F} = \frac{6}{v} - 1.$$

Now $4/F > 0$, so $v < 6$, and hence $v = 3$, 4 or 5, giving $F = 4$, 8 or 20 respectively. These three solutions correspond to the tessellations of S^2 formed by projecting onto S^2 the edges of a regular tetrahedron, octahedron or icosahedron respectively (see §2.13 for illustrations).

2.15 *PGL* (2, $\mathbb{C}$) as a Galois group

We have given an *analytic* representation of $PGL(2, \mathbb{C})$ as the group $\text{Aut}(\Sigma)$ of meromorphic bijections of Σ; we now give an *algebraic* representation of $PGL(2, \mathbb{C})$ as the Galois group Γ of the field extension $\mathbb{C}(z) \supset \mathbb{C}$, that is, as the group of all automorphisms of the field $\mathbb{C}(z)$ of rational functions in one variable z, which leave the field $\mathbb{C}$ of constant functions fixed. The following lemma shows that each $\gamma \in \Gamma$ is determined by its effect on the identity function $I : \Sigma \to \Sigma$, where we regard I as the rational function $z \in \mathbb{C}(z)$.

Lemma 2.15.1.   *Let* $\gamma \in \Gamma$ *and let* $\gamma(I) = g \in \mathbb{C}(z)$. *Then* $\deg(g) = 1$, *and* $\gamma(f) = f \circ g$ *for all* $f \in \mathbb{C}(z)$.

Proof.   Each $f \in \mathbb{C}(z)$ is a rational function $f : z \mapsto f(z) = \sum_{j=1}^{m} a_j z^j / \sum_{j=1}^{n} b_j z^j$, where $a_j, b_j \in \mathbb{C}$, so we have $f = \sum a_j I^j / \sum b_j I^j$, giving f as a rational function of I. Now γ is a field-automorphism of $\mathbb{C}(z)$, fixing the field $\mathbb{C}$ of constants, and hence fixing the coefficients a_j, b_j. Therefore

$$\gamma(f) = \gamma(\textstyle\sum a_j I^j) / \gamma(\textstyle\sum b_j I^j)$$
$$= \textstyle\sum \gamma(a_j I^j) / \sum \gamma(b_j I^j)$$
$$= \textstyle\sum a_j \gamma(I)^j / \sum b_j \gamma(I)^j$$
$$= \textstyle\sum a_j g^j / \sum b_j g^j,$$

where $g = \gamma(I)$, so that $\gamma(f) = f \circ g$ for all $f \in \mathbb{C}(z)$.

Since γ is an automorphism of $\mathbb{C}(z)$, γ is onto, so $I = \gamma(f) = f \circ g$ for some $f \in \mathbb{C}(z)$. Since I is one-to-one as a function $\Sigma \to \Sigma$, g must also be one-to-one, so $\deg(g) = 1$ by Corollary 1.4.3(ii). $\square$

Thus γ maps each rational function $f(z)$ to the rational function $f(g(z))$, where $g(z) = (az + b)/(cz + d)$ $(ad - bc \neq 0)$ depends only on γ. Now g lies in the group $PGL(2, \mathbb{C})$, so it has an inverse function $g^{-1} \in PGL(2, \mathbb{C})$.

Theorem 2.15.2.   *There is an isomorphism* $\theta : \Gamma \to PGL(2, \mathbb{C})$ *given by* $\theta(\gamma) = g^{-1}$, *where* $\gamma \in \Gamma$ *and* $g = \gamma(I)$.

Proof.   Using the lemma, we have seen that each $\gamma \in \Gamma$ determines an element $g^{-1} \in PGL(2, \mathbb{C})$, so let $\theta : \Gamma \to PGL(2, \mathbb{C})$ be given by $\gamma \mapsto g^{-1}$. If $\theta(\gamma_j) = g_j^{-1}$ for elements $\gamma_1, \gamma_2 \in \Gamma$, then each γ_j acts on $\mathbb{C}(z)$ by

$$\gamma_j : f \mapsto f \circ g_j,$$

so that the composition $\gamma_1 \gamma_2$ acts on $\mathbb{C}(z)$ by

$$\gamma_1 \gamma_2 : f \to (f \circ g_2) \circ g_1 = f \circ (g_2 \circ g_1),$$

and hence

$$\theta(\gamma_1\gamma_2) = (g_2 \circ g_1)^{-1} = g_1^{-1} \circ g_2^{-1} = \theta(\gamma_1) \circ \theta(\gamma_2).$$

Thus θ is a group-homomorphism. (Notice that the apparently more natural function $\gamma \mapsto g$ is *not* a homomorphism, since it reverses the order of products.)

If $\theta(\gamma) = g^{-1}$ is the identity element of $PGL(2, \mathbb{C})$, then γ acts on $\mathbb{C}(z)$ by $\gamma : f \mapsto f \circ g = f$, so γ is the identity automorphism of $\mathbb{C}(z)$; thus θ is one-to-one. If we choose any $h \in PGL(2, \mathbb{C})$, then $h = g^{-1}$ for some $g(= h^{-1})$ in $PGL(2, \mathbb{C})$, and the map $\gamma : \mathbb{C}(z) \to \mathbb{C}(z)$ given by $f \mapsto f \circ g$ is a field-automorphism of $\mathbb{C}(z)$ fixing $\mathbb{C}$: for example, γ sends $f_1 \pm f_2$ to $(f_1 \pm f_2) \circ g = (f_1 \circ g) \pm (f_2 \circ g)$, and similarly for products and quotients; the inverse of γ is given by $f \mapsto f \circ h$. Thus $\gamma \in \Gamma$ and $\theta(\gamma) = g^{-1} = h$, so θ is onto, and hence θ is an isomorphicm $\Gamma \cong PGL(2, \mathbb{C})$. $\square$

In fact, if F is *any* field, and $F(z)$ is the field of rational functions in one variable z with coefficients in F, then the Galois group of the extension $F(z) \supset F$ is isomorphic to $PGL(2, F)$. The proof is more difficult in the general case (see N. Jacobson, *Lectures in Abstract Algebra*, Vol. III, IV. 4), since a rational function g which is one-to-one need not have degree 1, so we cannot easily prove a result corresponding to Lemma 2.15.1: for example, $g(z) = z^3$ for $F = \mathbb{R}$, or $g(z) = z^p$ for $F = GF(p)$, the field of integers mod (p) where p is prime. When $F = \mathbb{C}$, however, we have the fundamental theorem of algebra (which lies behind Corollary 1.4.3 (ii)) to prove that g has degree 1.

EXERCISES

2A. Let z_0, z_1, z_2, z_3 be four distinct points in Σ. Show that there are precisely two values of k such that the ordered quadruple (z_0, z_1, z_2, z_3) can be mapped to $(1, -1, k, -k)$ by a Möbius transformation.

2B. Find the *type* (i.e. parabolic, hyperbolic, etc.) and the multiplier of each of the following Möbius transformations.

(i) $z \mapsto \dfrac{z + 1}{z + 3}$ (ii) $z \mapsto \dfrac{iz + 1}{z + 3i}$

(iii) $z \mapsto iz + 1$ (iv) $z \mapsto \dfrac{-z}{z + 4}$

2C. Let S and T be Möbius transformations with a common fixed-point. Prove that $STS^{-1}T^{-1}$ is parabolic or the identity. (Hint: assume the common fixed point is ∞. Why is this method valid?)

2D. Let S be a hyperbolic transformation with fixed-points p and q. Let T be a Möbius transformation which maps p to q. Prove that

(i) $STS^{-1}T^{-1}$ is hyperbolic,
(ii) $STST^{-1}$ is parabolic.

In the following three problems inversion is any transformation I_C where C is a circle in Σ. Thus it is inversion in a circle in $\mathbb{C}$ or reflection in a line in $\mathbb{C}$.

2E. (i) Show that the circle

$$a z \bar{z} + \bar{b} \bar{z} + b z + c = 0 \quad (a, c \in \mathbb{R})$$

does not intersect the real axis if and only if $(\operatorname{Re}(b))^2 < ac$.

(ii) Deduce that the composition of two inversions $I_{C_1} \circ I_{C_2}$ is hyperbolic if the circles C_1 and C_2 do not intersect. (Hint: It is enough to show this when one of the circles is the real axis – why?)

Show that the composition of the above two inversions is parabolic if C_1 and C_2 touch and elliptic if C_1 and C_2 intersect in two points.

2F. Show that if $\lambda > 1$ then U_λ is a composition of two inversions in concentric circles, if $\lambda = 1$ then U_λ is a composition of two reflections in parallel lines and if $|\lambda| = 1 \, (\lambda \neq 1)$ then U_λ is a composition of two inversions in intersecting lines. Deduce the converse of 2E (ii).

2G. Prove that if λ is not positive or if $|\lambda| \neq 1$ then U_λ is a composition of four inversions and hence show that every loxodromic transformation is a composition of four inversions.

Problems 2H–2L concern the isometric circle of a Möbius transformation. For more details and some uses of this circle see Ford [1951].

Let $T(z) = (az + b)/(cz + d) (a, b, c, d \in \mathbb{C}, ad - bc = 1)$. Its *isometric circle* $\mathscr{I}(T)$ is the circle $\{z \in \mathbb{C} \,|\, |cz + d| = 1\}$.

2H. Show that T maps $\mathscr{I}(T)$ onto $\mathscr{I}(T^{-1})$ and that if $z, w \in \mathscr{I}(T)$ then $|T(z) - T(w)| = |z - w|$. (Thus T maps $\mathscr{I}(T)$ *isometrically* onto $\mathscr{I}(T^{-1})$ – hence the name.) Show also that T maps the centre of $\mathscr{I}(T)$ to ∞. (Thus T maps the interior of $\mathscr{I}(T)$ to the exterior of $\mathscr{I}(T^{-1})$. We can also regard the interior of a circle as the region to the left of the circle as we travel around it in an anti-clockwise direction. As T preserves orientation, T maps $\mathscr{I}(T)$ with an anti-clockwise orientation to $\mathscr{I}(T^{-1})$ with a clockwise orientation. As T preserves distance on $\mathscr{I}(T)$ this means that once the image of one point of $\mathscr{I}(T)$ is known then the images of all points of $\mathscr{I}(T)$ are known.)

2I. Show that if T is hyperbolic then $\mathscr{I}(T)$ and $\mathscr{I}(T^{-1})$ are disjoint, if T is parabolic then $\mathscr{I}(T)$ and $\mathscr{I}(T^{-1})$ touch at one point and if T is elliptic with $T^2 \neq I$ then $\mathscr{I}(T)$ and $\mathscr{I}(T^{-1})$ intersect at two points. Show that $\mathscr{I}(T)$ and $\mathscr{I}(T^{-1})$ coincide if and only if $T^2 = I$.

2J. Suppose that $T^2 \neq I$. Let the straight line-segment joining the centres of $\mathscr{I}(T)$

and $\mathscr{I}(T^{-1})$ cut $\mathscr{I}(T)$ at z_1 and $\mathscr{I}(T^{-1})$ at z_2. Show that

$$z_1 + \frac{d}{c} = \frac{a}{c} - z_2 = \frac{a+d}{|a+d|c},$$

and deduce that if T is not loxodromic then $T(z_1) = z_2$. Prove that if T is not loxodromic then T is a composition of an inversion in $\mathscr{I}(T)$ followed by a reflection in the perpendicular bisector L of the straight line-segment joining the centres of $\mathscr{I}(T)$ and $\mathscr{I}(T^{-1})$. (No calculations are needed; just use the remarks made in 2H and Corollary 2.5.2.)

2K. If T is loxodromic show that

$$T(z_1) - \frac{a}{c} = e^{-2i\phi}\left(z_2 - \frac{a}{c}\right),$$

where $\phi = \arg(a+d)$. Deduce that T is a composition of an inversion in $\mathscr{I}(T)$ followed by a reflection in L followed by a rotation about the centre of $\mathscr{I}(T^{-1})$ through an angle of -2ϕ.

2L. If $T^2 = I$ prove that T is a composition of an inversion in $\mathscr{I}(T)$ with a suitable reflection.

2M. Let C be a circle in Σ passing through z_1, z_2, z_3. Prove that $I_C(z) = w$ if and only if

$$(w, z_1; z_2, z_3) = \overline{(z, z_1; z_2, z_3)}.$$

2N. If $\lambda \neq 0, \pm 1$ show that there are precisely two matrices of determinant 1 of the form

$$\begin{pmatrix} p & q \\ -q & 2-p \end{pmatrix}$$

$(p, q \in \mathbb{C})$ such that

$$\begin{pmatrix} \lambda & 0 \\ 0 & 1/\lambda \end{pmatrix}\begin{pmatrix} p & q \\ -q & 2-p \end{pmatrix}$$

has trace equal to 2. Deduce that every non-parabolic element of $PSL(2, \mathbb{C})$ is a product of two parabolic elements. (Thus $PSL(2, \mathbb{C})$ is generated by parabolic elements.)

2P. Let H be a normal subgroup of $PSL(2, \mathbb{C})$ containing more than just the identity element. Use the double transitivity of $PSL(2, \mathbb{C})$ on Σ to show that if H contains a non-parabolic element then it contains an element with two prescribed fixed points. Use Exercise 2C to show that H contains parabolic elements and deduce that $H = PSL(2, \mathbb{C})$. (Thus $PSL(2, \mathbb{C})$ is simple.)

2Q. Verify that a rotation of Σ is (represented by) an elliptic element of $PSL(2, \mathbb{C})$.

2R. Show that an elliptic element whose fixed-points are antipodal points of Σ is a rotation of Σ.

3
Elliptic functions

Having considered the sphere and its meromorphic functions in the first part of this book, we now turn our attention to another compact surface, the torus, and its meromorphic functions. These are the *elliptic* functions, which arise naturally as the doubly periodic meromorphic functions on $\mathbb{C}$. First we consider periodicity in general, and then, having obtained the torus as a quotient-space of $\mathbb{C}$, we consider the elementary properties of elliptic functions (regarded as functions from the torus to the Riemann sphere Σ). It is not a simple matter to construct non-constant elliptic functions, and for this purpose we have included sections on uniform and normal convergence and on infinite products (some readers may prefer to omit these sections at first reading, and refer back to them later, while others who are familiar with this type of analysis may be able to omit them completely). These techniques enable us to construct the Weierstrass functions $\wp(z)$, $\zeta(z)$ and $\sigma(z)$, and in the final sections of this chapter we use these functions to deduce further properties of elliptic functions in general.

3.1 Periods

Let f be a function defined on the complex plane $\mathbb{C}$. Then a complex number ω is called a *period* of f if

$$f(z + \omega) = f(z)$$

for all $z \in \mathbb{C}$, and f is called *periodic* if it has a period $\omega \neq 0$.

For example, $\sin z$ and $\cos z$ have period 2π, e^z has period $2\pi i$, and $\sin(2\pi z/\omega)$ has period ω for any complex number $\omega \neq 0$. Every function has period $\omega = 0$.

The set Ω_f of periods of a function f has two important properties: one algebraic, valid for all f, and one topological, valid for non-constant meromorphic functions f. These properties are given in Theorems 3.1.1 and 3.1.2.

Theorem 3.1.1. *Let Ω_f be the set of periods of a function f defined on $\mathbb{C}$; then Ω_f is a subgroup of the additive group $\mathbb{C}$.*

Proof. Let $\alpha, \beta \in \Omega_f$. Then $f(z + (\alpha + \beta)) = f((z + \alpha) + \beta) = f(z + \alpha) = f(z)$, and so $\alpha + \beta \in \Omega_f$. Moreover, $f(z - \alpha) = f((z - \alpha) + \alpha) = f(z)$, and so $-\alpha \in \Omega_f$. Finally, $f(z + 0) = f(z)$, and so $0 \in \Omega_f$. Thus Ω_f is a subgroup of $\mathbb{C}$. $\square$

A subset Δ of a topological space is called *discrete* if every $x \in \Delta$ has a neighbourhood U such that $U \cap \Delta = \{x\}$. For example

(i) the integers $\mathbb{Z}$ form a discrete subset of $\mathbb{R}$;
(ii) any finite subset of $\mathbb{R}^n$ is discrete;
(iii) $\{1/n \mid n \in \mathbb{Z}, n \neq 0\}$ is a discrete subset of $\mathbb{R}$.

However, $\{1/n \mid n \in \mathbb{Z}, n \neq 0\} \cup \{0\}$ is not discrete, since every neighbourhood of 0 contains real numbers of form $1/n$, $n \in \mathbb{Z}$.

Theorem 3.1.2. *Let Ω_f be the set of periods of a non-constant meromorphic function f defined on $\mathbb{C}$; then Ω_f is a discrete subset of $\mathbb{C}$.*

Proof. If Ω_f is not discrete then there exists $\omega \in \Omega_f$ such that every neighbourhood U of ω contains points of $\Omega_f \backslash \{\omega\}$. By taking these neighbourhoods to be discs, centred at ω, with radii approaching 0, we obtain a sequence of periods $\omega_n \neq \omega$ with $\lim_{n \to \infty} \omega_n = \omega$. Since $\omega, \omega_n \in \Omega_f$ we have $f(\omega) = f(0) = f(\omega_n)$. If $f(0) = \infty$ then f has a convergent sequence of poles, which is impossible by Theorem A.8. If $f(0) \neq \infty$ then the meromorphic function $g(z) = f(z) - f(0)$ has a convergent sequence of zeros, implying that g is identically zero (by Theorem A.8), and so f is constant. In either case, we have a contradiction. $\square$

To summarise, the set of periods of a non-constant meromorphic function is a discrete subgroup of $\mathbb{C}$. We now show that there are three types of discrete additive subgroups of $\mathbb{C}$, isomorphic to $\{0\}, \mathbb{Z}$ and $\mathbb{Z} \times \mathbb{Z}$ respectively.

Theorem 3.1.3. *Let Ω be a discrete subgroup of $\mathbb{C}$. Then one of the following holds:*

(i) $\Omega = \{0\}$;
(ii) $\Omega = \{n\omega_1 \mid n \in \mathbb{Z}\}$ *for some fixed $\omega_1 \in \mathbb{C} \backslash \{0\}$, and so Ω is isomorphic to $\mathbb{Z}$;*

(iii) $\Omega = \{m\omega_1 + n\omega_2 \,|\, m, n \in \mathbb{Z}\}$ *for some fixed* $\omega_1, \omega_2 \in \mathbb{C}$*, where* ω_1 *and* ω_2
are linearly independent over $\mathbb{R}$*, that is,* $\omega_1 \neq 0 \neq \omega_2$ *and* $\omega_1/\omega_2 \notin \mathbb{R}$*; in
this case,* Ω *is isomorphic to* $\mathbb{Z} \times \mathbb{Z}$.

Proof.   Suppose that $\Omega \neq \{0\}$. We first show that there exists $\omega_1 \in \Omega \setminus \{0\}$
with least value of $|\omega_1|$. Since Ω is discrete, there is some $\varepsilon > 0$ such that the
disc $|z| < \varepsilon$ contains no elements of $\Omega \setminus \{0\}$. It follows that for *any* $\omega \in \Omega$, the
disc $|z - \omega| < \varepsilon$ contains no elements of $\Omega \setminus \{\omega\}$: for if $z \in \Omega$ satisfies $0 <
|z - \omega| < \varepsilon$, then $z - \omega$ is an element of Ω (since Ω is a group) within distance
ε of 0 (as in Fig. 3.1), which is impossible. Thus each $\omega \in \Omega$ is the centre of a disc
of radius ε (independent of ω) containing no other elements of Ω. Therefore,
discs of radius $\frac{1}{2}\varepsilon$, centred on the elements of Ω, are disjoint from each other.
Now choose a disc $|z| < r$ sufficiently large to contain at least one element of

Fig. 3.1

Fig. 3.2

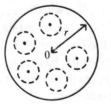

Fig. 3.3

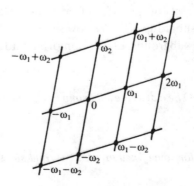

$\Omega \setminus \{0\}$ together with its surrounding disc of radius $\frac{1}{2}\varepsilon$; this disc has area πr^2, so, as in Fig. 3.2, it can contain only finitely many disjoint discs of radius $\frac{1}{2}\varepsilon$ (at most $4r^2\varepsilon^{-2}$, in fact); hence there are only finitely many elements of $\Omega \setminus \{0\}$ within distance r of 0, and out of this finite set we may choose ω_1 to have least modulus. (Of course, ω_1 is not unique: $-\omega_1$ will do equally well.)

Now let $L = \{\lambda\omega_1 | \lambda \in \mathbb{R}\}$ be the line through 0 and ω_1 in $\mathbb{C}$; then Ω contains $\{n\omega_1 | n \in \mathbb{Z}\}$, and this subgroup lies in L. First suppose that $\Omega \subseteq L$; we then claim that $\Omega = \{n\omega_1 | n \in \mathbb{Z}\}$, so that Ω is of type (ii). For if Ω contains some $\omega \neq n\omega_1$, then since $\Omega \subseteq L$ we have $\omega = \lambda\omega_1$ for some $\lambda \in \mathbb{R} \setminus \mathbb{Z}$, so $n < \lambda < n+1$ for some $n \in \mathbb{Z}$. Since Ω is a group containing ω and $n\omega_1$, it contains $\omega - n\omega_1 = (\lambda - n)\omega_1$. However, $0 \neq |(\lambda - n)\omega_1| < |\omega_1|$ which contradicts the minimality of $|\omega_1|$. Hence $\Omega = \{n\omega_1 | n \in \mathbb{Z}\}$, and we are in case (ii).

Now suppose that $\Omega \nsubseteq L$. By an argument similar to that used for ω_1, we can show that $\Omega \setminus L$ has an element ω_2 with least modulus. Then Ω contains the subgroup $\Lambda = \{m\omega_1 + n\omega_2 | m, n \in \mathbb{Z}\}$, and since $\omega_2 \notin L$, ω_1 and ω_2 are linearly independent over $\mathbb{R}$, so Λ consists of the vertices of a tessellation of $\mathbb{C}$ by congruent parallelograms, as shown in Fig. 3.3.

We now show that $\Omega = \Lambda$. If this is not the case, then there exists $\omega \in \Omega$ with $\omega \neq m\omega_1 + n\omega_2$ for any $m, n \in \mathbb{Z}$. Let $\omega = \lambda\omega_1 + \mu\omega_2$ with $\lambda, \mu \in \mathbb{R}$; then by adding or subtracting suitable multiples of ω_1 and ω_2 we may assume that $|\lambda| \leqslant \frac{1}{2}$, $|\mu| \leqslant \frac{1}{2}$. If $\mu = 0$ then $\omega = \lambda\omega_1 \in L$, with $|\omega| = |\lambda\omega_1| < |\omega_1|$; by minimality of $|\omega_1|$ we have $\omega = 0$ and hence $\omega \in \Lambda$, against our assumption. If $\lambda = 0$ then $\omega = \mu\omega_2$, and again $\omega = 0$, this time by minimality of $|\omega_2|$. Hence $\lambda\omega_1$ and $\mu\omega_2$ are non-zero and therefore linearly independent over $\mathbb{R}$, giving

$$|\omega| < |\lambda\omega_1| + |\mu\omega_2| \leqslant \tfrac{1}{2}|\omega_1| + \tfrac{1}{2}|\omega_2| \leqslant \tfrac{1}{2}|\omega_2| + \tfrac{1}{2}|\omega_2| = |\omega_2|,$$

the first inequality being strict because $\lambda\omega_1$ and $\mu\omega_2$ are linearly independent. Now $\omega \in \Omega \setminus L$ (since $\mu \neq 0$), so by minimality of $|\omega_2|$ we have $\omega = 0$, contradicting the fact that $\lambda, \mu \neq 0$. Thus $\omega \in \Lambda$, so $\Omega = \Lambda$ and we are in case (iii). □

Definition. If a function f has its set Ω_f of periods of type (ii), then f is *simply periodic*; if Ω_f is of type (iii), then f is *doubly periodic*. Groups Ω of type (iii) are called *lattices*, and any pair ω_1, ω_2 such that $\Omega = \{m\omega_1 + n\omega_2 | m, n \in \mathbb{Z}\}$ is called a *basis* for the lattice. The above results show that periodic functions are either simply or doubly periodic (apart from constant functions, where $\Omega = \mathbb{C}$, and non-meromorphic functions, where Ω can be any subgroup of $\mathbb{C}$, not necessarily discrete). Our main concern in this chapter will be with doubly periodic functions; however, we first look at some generalisations of the ideas in this section, and then we briefly consider simply periodic functions.

3.2 Topological groups

Definition. A *topological group* is a topological space G which is also a group, in which the group multiplication and the taking of inverses are continuous operations. More precisely, the maps

$$m: G \times G \to G, \text{ defined by } m(g, h) = gh,$$
$$i: G \to G, \qquad \text{defined by } i(g) = g^{-1},$$

are continuous.

Thus $\mathbb{C}$, with its additive group structure, is a topological group, since the group operations $m(z, w) = z + w$ and $i(z) = -z$ are continuous. Other important examples of topological groups are:

(i) The circle $S^1 = \{z \in \mathbb{C} \,|\, |z| = 1\}$, with multiplication of complex numbers as the group operation.

(ii) $GL(n, \mathbb{C})$, with matrix multiplication as the group operation. We obtain the topology on this group by considering the $n \times n$ matrix (a_{ij}) to be the point $(a_{11}, a_{12}, \ldots, a_{1n}, a_{21}, \ldots, a_{2n}, \ldots, a_{nn})$ in $\mathbb{C}^{n^2}$; the operations m and i of this group may be expressed in terms of rational functions of the coordinates a_{ij}, so they are continuous. Similarly, $GL(n, \mathbb{R})$ is a topological group.

(iii) $PGL(n, \mathbb{C})$, with multiplication of the cosets of $\{\lambda I \,|\, \lambda \neq 0\}$ as the group operation (see §2.2). In this group, the matrix $A \in GL(n, \mathbb{C})$ is identified with λA ($\lambda \in \mathbb{C} \setminus \{0\}$), so we topologise the group by identifying the point $(a_{11}, \ldots, a_{nn})$ in $\mathbb{C}^{n^2}$ with the point $(\lambda a_{11}, \ldots, \lambda a_{nn})$ and using the identification topology. Similarly, $PGL(n, \mathbb{R})$ is a topological group.

In any topological group G, the map m_g defined by

$$m_g(x) = xg \quad (x, g \in G)$$

is a continuous bijection with a continuous inverse $m_{(g^{-1})}$. Thus m_g is a homeomorphism $G \to G$, known as *right translation*. If $x, y \in G$ then

$$m_{(x^{-1}y)}(x) = y,$$

and so there is a homeomorphism of G taking any given point to any other. (In other words, the group of homeomorphisms of G is transitive; see §2.5.) This means that any one point of G looks topologically like any other point of G; in the case of $\mathbb{C}$ and S^1 this is visually apparent. In particular, each neighbourhood of a given point in G is homeomorphic to a neighbourhood of the identity $e \in G$, so we can define a *discrete subgroup* Ω of G to be a subgroup with the property that there is a neighbourhood U of e in G such that $U \cap \Omega = \{e\}$. (This idea of translating neighbourhoods from an arbitrary point back to the identity element has already been used at the beginning of the proof of Theorem 3.1.3.)

An extension of Theorem 3.1.3 shows that the discrete sub-groups of $\mathbb{R}^n$ are isomorphic to $\{0\}$ or to $\mathbb{Z}^m$ where $1 \leqslant m \leqslant n$. In general, however, it is difficult to describe the discrete subgroups Ω of an arbitrary topological group G; for later applications (especially in Chapter 5 where we consider $G = PSL(2, \mathbb{R})$) we will now show that Ω meets each compact subset of G in only finitely many points.

Let $\mathcal{N}$ denote the collection of all open subsets of G containing e. If $x \in G$ then $\{m_x(U) | U \in \mathcal{N}\}$ is the collection of all open sets containing x, so the open sets in G are determined by the elements of $\mathcal{N}$. If $U, V \in \mathcal{N}$ then $UV \in \mathcal{N}$ and $U^{-1} \in \mathcal{N}$, where $UV = \{uv | u \in U, v \in V\}$ and $U^{-1} = \{u^{-1} | u \in U\}$. If $U = U^{-1}$ then U is called a *symmetric* open set.

Lemma 3.2.1. *If $U \in \mathcal{N}$ then there exists a symmetric open set $V \in \mathcal{N}$ such that $VV \subseteq U$.*

Proof. Let W denote $m^{-1}(U) = \{(x, y) \in G \times G | xy \in U\}$. Since $m: G \times G \to G$ is continuous and $(e, e) \in W$, it follows that W is an open neighbourhood of (e, e) in $G \times G$, so by definition of the product topology, there exist $V_1, V_2 \in \mathcal{N}$ such that $V_1 \times V_2 \subseteq W$ and hence $V_1 V_2 \subseteq U$. Now let $V_3 = V_1 \cap V_2$; then $V_3 V_3 \subseteq U$, so if we put $V = V_3 \cap V_3^{-1}$ then $V \in \mathcal{N}$, V is symmetric, and $VV \subseteq U$. $\square$

Theorem 3.2.2. *If Ω is a discrete subgroup of a topological group G, and if K is a compact subset of G, then $\Omega \cap K$ is finite.*

Proof. Since Ω is discrete there exists $U \in \mathcal{N}$ such that $U \cap \Omega = \{e\}$, and by Lemma 3.2.1 there exists a symmetric $V \in \mathcal{N}$ such that $VV \subseteq U$. The open sets $gV (g \in G)$ cover G, and hence cover K, so by compactness there is a finite subcover, $g_1 V \cup \ldots \cup g_n V \supseteq K$. We now show that for each $i = 1, \ldots, n$, $|\Omega \cap g_i V| \leqslant 1$. For suppose that $h_1, h_2 \in \Omega \cap g_i V$; then there exist $v_1, v_2 \in V$ such that $h_j = g_i v_j$ $(j = 1, 2)$ and hence $h_1^{-1} h_2 = v_1^{-1} g_i^{-1} . g_i v_2 = v_1^{-1} v_2 \in V^{-1} V = VV \subseteq U$, so that $h_1 = h_2$ since $h_1^{-1} h_2 \in U \cap \Omega = \{e\}$. Thus $|\Omega \cap K| \leqslant n$, as required. $\square$

Corollary 3.2.3. *In a compact topological group, every discrete subgroup is finite.* □

For example, in §2.12 we showed that the rotation group $\text{Rot}(\Sigma)$ is homeomorphic to $P_3(\mathbb{R})$, the quotient of S^3 by the antipodal map. Since S^3 is compact, so is $\text{Rot}(\Sigma)$, so it follows that the discrete groups of rotations of Σ are just the finite groups listed in Theorem 2.13.5.

We could have used Theorem 3.2.2 to shorten the proof of Theorem 3.1.3 (see Exercise 3B), but we preferred to give a more elementary proof, using the Euclidean metric on $\mathbb{R}^2$.

In Theorem 3.2.2, the condition that Ω is a subgroup is necessary: for example, let $G = \mathbb{R}$ (under addition), $\Omega = \{1/n \mid n = 1, 2, 3, \ldots\}$, and $K = [0, 1]$; then Ω is a discrete subset of G (though not a subgroup), K is compact, and $\Omega \cap K$ *is* infinite.

3.3 Simply periodic functions

In this section we will show that a simply periodic meromorphic function can be expressed in terms of the standard exponential and trigonometric functions. (In fact, all we require of the set of periods is that it should contain a subgroup isomorphic to $\mathbb{Z}$, so our results are also valid for doubly periodic functions; however, we shall see later on that much stronger results are available for such functions.)

If f is simply periodic with set of periods $\Omega_f = \{n\omega_1 \mid n \in \mathbb{Z}\}$, then by replacing z by $\omega_1 z$ we can assume that $\Omega_f = \mathbb{Z}$ (for example, replacing $\sin z$ by $\sin 2\pi z$, or e^z by $e^{2\pi i z}$), so we have

$$f(z) = f(z + n) \quad \text{for all} \quad n \in \mathbb{Z}.$$

We define complex numbers z_1, z_2 to be *congruent* mod $\mathbb{Z}$ if and only if $z_1 - z_2 \in \mathbb{Z}$. This defines an equivalence relation on $\mathbb{C}$, in which the equivalence classes are the cosets of $\mathbb{Z}$, and f takes the same value at congruent points. Each complex number is congruent to precisely one point in the infinite vertical strip $S = \{z \in \mathbb{C} \mid 0 \leqslant \text{Re}(z) < 1\}$, so the behaviour

Fig. 3.4

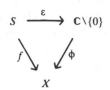

of f on $\mathbb{C}$ is determined by its behaviour on S, this behaviour being repeated on each parallel strip $S + n, n \in \mathbb{Z}$.

The function $\varepsilon : z \mapsto \zeta = e^{2\pi i z}$ has $\mathbb{Z}$ as its set of periods, and it is a bijection between S and $\mathbb{C} \backslash \{0\}$. We define

$$\phi(\zeta) = f(z) = (f \circ \varepsilon^{-1})(\zeta) = f\left(\frac{1}{2\pi i} \log \zeta\right),$$

so that if f is a simply periodic function from $\mathbb{C}$ to a set X, then ϕ is a function from $\mathbb{C} \backslash \{0\}$ to X (see Fig. 3.4). Notice that although $\log \zeta$ is not a single-valued function of ζ, successive values of $\log \zeta$ differ by $2\pi i$, so different values of $(1/2\pi i) \log \zeta$ differ by integers, and since these integers are periods of $f, \phi(\zeta) = f((1/2\pi i) \log \zeta)$ is a single-valued function of ζ: the periodicity of f cancels the many-valued nature of $\log \zeta$. In general, $\phi(\zeta)$ is a simpler function than $f(z)$. For instance, if $f(z) = \sin 2\pi z$ then

$$\phi(\zeta) = f(z) = \sin 2\pi z = \frac{1}{2i}\left(e^{2\pi i z} - e^{-2\pi i z}\right) = \frac{1}{2i}\left(\zeta - \frac{1}{\zeta}\right),$$

and similarly if $f(z) = \cos 2\pi z$ then $\phi(\zeta) = \frac{1}{2}(\zeta + \zeta^{-1})$.

Conversely, if $\phi : \mathbb{C} \backslash \{0\} \to X$ is any function, then we obtain a simply periodic function $f = \phi \circ \varepsilon : \mathbb{C} \to X$ given by

$$f(z) = \phi(\zeta) = \phi(e^{2\pi i z}).$$

Thus the functions $f : \mathbb{C} \to X$ which are periodic with respect to $\mathbb{Z}$ are precisely those of the form $f = \phi \circ \varepsilon$, that is, the functions of $\zeta = e^{2\pi i z}$.

In a sufficiently small neighbourhood of each point in $\mathbb{C} \backslash \{0\}$, there is a single-valued analytic branch of $\log \zeta$, so if f is a simply periodic meromorphic function $\mathbb{C} \to \Sigma$ then ϕ is a meromorphic function $\mathbb{C} \backslash \{0\} \to \Sigma$, with poles in one-to-one correspondence with the congruence classes of poles of f on $\mathbb{C}$. (Of course, ϕ may have singularities at 0 and ∞, since $\log \zeta$ does.) For example, if $f(z) = \tan \pi z$, with a single congruence class of poles at $z = n + \frac{1}{2}$ $(n \in \mathbb{Z})$, then $\phi(\zeta) = -i(\zeta - 1)/(\zeta + 1)$ with a single pole at $\zeta = -1 = e^{2\pi i (n + 1/2)}$. Conversely, if ϕ is meromorphic then so if f, since $f = \phi \circ \varepsilon$ and ε is analytic.

We sum up these results in

Theorem 3.3.1. *The functions f for which $\mathbb{Z} \subseteq \Omega_f$ are the single-valued functions $\phi(\zeta)$ of $\zeta = e^{2\pi i z}$; f is meromorphic on $\mathbb{C}$ if and only if ϕ is meromorphic on $\mathbb{C} \backslash \{0\}$, and the congruence classes of poles of f on $\mathbb{C}$ are in one-to-one correspondence with the poles of ϕ on $\mathbb{C} \backslash \{0\}$.* $\square$

If we now assume that f is meromorphic then its poles are discrete, so by an argument similar to that at the beginning of the proof of Theorem 3.1.3 we can find a rectangle $R = \{z | y_1 < \operatorname{Im}(z) < y_2, 0 \leqslant \operatorname{Re}(z) < 1\}$ within the infinite strip S such that R contains no poles of f. Now the function $\varepsilon : z \mapsto \zeta = e^{2\pi iz}$ maps the edges $\{x + iy_j | 0 \leqslant x < 1\}$ of R to $\{e^{-2\pi y_j}.e^{2\pi ix} | 0 \leqslant x < 1\}$ for $j = 1, 2$, and these are circles of radii $r_j = e^{-2\pi y_j}$ in the ζ-plane, so R is mapped by ε to an annular region $\varepsilon(R)$ given by $r_2 < |\zeta| < r_1$, within which $\phi(\zeta)$ is analytic (see Fig. 3.5).

It follows that $\phi(\zeta)$ has a unique Laurent expansion

$$\phi(\zeta) = \sum_{n=-\infty}^{\infty} a_n \zeta^n$$

valid for $r_2 < |\zeta| < r_1$ (at least), so that $f(z)$ has an expansion

$$f(z) = \sum_{n=-\infty}^{\infty} a_n e^{2\pi niz},$$

valid for $y_1 < \operatorname{Im}(z) < y_2$. Putting $e^{2\pi niz} = \cos 2\pi nz + i \sin 2\pi nz$, we obtain the Fourier series (J.B.J. Fourier, 1768–1830)

$$f(z) = a_0 + \sum_{n=1}^{\infty} (A_n \cos 2\pi nz + B_n \sin 2\pi nz),$$

where $A_n = a_n + a_{-n}$ and $B_n = i(a_n - a_{-n})$ for $n \geqslant 1$. This series is valid in the horizontal strip $y_1 < \operatorname{Im}(z) < y_2$ consisting of R and its translates $R + m$, $m \in \mathbb{Z}$. However, different choices for the rectangle R may give rise to different Fourier series for f, valid on disjoint horizontal strips (see Exercise 3D).

Fig. 3.5

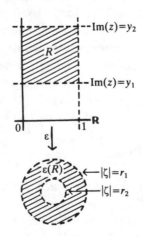

3.4 Lattices and fundamental regions

We now turn to the study of doubly periodic functions, starting with a closer look at some of the algebraic and geometric ideas involved. A group Ω of type (iii) in Theorem 3.1.3 is called a lattice, and is denoted by $\Omega(\omega_1, \omega_2)$ where $\{\omega_1, \omega_2\}$ is a basis for Ω, that is, a pair of generators for Ω. There are other bases for Ω besides $\{\omega_1, \omega_2\}$: for instance $\{\omega_1, \omega_1 + \omega_2\}$ is also a basis, for if $\omega \in \Omega(\omega_1, \omega_2)$ then

$$\omega = m\omega_1 + n\omega_2 = (m - n)\omega_1 + n(\omega_1 + \omega_2),$$

with $m - n, n \in \mathbb{Z}$. In general, if $\omega_1', \omega_2' \in \Omega(\omega_1, \omega_2)$ then

$$\begin{aligned} \omega_2' &= a\omega_2 + b\omega_1, \\ \omega_1' &= c\omega_2 + d\omega_1, \end{aligned} \tag{3.4.1}$$

where a, b, c, d are integers.

Theorem 3.4.2. *Equations* (3.4.1) *define a basis* $\{\omega_1', \omega_2'\}$ *for* $\Omega(\omega_1, \omega_2)$ *if and only if* $ad - bc = \pm 1$.

Proof. It is convenient to write (3.4.1) using matrix notation

$$\begin{pmatrix} \omega_2' \\ \omega_1' \end{pmatrix} = A \begin{pmatrix} \omega_2 \\ \omega_1 \end{pmatrix}, \tag{3.4.3}$$

where

$$A = \begin{pmatrix} a & b \\ c & d \end{pmatrix}.$$

If $ad - bc = \pm 1$, then A^{-1} has integer coefficients, and we have

$$\begin{pmatrix} \omega_2 \\ \omega_1 \end{pmatrix} = A^{-1} \begin{pmatrix} \omega_2' \\ \omega_1' \end{pmatrix} = \pm \begin{pmatrix} d & -b \\ -c & a \end{pmatrix} \begin{pmatrix} \omega_2' \\ \omega_1' \end{pmatrix}.$$

Thus $\omega_1, \omega_2 \in \Omega(\omega_1', \omega_2')$ and hence $\Omega(\omega_1, \omega_2) \subseteq \Omega(\omega_1', \omega_2')$. The reverse inclusion is obvious, so $\Omega(\omega_1, \omega_2) = \Omega(\omega_1', \omega_2')$ and hence $\{\omega_1', \omega_2'\}$ is a basis.

Conversely, suppose that equation (3.4.3) defines a basis $\{\omega_1', \omega_2'\}$ for $\Omega(\omega_1, \omega_2)$. Expressing the elements ω_1, ω_2 in terms of this basis, we have

$$\begin{pmatrix} \omega_2 \\ \omega_1 \end{pmatrix} = B \begin{pmatrix} \omega_2' \\ \omega_1' \end{pmatrix}$$

for some matrix B with integer coefficients, so

$$\begin{pmatrix} \omega_2 \\ \omega_1 \end{pmatrix} = BA \begin{pmatrix} \omega_2 \\ \omega_1 \end{pmatrix}. \tag{3.4.4}$$

Now $\mathbb{C}$ is a 2-dimensional vector space over $\mathbb{R}$, and since ω_1 and ω_2 are linearly independent over $\mathbb{R}$ they form a basis for $\mathbb{C}$. It follows from (3.4.4) that the matrix BA induces the identity linear transformation of $\mathbb{C}$, so $BA = I$ and hence

$$\det(B).\det(A) = 1.$$

Since A and B have integer coefficients, their determinants are integers, so $\det(A) = \pm 1$, that is, $ad - bc = \pm 1$. $\square$

It is easily seen that there are infinitely many sets of integers a, b, c, d satisfying $ad - bc = \pm 1$, so any lattice Ω has infinitely many bases. Exercise 3E gives necessary and sufficient conditions for an element $\omega \in \Omega$ to be contained in a basis.

FUNDAMENTAL REGIONS

Given a lattice Ω, we define $z_1, z_2 \in \mathbb{C}$ to be *congruent* mod Ω, written $z_1 \sim z_2$, if $z_1 - z_2 \in \Omega$. Congruence mod Ω is easily seen to be an equivalence relation on $\mathbb{C}$, and the equivalence classes are the cosets $z + \Omega$ of Ω in the additive group $\mathbb{C}$. Alternatively, we may regard Ω as acting on $\mathbb{C}$ as a transformation group, each $\omega \in \Omega$ inducing the translation

$$t_\omega : z \mapsto z + \omega$$

of $\mathbb{C}$; since

$$t_{(\omega_1 + \omega_2)} = t_{\omega_1} \circ t_{\omega_2},$$

we have a group isomorphism $\Omega \cong \{t_\omega | \omega \in \Omega\}$. Then two points $z_1, z_2 \in \mathbb{C}$ are congruent mod Ω if and only if they lie in the same orbit under this action of Ω.

A closed, connected subset P of $\mathbb{C}$ is defined to be a *fundamental region* for Ω if

(i) for each $z \in \mathbb{C}$, P contains at least one point in the same Ω-orbit as z (i.e. every point $z \in \mathbb{C}$ is congruent to some point in P);

(ii) no two points in the interior of P are in the same Ω-orbit (i.e. no pair of points in the interior of P are congruent).

If, as is usually the case, P is also a Euclidean polygon, with a finite number of sides, then we call P a *fundamental polygon* for Ω; in particular, if P is a parallelogram, then it is called a *fundamental parallelogram* for Ω.

For example, the parallelogram P, shown in Fig. 3.6, with vertices $0, \omega_1, \omega_2, \omega_1 + \omega_2$, is a fundamental parallelogram for the lattice $\Omega(\omega_1, \omega_2)$.

Conditions (i) and (ii) ensure that if P is any fundamental region for a lattice Ω, then P and its images under the action of Ω (that is, its translates $P + \omega, \omega \in \Omega$) cover the plane $\mathbb{C}$ completely, overlapping only at their boundaries; this type of covering is known as a *tessellation* of $\mathbb{C}$, and the diagram shows the tessellation arising from a fundamental parallelogram. By using Theorem 3.4.1 we can obtain fundamental parallelograms of different shapes, and hence obtain different tessellations of $\mathbb{C}$.

If P is any fundamental region for Ω, then for fixed $t \in \mathbb{C}$, the set

$$P + t = \{z + t \,|\, z \in P\}$$

is also a fundamental region. This is useful when we need to find a fundamental region containing or avoiding certain specified points; for example, we can always find a fundamental parallelogram for Ω with 0 in its interior.

We can obtain fundamental regions which are not parallelograms, nor even polygons, by the following procedure. Let P be the fundamental parallelogram for $\Omega(\omega_1, \omega_2)$ with vertices $0, \omega_1, \omega_2, \omega_1 + \omega_2$. The transformation $z \mapsto z + \omega_2$ maps one side of P to the opposite side. If we cut out a

Fig. 3.6

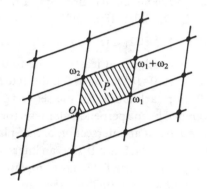

Fig. 3.7

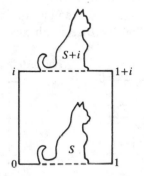

section S of P which intersects this side and glue on $S + \omega_2$ at the opposite side, then we still have a fundamental region $(P \backslash S) \cup (S + \omega_2)$. This is illustrated in Fig. 3.7 where $\Omega = \Omega(1, i)$. Similarly, we can use the transformations $z \mapsto z - \omega_2$ and $z \mapsto z \pm \omega_1$. By performing several of these cutting and gluing operations, we can obtain a fundamental region which is the union of two squares, as shown in Fig. 3.8; this gives a neat proof of Pythagoras' theorem (see the comment after Theorem 3.4.6).

Fig. 3.8

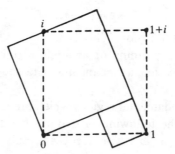

As a final example of a fundamental region for a lattice Ω, we consider the *Dirichlet region* (G.P.L. Dirichlet, 1805–59)

$$D(\Omega) = \{z \in \mathbb{C} \mid |z| \leqslant |z - \omega| \text{ for all } \omega \in \Omega\}.$$

This is the set of points which are at least as close to 0 as they are to any other lattice-point. Clearly, $0 \in D(\Omega)$, and as Ω is discrete $D(\Omega)$ contains some neighbourhood of 0. For each $\omega \in \Omega \backslash \{0\}$ the set $\{z \in \mathbb{C} \mid |z| \leqslant |z - \omega|\}$ is a closed half-plane bounded by the perpendicular bisector of the line segment joining 0 and ω. Thus $D(\Omega)$ is an intersection of half-planes, each of which is convex, and so $D(\Omega)$ is convex. Since $D(\Omega)$ is an intersection of just finitely many half-planes (why?), and since $D(\Omega)$ has non-empty interior (containing 0 for example) it follows that $D(\Omega)$ is a polygon. For this reason the Dirichlet region is often called the *Dirichlet polygon*.

Figs. 3.9 and 3.10 illustrate two examples of Dirichlet polygons. In the most general case, as in Fig. 3.9, $D(\Omega)$ is a hexagon with its opposite sides

Fig. 3.9   **Fig. 3.10**

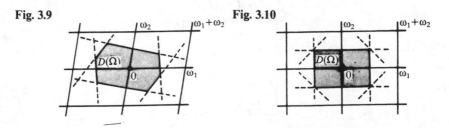

parallel and of equal lengths, but when Ω is 'rectangular', that is, Ω has a pair of perpendicular generators, as in Fig. 3.10, then $D(\Omega)$ is a rectangle.

Theorem 3.4.5. *$D(\Omega)$ is a fundamental region for Ω.*

Proof. Let $z_1 \in \mathbb{C}$ and let z_0 be a point in the Ω-orbit of z_1 with smallest modulus; such a point exists by the argument used at the beginning of the proof of Theorem 3.1.3. Then $|z_0| \leqslant |z_0 - \omega|$ for all $\omega \in \Omega$, and so $z_0 \in D(\Omega)$. Thus $D(\Omega)$ contains at least one point from every Ω-orbit.

Next we show that if z_1, z_2 are in the interior of $D(\Omega)$ then they cannot lie in the same Ω-orbit. If $|z| = |z - \omega|$ for some $\omega \in \Omega \backslash \{0\}$, then either $z \notin D(\Omega)$ or else z lies on the boundary of $D(\Omega)$; hence if z is in the interior of $D(\Omega)$ then $|z| < |z - \omega|$ for all $\omega \in \Omega \backslash \{0\}$. If two interior points z_1, z_2 lie in the same Ω-orbit then this implies that $|z_1| < |z_2|$ and $|z_2| < |z_1|$, a contradiction. Thus the interior of $D(\Omega)$ contains at most one point in each Ω-orbit.

Being an intersection of closed half-planes, $D(\Omega)$ is closed. Since $D(\Omega)$ is convex, it is path-connected and hence connected. Thus $D(\Omega)$ is a fundamental region. $\square$

We have seen that a fundamental region for a lattice is not unique. However, as the next theorem shows, its area is unique, and may therefore be regarded as a function of the lattice alone. Even though this result is true for any 'reasonably well-behaved' fundamental regions, in order to avoid having to consider pathological cases we will just prove it for fundamental polygons.

For any measurable set $X \subseteq \mathbb{C}$, let $\mu(X)$ be the area, or measure, of X. For notational convenience, we will write $\omega(X)$ for $X + \omega$; since the translation $z \mapsto z + \omega$ is an isometry of $\mathbb{C}$, we have $\mu(\omega(X)) = \mu(X)$.

Theorem 3.4.6. *Let P_1 and P_2 be fundamental polygons for a lattice Ω. Then $\mu(P_1) = \mu(P_2)$.*

Proof. If Q_j is the interior of $P_j (j = 1, 2)$ then $\mu(P_j) = \mu(Q_j)$. Now

$$P_1 \supseteq P_1 \cap \bigcup_{\omega \in \Omega} \omega(Q_2) = \bigcup_{\omega \in \Omega} (P_1 \cap \omega(Q_2)).$$

As Q_2 is the interior of a fundamental region, the sets $P_1 \cap \omega(Q_2)$ are disjoint, and hence

$$\mu(P_1) \geqslant \sum_{\omega \in \Omega} \mu(P_1 \cap \omega(Q_2))$$

$$= \sum_{\omega \in \Omega} \mu((-\omega)(P_1) \cap Q_2)$$

$$= \sum_{\omega \in \Omega} \mu(\omega(P_1) \cap Q_2),$$

since ω ranges over Ω as $-\omega$ does. Now as P_1 is a fundamental region,

$$\bigcup_{\omega \in \Omega} \omega(P_1) = \mathbb{C},$$

and hence

$$\bigcup_{\omega \in \Omega} (\omega(P_1) \cap Q_2) = Q_2.$$

Hence

$$\sum_{\omega \in \Omega} \mu(\omega(P_1) \cap Q_2) \geqslant \mu(Q_2) = \mu(P_2),$$

giving

$$\mu(P_1) \geqslant \mu(P_2).$$

Interchanging P_1 and P_2 we have $\mu(P_2) \geqslant \mu(P_1)$, and hence $\mu(P_1) = \mu(P_2)$. □

Applying this result to the fundamental regions in Fig. 3.8, we get a proof of Pythagoras' theorem.

3.5 The torus

If a function f is doubly periodic with respect to a lattice $\Omega = \Omega(\omega_1, \omega_2)$, then its behaviour on $\mathbb{C}$ is determined by its behaviour on a fundamental region P for Ω, which we can take to be the parallelogram with vertices 0, ω_1, ω_2, $\omega_1 + \omega_2$; this behaviour is then repeated on all translates $P + \omega$ ($\omega \in \Omega$). Hence we can regard f as a function defined on P, and since f takes the same values on congruent boundary points we may as well identify them and regard f as a function defined on the resulting space T, which is known as a *torus* (see Fig. 3.11). Conversely, any function defined on T may be regarded as a doubly periodic function on $\mathbb{C}$.

Fig. 3.11

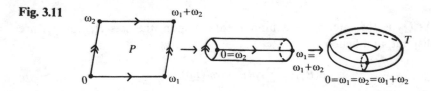

By the definition of a fundamental region, we see that for every Ω-orbit on $\mathbb{C}$ there is just one point of T, and vice-versa, so we can think of T as the set of Ω-orbits, that is, as the set $\mathbb{C}/\Omega$ of cosets of Ω in $\mathbb{C}$. Since Ω is a normal subgroup of the additive group $\mathbb{C}$ (normal since $\mathbb{C}$ is abelian), the quotient set $T = \mathbb{C}/\Omega$ has the structure of a group (also abelian). Since there is a continuous function from the closed bounded set P onto T (given by identifying boundary points), it follows that T is compact.

This construction of T from the action of Ω on $\mathbb{C}$ may be generalised as follows. Let X be a topological space and G a group of homeomorphisms of X. Then the action of G on X breaks X up into G-orbits. We denote the G-orbit of x by $[x]_G$, so that $y \in [x]_G$ if and only if $g(x) = y$ for some $g \in G$. The set of G-orbits is denoted by X/G and is called the *orbit-space* (or *quotient-space*) of X by G. We define the projection map $p: X \to X/G$ by $p(x) = [x]_G$; then we topologise X/G by defining a set $V \subseteq X/G$ to be open if and only if $p^{-1}(V)$ is open in X. With this definition, p is clearly continuous; p is also open (see §1.5) since if U is open in X then

$$p^{-1}(p(U)) = \bigcup_{g \in G} g(U)$$

is also open (as each $g \in G$ is a homeomorphism), and thus $p(U)$ is open. For example, if $\mathbb{Z}$ acts on $\mathbb{R}$ by translation ($n(x) = x + n$ for all $x \in \mathbb{R}$, $n \in \mathbb{Z}$) then $\mathbb{R}/\mathbb{Z}$ is homeomorphic to the circle S^1; the quotient of the sphere S^2 by the group G generated by the antipodal map $(x_1, x_2, x_3) \mapsto (-x_1, -x_2, -x_3)$ is homeomorphic to the real projective plane.

Returning to the torus $T = \mathbb{C}/\Omega$, we see that for each point $[z] = [z]_\Omega \in T$, $p^{-1}([z])$ is the Ω-orbit $[z] = z + \Omega$ of z, and is therefore discrete. Let d be the smallest distance between any two points of $p^{-1}([z])$ (we have $d > 0$ by the argument in Theorem 3.1.3), and let U be an open disc of radius at most $d/2$, centred on any point in $p^{-1}([z])$, for example on z itself. Then U contains at most one point from each Ω-orbit, so if we define $V = p(U)$ then the map $p: U \to V$ is bijective, open and continuous, and is therefore a homeomorphism. Thus every point $[z] \in T$ has a neighbourhood V homeomorphic to an open set in $\mathbb{C}$. Such a space is called a *surface*; the torus is the simplest

Fig. 3.12

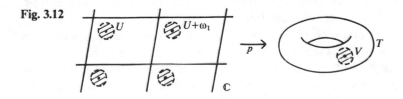

compact surface other than the sphere. (See Chapter 4 for further details about surfaces.)

Note that $p^{-1}(V)$ consists of disjoint open sets of the form $U + \omega$ ($\omega\in\Omega$), each of which is mapped homeomorphically onto V by p (see Fig. 3.12). Thus, in the terminology of §1.5, $\mathbb{C}$ is a covering space of T, and p is a covering map. There are infinitely many sheets, and no branch-points.

3.6 General properties of elliptic functions

Definition. A meromorphic function $f:\mathbb{C}\to\Sigma$ is *elliptic* with respect to a lattice $\Omega\subseteq\mathbb{C}$ if f is doubly periodic with respect to Ω, that is, if

$$f(z + \omega) = f(z) \quad \text{for all } z\in\mathbb{C}, \omega\in\Omega,$$

so that each $\omega\in\Omega$ is a period of f.

The connection between doubly periodic functions and the ellipse is tenuous, but historically interesting. From the second half of the seventeenth century onwards, much attention was devoted to certain integrals, which, it seemed, could not be evaluated using so-called 'elementary' functions. These were known as *elliptic integrals*, since they included the integral

$$\int\sqrt{\left(\frac{a^2 - ex^2}{a^2 - x^2}\right)}\,dx,$$

giving the circumference of the ellipse $(x^2/a^2) + (y^2/b^2) = 1$, where $e = 1 - (b^2/a^2)$. In the late 1790s, Gauss noticed that, just as the inverse functions of the integrals

$$\int\frac{dx}{1 + x^2} \quad \text{and} \quad \int\frac{dx}{\sqrt{(1 - x^2)}},$$

give the *simply* periodic trigonometric functions $\tan x$ and $\sin x$, the inverse functions of certain elliptic integrals, such as

$$\int\frac{dx}{\sqrt{(1 - x^4)}},$$

give *doubly* periodic functions. These ideas were unpublished, and were rediscovered in the 1820s by Abel and Jacobi, who gave the name 'elliptic' to these doubly periodic functions.

If f is elliptic with respect to Ω, then we may regard f as a function $f:T\to\Sigma$, where T is the torus $T = \mathbb{C}/\Omega$. When we considered meromorphic functions $f:\Sigma\to\Sigma$, the compactness of the domain Σ allowed us to use Liouville's theorem to show that if f is analytic then f is constant: the proofs of Theorems 1.3.3 and 1.3.4 are based on this idea. Here the domain T is compact, so we can prove similar results for elliptic functions; we shall see that elliptic functions are related to the torus in the same way that rational functions are related to the sphere.

So far, the only elliptic functions we have met are the constant functions, and it is a substantial problem to construct non-constant elliptic functions. Before doing this, in §3.9, we will examine some of the elementary properties which elliptic functions must possess.

Suppose that f is elliptic with respect to a lattice Ω, that $c \in \Sigma$, and that f is not identically equal to c. Then the solutions of $f(z) = c$ are isolated, and each solution has finite multiplicity, congruent solutions having the same multiplicity. Since the solutions are isolated, any fundamental polygon P for Ω contains only finitely many solutions (since P is compact), and, by replacing P by $P + t\,(t \in \mathbb{C})$ if necessary, we may assume that there are no solutions on the boundary ∂P of P. Let the solutions within P be $z = z_1, \ldots, z_r$, with multiplicities $k_1, \ldots, k_r$, and let $N = k_1 + \ldots + k_r$. Then we say that 'there are N solutions of $f(z) = c$'. Since $z_1, \ldots, z_r$ are representatives of the congruence classes of solutions of $f(z) = c$, for $z \in \mathbb{C}$, we can think of N as the sum of the multiplicities of the solutions of $f([z]) = c$, where $[z] \in T = \mathbb{C}/\Omega$.

With this in mind, we define the *order* ord(f) of an elliptic function f to be the number of solutions of $f(z) = \infty$, that is, the sum of the orders of the congruence classes of poles of f. (This is analogous to the degree deg(f) of a rational function f, equal to the number of solutions of $f(z) = \infty$, counting multiplicities, by Theorem 1.4.2.) For the rest of this section, we assume that f is elliptic with respect to Ω, that ord$(f) = N$, and that P is a fundamental parallelogram for Ω with vertices t, $t + \omega_1$, $t + \omega_2$, $t + \omega_1 + \omega_2$, where $\{\omega_1, \omega_2\}$ is a basis for Ω, and t is chosen so that ∂P contains no zeros or poles of f (see Fig. 3.13).

Fig. 3.13

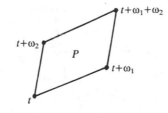

Theorem 3.6.1. *f is constant if and only if $N = 0$. (Thus an analytic elliptic function must be constant.)*

Proof. If f is constant and meromorphic, then it has no poles in $\mathbb{C}$, so $N = 0$. Conversely, suppose that $N = 0$. Then f has no poles, so f is analytic on $\mathbb{C}$. Now P is compact, and f is continuous, so $f(P)$ is a compact subset of $\mathbb{C}$ and is therefore bounded. Since $f(\mathbb{C}) = f(P)$, it follows that f is bounded on $\mathbb{C}$, so Liouville's theorem implies that f, being analytic and bounded, must be constant. (Equivalently, we can use the compactness of $T = \mathbb{C}/\Omega$ to prove that f is bounded, since $f(\mathbb{C}) = f(T)$.) $\square$

Theorem 3.6.2. *The sum of the residues of f within P is zero.*

Proof. Since f is meromorphic, and is analytic on ∂P, $(1/2\pi i)\int_{\partial P} f(z)\,dz$ is equal to the sum of the residues within P.

Now let Γ_1, Γ_2, Γ_3 and Γ_4 be the sides of P from t to $t + \omega_1$, $t + \omega_1$ to $t + \omega_1 + \omega_2$, $t + \omega_1 + \omega_2$ to $t + \omega_2$, and $t + \omega_2$ to t respectively, so that

$$\int_{\partial P} f(z)\,dz = \sum_{j=1}^{4} \int_{\Gamma_j} f(z)\,dz,$$

where the direction of integration along Γ_j is consistent with the positive (i.e. anti-clockwise) orientation of ∂P, as in Fig. 3.14. Now

$$\int_{\Gamma_3} f(z)\,dz = \int_{\Gamma_3} f(z + \omega_2)\,dz \quad \text{(since ω_2 is a period of f)}$$

$$= -\int_{\Gamma_1 + \omega_2} f(z + \omega_2)\,d(z + \omega_2)$$

$$\text{(since $\Gamma_3 = \Gamma_1 + \omega_2$ with the reverse orientation)}$$

$$= -\int_{\Gamma_1} f(z)\,dz \quad \text{(substituting $z - \omega_2$ for z).}$$

Similarly $\int_{\Gamma_4} f(z)\,dz = -\int_{\Gamma_2} f(z)\,dz$ since ω_1 is a period, so $\int_{\partial P} f(z)\,dz = 0$ and hence the sum of the residues is zero. $\square$

Fig. 3.14

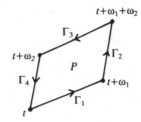

Corollary 3.6.3. *There are no elliptic functions of order $N = 1$.*

Proof. If f were elliptic of order 1, it would have a single pole of order 1 in P, say at $z = a \in P$, so

$$f(z) = \sum_{j=-1}^{\infty} a_j (z-a)^j \quad \text{near } z = a,$$

with $a_{-1} \neq 0$. Thus the sum of the residues of f within P is equal to a_{-1}, which is non-zero, contradicting Theorem 3.6.2. $\square$

Later, we shall construct an elliptic function $\wp$ of order 2, and then show that all other elliptic functions may be expressed in terms of $\wp$, just as all simply periodic functions may be expressed in terms of $e^{2\pi i z}$.

Theorem 3.6.4. *If f has order $N > 0$ then f takes each value $c \in \Sigma$ exactly N times.*

Proof. This is the definition of N if $c = \infty$, so we may assume that $c \in \mathbb{C}$. Replacing f by $f - c$ (which has the same order as f), we may assume that $c = 0$. Now f'/f is meromorphic, and since ∂P contains no poles or zeros of f, f'/f is analytic on ∂P. We may therefore integrate f'/f around ∂P. Since f is elliptic, so is f' and hence so is f'/f, so applying the argument used in the proof of Theorem 3.6.2, we see that

$$\int_{\partial P} \frac{f'(z)}{f(z)} \, dz = 0,$$

and hence the sum of the residues of f'/f within P must be zero.

Now f'/f has poles at the zeros and poles of f, and nowhere else. Suppose that f has a zero of multiplicity k at $z = a \in P$, so that $f(z) = (z-a)^k g(z)$ near $z = a$, where g is analytic and $g(a) \neq 0$. Then $f'(z) = k(z-a)^{k-1} g(z) + (z-a)^k g'(z)$ near $z = a$, and so

$$\frac{f'(z)}{f(z)} = \frac{k}{z-a} + \frac{g'(z)}{g(z)}$$

near $z = a$, so that f'/f has residue k at $z = a$. A similar argument, with $f(z) = (z-a)^{-k} g(z)$, shows that f'/f has residue $-k$ at each pole of multiplicity k of f. Since the sum of the residues of f'/f is zero, the number of zeros of f must equal the number of poles, counting multiplicities, so the equation $f(z) = 0$ has N solutions, as required. $\square$

Compare the above result with Theorem 1.4.2 for rational functions. Similarly, there are analogues of Theorems 1.3.3 and 1.3.4:

Theorem 3.6.5. *Let f and g be elliptic functions with respect to Ω, with* poles at the same points in $\mathbb{C}$, and with the same principal parts at these points. Then $f(z) = g(z) + c$ for some constant c.

Proof. The function $f - g$ is elliptic, and has order 0 since it has no poles, so $f - g$ is constant by Theorem 3.6.1. $\square$

Theorem 3.6.6. *Let f and g be elliptic functions with respect to Ω, with zeros and poles of the same orders at the same points of $\mathbb{C}$. Then $f(z) = cg(z)$ for some constant $c \neq 0$.*

Proof. Replace $f - g$ by f/g in the proof of Theorem 3.6.5. $\square$

A rational function $f : \Sigma \to \Sigma$, which is not identically zero, must have finitely many zeros (say at $a_1, \ldots, a_r$ with multiplicities $k_1, \ldots, k_r$) and finitely many poles (say at $b_1, \ldots, b_s$ with multiplicities $l_1, \ldots, l_s$); conversely, given any choice of points $a_1, \ldots, a_r, b_1, \ldots, b_s \in \Sigma$ and multiplicities $k_1, \ldots, k_r$, $l_1, \ldots, l_s \geqslant 1$, there exists a rational function f with these zeros and poles, with these multiplicities, provided

(i) $k_1 + \ldots + k_r = l_1 + \ldots + l_s$ (both must equal the degree of f), and
(ii) the sets $\{a_1, \ldots, a_r\}$ and $\{b_1, \ldots, b_s\}$ are disjoint (zeros and poles cannot coincide):

we take $f(z) = \prod_j (z - a_j)^{k_j} / \prod_j (z - b_j)^{l_j}$, where these products range over all j such that $a_j, b_j \in \mathbb{C}$, but exclude factors where $a_j = \infty$ or $b_j = \infty$.

If an elliptic function $f : \mathbb{C} \to \Sigma$ is to have its zeros and poles at the congruence classes $[a_1], \ldots, [a_r]$ and $[b_1], \ldots, [b_s]$ with multiplicities $k_1, \ldots, k_r$ and $l_1, \ldots, l_s$, then condition (i) is necessary by Theorem 3.6.4, and corresponding to (ii) we have the necessary condition

(ii)' the sets $[a_1] \cup \ldots \cup [a_r]$ and $[b_1] \cup \ldots \cup [b_s]$ are disjoint.

The next result shows that, in contrast with the situation for rational functions, these conditions are not *sufficient* for the existence of f.

Theorem 3.6.7. *Let the congruence classes of zeros and poles of an elliptic function f be $[a_1], \ldots, [a_r]$ and $[b_1], \ldots, [b_s]$, with multiplicities $k_1, \ldots, k_r$ and $l_1, \ldots, l_s$. Then*

$$\sum_{j=1}^{r} k_j a_j \sim \sum_{j=1}^{s} l_j b_j \bmod \Omega.$$

Proof. As usual, let P be a fundamental parallelogram for Ω, chosen so that f has no zeros or poles on ∂P. The effect of replacing any a_j or b_j by a congruent point is to add an element of Ω to $\sum k_j a_j - \sum l_j b_j$, and this does not affect the conclusion of the theorem, so we may assume that a_j, $b_j \in P$ for all j.

First we prove that

$$\sum k_j a_j - \sum l_j b_j = \frac{1}{2\pi i} \int_{\partial P} \frac{z f'(z)}{f(z)} \, dz.$$

The poles of $z f'/f$ are at the zeros and poles of f, and if f has a zero of multiplicity k at $z = a$, then $f(z) = (z - a)^k g(z)$ near $z = a$, with g analytic and $g(a) \neq 0$. Then

$$\frac{z f'(z)}{f(z)} = \frac{z}{(z - a)^k g(z)} (k(z - a)^{k-1} g(z) + (z - a)^k g'(z))$$

$$= \frac{kz}{z - a} + \frac{z g'(z)}{g(z)}$$

near $z = a$, with $z g'/g$ analytic at a, so $z f'/f$ has residue ka at $z = a$. Similarly, if f has a pole of multiplicity l at $z = b$, then $z f'/f$ has residue $-lb$ at $z = b$. Now the zeros and poles of f within P are at $a_1, \ldots, a_r$ and $b_1, \ldots, b_s$ with multiplicities $k_1, \ldots, k_r$ and $l_1, \ldots, l_s$, so

$$\frac{1}{2\pi i} \int_{\partial P} \frac{z f'(z)}{f(z)} \, dz,$$

which is equal to the sum of the residues of $z f'/f$, takes the value $\sum k_j a_j - \sum l_j b_j$.

If we label the sides of P as in the proof of Theorem 3.6.2, then

$$\int_{\Gamma_2} \frac{z f'(z)}{f(z)} \, dz = \int_{\Gamma_2} \frac{(z - \omega_1) f'(z)}{f(z)} \, dz + \int_{\Gamma_2} \frac{\omega_1 f'(z)}{f(z)} \, dz$$

$$= - \int_{\Gamma_4} \frac{z f'(z + \omega_1)}{f(z + \omega_1)} \, dz + \omega_1 [\log f(z)]_{\Gamma_2}$$

$$= - \int_{\Gamma_4} \frac{z f'(z)}{f(z)} \, dz + 2\pi n_1 i \omega_1$$

for some $n_1 \in \mathbb{Z}$, using the facts that Γ_4 is just the path $\Gamma_2 - \omega_1$, with reverse orientation, that f and f' are periodic, and that $\log f(z)$ changes its value by an integer multiple of $2\pi i$ as z travels along Γ_2 from $t + \omega_1$ to $t + \omega_1 + \omega_2$ (since $f(t + \omega_1) = f(t + \omega_1 + \omega_2)$).

Similarly, we have

$$\int_{\Gamma_1} \frac{z f'(z)}{f(z)} \, dz = - \int_{\Gamma_3} \frac{z f'(z)}{f(z)} \, dz + 2\pi n_2 i \omega_2$$

for some $n_2 \in \mathbb{Z}$, and hence

$$\Sigma k_j a_j - \Sigma l_j b_j = \frac{1}{2\pi i} \int_{\partial P} \frac{z f'(z)}{f(z)} dz$$

$$= \frac{1}{2\pi i} \sum_{j=1}^{4} \int_{\Gamma_j} \frac{z f'(z)}{f(z)} dz$$

$$= \frac{1}{2\pi i} (2\pi n_1 i \omega_1 + 2\pi n_2 i \omega_2)$$

$$= n_1 \omega_1 + n_2 \omega_2,$$

which is an element of Ω, as required. $\square$

When we come to construct elliptic functions, we shall see that conditions (i) and (ii)′, together with the condition

(iii) $\Sigma k_j a_j \sim \Sigma l_j b_j \bmod \Omega$,

are sufficient for the existence of an elliptic function f with prescribed zeros and poles.

3.7 Uniform and normal convergence

In §3.9 we shall construct elliptic functions explicitly, using infinite series and products. To illustrate some of the ideas involved let us briefly consider *simply* periodic functions. Suppose that we wanted to construct such a function $F(z)$ without assuming knowledge of the exponential or trigonometric functions. We might try defining $F(z)$ by means of an infinite series

$$F(z) = \sum_{n=-\infty}^{\infty} f(z-n),$$

where f is chosen so that this series converges at z. By this series we mean

$$F(z) = \sum_{n=0}^{-\infty} f(z-n) + \sum_{n=1}^{\infty} f(z-n)$$

$$= \sum_{n=-1}^{-\infty} f(z-n) + \sum_{n=0}^{\infty} f(z-n)$$

$$= \sum_{m=0}^{-\infty} f(z+1-m) + \sum_{m=1}^{\infty} f(z+1-m) \qquad (m = n+1)$$

$$= F(z+1),$$

so that F is a function of period 1. For example, we shall see later that $\sum_{n=-\infty}^{\infty} (z-n)^{-2}$ represents the simply periodic meromorphic function $\pi^2 \operatorname{cosec}^2 \pi z$.

We can use the same technique to construct a *doubly* periodic function with respect to a lattice Ω, by defining

$$F(z) = \sum_{\omega \in \Omega} f(z - \omega).$$

Now the order of summation over Ω is not so apparent, and we may need to rearrange this series in order to prove periodicity. However, if the series is *absolutely* convergent then its sum is independent of the order of summation and so this procedure is justified.

Another difficulty is to ensure that the simply or doubly periodic function F is meromorphic; for this we require f to be meromorphic and the series defining F to be *uniformly* convergent, since this allows term-by-term differentiation. After considering these problems in a little more detail, we will therefore introduce the concept of *normal* convergence, since this implies both absolute and uniform convergence.

The theory of infinite series usually deals with summation of terms which are indexed by the set $\mathbb{N}$ of non-negative integers. However, as indicated above, we will need to consider series (and also products) whose terms are indexed by the set $\mathbb{Z}$ of all integers, or by a lattice Ω. Now each of these indexing sets $\Lambda(=\mathbb{Z}$ or $\Omega)$ is countably infinite, that is, there is a bijection $n \mapsto \lambda_n$ between $\mathbb{N}$ and Λ, so that the elements of Λ may be arranged in a sequence. For example,

$$0, \quad 1, \quad -1, \quad 2, \quad -2, \quad 3, \quad -3, \ldots$$

gives an ordering of $\mathbb{Z}$, and we will show in §3.9 how the elements of a lattice Ω may be ordered.

If Λ is *any* countably infinite set, with a particular ordering $\lambda_0, \lambda_1, \lambda_2, \ldots$, then by $\sum_{\lambda \in \Lambda} a_\lambda$ we will mean $\lim_{n \to \infty} \sum_{j=0}^{n} a_{\lambda_j}$ provided this limit exists. In general, the value (and even the existence) of this limit will depend on the particular ordering we choose for Λ, but if the series is absolutely convergent (that is, if $\lim_{n \to \infty} \sum_{j=0}^{n} |a_{\lambda_j}|$ exists), then $\sum_{\lambda \in \Lambda} a_\lambda$ is independent of the ordering of Λ. For example, if we choose the ordering $0, 1, -1, 2, -2, \ldots$ of $\Lambda = \mathbb{Z}$, then the sum of an absolutely convergent series $\sum_{\lambda \in \Lambda} a_\lambda$ is easily seen to be $\sum_{n=0}^{-\infty} a_n + \sum_{n=1}^{\infty} a_n$; however, if $\sum_{\lambda \in \mathbb{Z}} a_\lambda$ is conditionally convergent, then neither $\sum_{n=0}^{-\infty} a_n$ nor $\sum_{n=1}^{\infty} a_n$ need converge.

We can simplify the notation by writing b_j for a_{λ_j}. Then it is easy to check that our definition of $\sum_{\lambda \in \Lambda} a_\lambda$ coincides with the traditional definition of $\sum_{j=0}^{\infty} b_j$, so in effect we have replaced Λ by $\mathbb{N}$ as an indexing set, where $\mathbb{N}$ is ordered in the usual way. We will therefore state our results in this section for series of terms indexed by $\mathbb{N}$, but the reader should bear in mind that these results generalise in a straightforward way to all countably infinite indexing sets provided we have absolute convergence.

Definition. Let (u_n) be a sequence of functions $u_n : E \to \mathbb{C}$, defined on some set E; then u_n *converges uniformly* to a function $u : E \to \mathbb{C}$ if, for every $\varepsilon > 0$ there exists $n_0 \in \mathbb{N}$ (depending possibly on ε but not on z) such that $|u_n(z) - u(z)| < \varepsilon$ for all $n > n_0$ and for all $z \in E$.

For example, if $E_k = \{z \in \mathbb{C} \,|\, |z| < k\}$, then z^n converges uniformly to 0 on E_k provided $k < 1$ (since $n > \ln(\varepsilon)/\ln(k)$ implies $|z^n| < \varepsilon$ for all $z \in E_k$). However, if $k = 1$, then although z^n still converges to 0 on E_1, convergence is no longer uniform, since for each fixed n we have $\lim_{z \to 1} z^n = 1$ and hence there is no n satisfying $|z^n| < \varepsilon = \frac{1}{2}$ for all $z \in E_1$. Nevertheless, each compact subset $K \subseteq E_1$ is contained in some E_k with $k < 1$, so z^n converges uniformly on all compact subsets of E_1; this property, which we now define precisely, turns out to be sufficient to prove the results we need.

Definition. Let R be a region in $\mathbb{C}$, and let (u_n) be a sequence of functions $u_n : R \to \mathbb{C}$; then (u_n) *converges uniformly on all compact subsets* of R if, for each compact $K \subseteq R$, the sequence of restrictions $(u_n | K)$ converges uniformly on K. Since K can be covered by the interiors of finitely many closed discs $D \subseteq R$, it is easily seen that this condition is equivalent to the property that $(u_n | D)$ converges uniformly on D for each closed disc $D \subseteq R$.

Then we have the following important theorem:

Theorem 3.7.1. *Let (u_n) be a sequence of analytic functions on a region $R \subseteq \mathbb{C}$, uniformly convergent to a function u on all compact subsets of R. Then u is analytic on R, and the sequence of derivatives (u'_n) converges uniformly to u' on all compact subsets of R.*

Proof (outline). As remarked we just need to prove this result for a closed disc $D \subseteq R$ which we will suppose has centre z_0 and radius r. We use the following well-known and easily proved properties of uniform convergence: (i) the limit function u is continuous in D, (ii) $\lim_{n \to \infty} \int_\gamma u_n = \int_\gamma u$, for all closed curves γ in D.

Now as u_n is analytic, Cauchy's theorem implies that $\int_\gamma u_n = 0$, and hence by (ii) $\int_\gamma u = 0$ for all closed curves γ in D. Hence by Morera's theorem (Theorem A.5), u is analytic in D.

To prove that $u'_n \to u'$ we use the Cauchy integral formula (Theorem A.3). This implies that if δ is a circle with centre z_0 and radius $\rho > r$ whose interior is contained in R then

$$|u_n'(z) - u'(z)| = \left| \frac{1}{2\pi i} \int_\delta \frac{u_n(w) - u(w)}{(w - z)^2} dw \right|$$

and standard estimates now show that u_n' converges to u'. □

We can extend this result from sequences to series in the usual way. We say that $\sum_{n=0}^\infty u_n(z)$ *converges uniformly* to $u(z)$ on a set E if the sequence of partial sums $\sum_{n=0}^m u_n(z)$ converges uniformly to $u(z)$ on E as $m \to \infty$. We immediately deduce:

Corollary 3.7.2. *Let (u_n) be a sequence of analytic functions on a region $R \subseteq \mathbb{C}$; if $\sum_{n=0}^\infty u_n(z)$ is uniformly convergent to $u(z)$ on all compact subsets of R, then $u(z)$ is analytic on R and $\sum_{n=0}^\infty u_n'(z)$ is uniformly convergent to $u'(z)$ on all compact subsets of R.* □

For example, $\sum_{n=0}^\infty z^n$ is uniformly convergent to $(1 - z)^{-1}$ on all compact subsets of E_1 (though not on E_1 itself), and differentiating term-by-term, $\sum_{n=1}^\infty nz^{n-1}$ converges uniformly to $(d/dz)(1 - z)^{-1} = (1 - z)^{-2}$ on all compact subsets of E_1.

In order to apply Corollary 3.7.2, we need to be able to prove uniform convergence of $\sum_{n=0}^\infty u_n(z)$ on compact subsets. This can often be done using

Theorem 3.7.3 *(Weierstrass' M-test). Let (u_n) be a sequence of functions $u_n : E \to \mathbb{C}$, defined on some set E, such that*

(i) *for each $n \in \mathbb{N}$ there exists $M_n \in \mathbb{R}$ satisfying $|u_n(z)| \leqslant M_n$ for all $z \in E$,*
(ii) *$\sum_{n=0}^\infty M_n$ converges.*

Then $\sum_{n=0}^\infty u_n(z)$ converges uniformly on E, and converges absolutely for each $z \in E$.

Proof. Absolute convergence follows immediately from (i) and (ii) by the comparison test. For a proof of uniform convergence, using Cauchy's criterion, see Apostol [1963]. □

A useful method of applying Weierstrass' M-test is to use *normal convergence*. We define the *norm* $\|f\| = \|f\|_E$ of a function $f : E \to \mathbb{C}$ to be $\sup_{z \in E} |f(z)|$, provided f is bounded, and we say that a series of functions $\sum u_n(z)$ is *normally convergent* on E if

(i) each u_n is bounded on E,
(ii) the series $\sum \|u_n\|$ converges.

By taking $M_n = \|u_n\|$ in Theorem 3.7.3, we see that normal convergence implies both uniform and absolute convergence of $\sum u_n$ on E.

For example, suppose that each function u_n is analytic on some region R; then $|u_n(z)|$ is continuous, so $\|u_n\|_K$ exists for each compact $K \subseteq R$. The series $\sum \|u_n\|_K$ consists of non-negative real numbers, so we may be able to prove that it converges by using elementary tests, such as the comparison, ratio or integral tests; if so, then $\sum u_n$ is normally and hence uniformly and absolutely convergent on compact subsets of R, so it represents an analytic function and may be differentiated term-by-term.

We also need to deal with series of *meromorphic* functions. Suppose that (u_n) is a sequence of meromorphic functions on a region R, and that for each compact subset $K \subseteq R$ there exists $N_K \in \mathbb{N}$ such that

(i) $u_n(z)$ has no poles (and is thus analytic) in K for $n > N_K$,

(ii) $\sum_{n > N_K} u_n(z)$ is uniformly convergent on K.

Then we say that $\sum u_n(z)$ *converges uniformly on all compact subsets* of R. Since $\sum_{n \leq N_K} u_n(z)$ is meromorphic on the interior $\overset{\circ}{K}$ of K (being a sum of finitely many meromorphic functions), and since $\sum_{n > N_K} u_n(z)$ is analytic on $\overset{\circ}{K}$ (being a uniformly convergent series of analytic functions), the function

$$\sum_{n=0}^{\infty} u_n(z) = \sum_{n \leq N_K} u_n(z) + \sum_{n > N_K} u_n(z)$$

is meromorphic on $\overset{\circ}{K}$, its poles being included among the poles of the functions $u_n(z)$ for $n \leq N_K$. (It is easily seen that, for any given z, the value of $\sum u_n(z)$ is independent of the choice of K or of N_K.) Since each point $z \in R$ has a neighbourhood with compact closure $K \subseteq R$, $\sum u_n(z)$ is meromorphic on R.

From Corollary 3.7.2 and the above argument we have

Theorem 3.7.4.   *Let $\sum u_n(z)$ be a series of meromorphic functions on a region $R \subseteq \mathbb{C}$, uniformly convergent to $u(z)$ on all compact subsets of R. Then $u(z)$ is meromorphic on R, and the series $\sum u'_n(z)$ converges uniformly to $u'(z)$ on all compact subsets of R.* $\square$

For example, consider the series $\sum_{n=-\infty}^{\infty} (z-n)^{-2}$; here we are summing over $\mathbb{Z}$ rather than $\mathbb{N}$, but this is no problem provided we have absolute convergence. Each compact set $K \subseteq \mathbb{C}$ is bounded, so all but finitely many of the meromorphic functions $(z-n)^{-2}$ are analytic on K. By comparing $\sum \|(z-n)^{-2}\|_K$ with the convergent series $\sum n^{-2}$ it is not difficult to show that $\sum (z-n)^{-2}$ converges normally, and hence uniformly and absolutely, on all compact subsets of $\mathbb{C}$, so this series represents a simply periodic meromorphic function on $\mathbb{C}$, with double poles at each $n \in \mathbb{Z}$.

3.8 Infinite products

In order to construct elliptic functions, we shall consider certain infinite products of analytic functions. The terms of such a product will be indexed by a lattice Ω, but as in the case of infinite series, it is sufficient to state general results for products indexed by $\mathbb{N}$.

We first consider products of non-zero complex numbers; later, when we consider products of *functions*, we will consider products in which some factors may be zero. Let (b_n) be a sequence of non-zero complex numbers, and let $p_n = b_0 b_1 \ldots b_n$. Then $\prod_{n=0}^{\infty} b_n$ *converges to* p, written $\prod_{n=0}^{\infty} b_n = p$, if $p \in \mathbb{C}$ and

(i) $\lim_{n \to \infty} p_n = p$,

(ii) $p \neq 0$.

If $\prod_{n=0}^{\infty} b_n$ converges, then $\lim_{n \to \infty} b_n = \lim_{n \to \infty} p_n / p_{n-1} = \lim_{n \to \infty} p_n / \lim_{n \to \infty} p_{n-1} = 1$, using condition (ii). Thus, putting $b_n = 1 + c_n$ we see that the convergence of $\prod_{n=0}^{\infty} (1 + c_n)$ implies that $\lim_{n \to \infty} c_n = 0$.

We can convert infinite products into infinite series by using logarithms. For complex numbers, $\log(z)$ is not a uniquely defined function of z, distinct values differing by multiplies of $2\pi i$. We therefore introduce the *principal value* of $\log(z)$, defined to be

$$\text{Log}(z) = \ln(|z|) + i \arg(z)$$

for each $z \neq 0$, where $-\pi < \arg(z) \leq \pi$ and $\ln(|z|)$ is the unique real value of $\log(|z|)$. We cannot assume that $\text{Log}(ab) = \text{Log}(a) + \text{Log}(b)$, since the two sides may differ by $\pm 2\pi i$. However, as $\exp(2\pi i) = 1$ we have

$$ab = \exp(\text{Log}(ab)) = \exp(\text{Log}(a) + \text{Log}(b))$$

for all $a, b \neq 0$.

Theorem 3.8.1. *If $b_n \neq 0$ for all n, then $\prod_{n=0}^{\infty} b_n$ converges if and only if $\sum_{n=0}^{\infty} \text{Log}(b_n)$ converges, in which case $\prod_{n=0}^{\infty} b_n = \exp \sum_{n=0}^{\infty} \text{Log}(b_n)$.*

Proof. Suppose that $\sum_{n=0}^{\infty} \text{Log}(b_n)$ converges to w. As the exponential function is continuous, we have

$$\exp(w) = \exp\left(\lim_{n \to \infty} \sum_{k=0}^{n} \text{Log}(b_k) \right)$$

$$= \lim_{n \to \infty} \left(\exp \sum_{k=0}^{n} \text{Log}(b_k) \right)$$

$$= \lim_{n \to \infty} \left(\prod_{k=0}^{n} b_k \right).$$

Since $\exp(w) \neq 0$, this shows that $\prod_{n=0}^{\infty} b_n$ converges to $\exp(w) = \exp \sum_{n=0}^{\infty} \text{Log}(b_n)$.

Conversely, suppose that $\prod_{n=0}^{\infty} b_n$ converges to p (so $p \neq 0$). Thus $\lim_{n \to \infty} p_n = p$ where $p_n = \prod_{k=0}^{n} b_k$. Putting

$$s_n = \sum_{k=0}^{n} \text{Log}(b_k),$$

we have

$$s_n = \text{Log}(p_n) + 2\pi i q_n,$$

where $q_n \in \mathbb{Z}$ for each n. We need to show that q_n is constant for all sufficiently large n. We have

$$2\pi i(q_{n+1} - q_n) = s_{n+1} - s_n + \text{Log}(p_n) - \text{Log}(p_{n+1})$$
$$= \text{Log}(b_{n+1}) + \text{Log}(p_n) - \text{Log}(p_{n+1})$$
$$= \ln|b_{n+1}| + \ln|p_n| - \ln|p_{n+1}|$$
$$+ i(\arg(b_{n+1}) + \arg(p_n) - \arg(p_{n+1})).$$

Equating imaginary parts, we have

$$|q_{n+1} - q_n| = \frac{1}{2\pi} |\arg(b_{n+1}) + (\arg p_n - \arg p) + (\arg p - \arg p_{n+1})|.$$

As $n \to \infty$ we have $b_{n+1} \to 1$ and $p_n, p_{n+1} \to p$, so for sufficiently large n we can make each of $|\arg(b_{n+1})|, |\arg p_n - \arg p|$ and $|\arg p - \arg p_{n+1}|$ less than $\frac{2}{3}\pi$. This gives $|q_{n+1} - q_n| < 1$, so $q_{n+1} = q_n$ since $q_{n+1} - q_n \in \mathbb{Z}$. Hence q_n is constant, say $q_n = q$, for all sufficiently large n, giving

$$\lim_{n \to \infty} s_n = \lim_{n \to \infty} (\text{Log}(p_n) + 2\pi i q_n)$$
$$= \text{Log}(p) + 2\pi i q,$$

so $\sum_{n=0}^{\infty} \text{Log}(b_n)$ converges to $\text{Log}(p) + 2\pi i q$, and hence $\prod_{n=0}^{\infty} b_n = p = \exp \sum_{n=0}^{\infty} \text{Log}(b_n)$. $\square$

We say that an infinite product $\prod_{n=0}^{\infty} b_n$ of non-zero terms *converges absolutely* if the series $\sum_{n=0}^{\infty} \text{Log}(b_n)$ converges absolutely. An important property of absolutely convergent series is that the order in which we sum the terms can be altered without affecting the convergence of the series or changing its sum, so that, in a sense, the commutative law applies to absolutely convergent series. Since we can obtain all infinite products of non-zero complex numbers by applying the exponential function to the appropriate infinite series, it follows that the terms of an absolutely convergent infinite product may be rearranged without affecting the convergence or the value of the product. We have a simple criterion for absolute convergence of infinite products:

Theorem 3.8.2. *If* $1 + c_n \neq 0$ *for all* n, *then* $\prod_{n=0}^{\infty}(1 + c_n)$ *converges absolutely if and only if* $\sum_{n=0}^{\infty} c_n$ *converges absolutely, that is, if and only if* $\sum_{n=0}^{\infty} |c_n|$ *converges.*

Proof. As the derivative of $\mathrm{Log}\,(1 + z)$ at $z = 0$ is 1, we have

$$\lim_{z \to 0} \frac{\mathrm{Log}(1 + z)}{z} = 1,$$

and so for sufficiently small $|z|$ we have

$$\tfrac{1}{2}|z| \leqslant |\mathrm{Log}\,(1 + z)| \leqslant 2|z|. \tag{3.8.3}$$

Now if $\sum_{n=0}^{\infty} |c_n|$ converges, then $|c_n| \to 0$ as $n \to \infty$, so $|\mathrm{Log}\,(1 + c_n)| \leqslant 2|c_n|$ for all sufficiently large n, and hence $\sum_{n=0}^{\infty} |\mathrm{Log}\,(1 + c_n)|$ converges by the comparison test, that is, $\prod_{n=0}^{\infty}(1 + c_n)$ is absolutely convergent.

Similarly, if $\sum_{n=0}^{\infty} |\mathrm{Log}(1 + c_n)|$ converges, then $\mathrm{Log}(1 + c_n) \to 0$ as $n \to \infty$, so $c_n \to 0$ as $n \to \infty$, and hence $\tfrac{1}{2}|c_n| \leqslant |\mathrm{Log}\,(1 + c_n)|$ for all sufficiently large n; by the comparison test, it follows that $\sum_{n=0}^{\infty} |c_n|$ converges. $\square$

So far, we have insisted that the terms b_n in an infinite product should be non-zero, so that we can take logarithms. However, in order to deal with products of functions we need to allow some terms to take the value zero, so we extend the definition of an infinite product as follows.

Definition. Let (b_n) be a sequence of complex numbers. Then $\prod_{n=0}^{\infty} b_n$ *converges* if there exists $N \in \mathbb{N}$ such that

(i) $b_n \neq 0$ for all $n \geqslant N$,
(ii) $\prod_{n=N}^{\infty} b_n$ converges to a non-zero complex number (in the sense that $b_N b_{N+1} \ldots b_m \to p \neq 0$ as $m \to \infty$).

We define the value of $\prod_{n=0}^{\infty} b_n$ to be

$$(b_0 b_1 \ldots b_{N-1}) . \prod_{n=N}^{\infty} b_n = b_0 b_1 \ldots b_{N-1} p.$$

It is easily seen that this is independent of the choice of N, and that $\prod_{n=0}^{\infty} b_n = 0$ if and only if some $b_n = 0$.

Definition. Let (f_n) be a sequence of functions $f_n : E \to \mathbb{C}$. Then $\prod_{n=0}^{\infty} f_n(z)$ *converges normally* on E if

(i) $f_n(z) \to 1$ uniformly on E (so that $\| f_n - 1 \| < 1$ and hence $\mathrm{Log}\,(f_n)$ is well defined for large n, say for $n \geqslant N$),

(ii) $\sum_{n \geqslant N} \mathrm{Log}\, f_n(z)$ is normally convergent to a function $w(z)$ on E.

Then we write

$$\prod_{n=0}^{\infty} f_n(z) = f_0(z) \ldots f_{N-1}(z) . \prod_{n \geqslant N} f_n(z)$$
$$= f_0(z) \ldots f_{N-1}(z) \exp(w(z));$$

clearly this is independent of the choice of N.

The following result shows that normal convergence of an infinite product is equivalent to normal convergence of an appropriate series.

Theorem 3.8.4. *Let $f_n = 1 + F_n$ be a sequence of functions $f_n : E \to \mathbb{C}$. Then $\prod_{n=0}^{\infty} f_n$ is normally convergent on E if and only if $\sum_{n=0}^{\infty} F_n$ is normally convergent on E.*

Proof. Suppose that $\prod_{n=0}^{\infty} f_n$ converges normally on E, so that conditions (i) and (ii) are satisfied. By (i), $\| F_n \| \to 0$ as $n \to \infty$, so for all sufficiently large n, say $n \geqslant N$, the function $\mathrm{Log}\,(1 + F_n) = \mathrm{Log}\,(f_n)$ is well defined on E and satisfies $\frac{1}{2}|F_n(z)| \leqslant |\mathrm{Log}\,(f_n(z))|$ for all $z \in E$, using (3.8.3). Hence $\frac{1}{2}\| F_n \| \leqslant \| \mathrm{Log}\,(f_n) \|$ for all $n \geqslant N$; since $\sum \| \mathrm{Log}\, f_n \|$ converges, the comparison test implies that $\sum \| F_n \|$ converges, so $\sum F_n$ is normally convergent on E.

Conversely, suppose that $\sum \| F_n \|$ converges. Then $\| F_n \| \to 0$ as $n \to \infty$, so $f_n \to 1$ uniformly on E, giving (i). As above, $\mathrm{Log}\,(1 + F_n) = \mathrm{Log}\,(f_n)$ is well defined on E for large n and satisfies $|\mathrm{Log}\,(f_n(z))| \leqslant 2|F_n(z)|$ for all $z \in E$, by (3.8.3). Hence $\| \mathrm{Log}\,(f_n) \| \leqslant 2\| F_n \|$ for large n, so the comparison test implies that $\sum \| \mathrm{Log}\,(f_n) \|$ converges, giving (ii). $\square$

Corollary 3.8.5. *If $\prod_{n=0}^{\infty} f_n$ is normally convergent on E, then $\prod_{n=0}^{\infty} f_n$ is absolutely convergent on E.*

Proof. If $\prod f_n$ converges normally on E then so does $\sum F_n$ by Theorem 3.8.4, where $f_n = 1 + F_n$. Hence $\sum F_n(z)$ is absolutely convergent for each

$z \in E$, so by Theorem 3.8.2, $\prod(1 + F_n(z)) = \prod f_n(z)$ is absolutely convergent for each $z \in E$. □

Theorem 3.8.6. *Let (f_n) be a sequence of analytic functions on a region $R \subseteq \mathbb{C}$, and suppose that $\prod_{n=0}^{\infty} f_n$ converges normally on all compact subsets of R. Then the function $f = \prod_{n=0}^{\infty} f_n$ is analytic on R.*

Proof. Given any compact $K \subseteq R$, the product $\prod f_n$ converges normally on K, so there exists $N = N_K \in \mathbb{N}$ such that $\| f_n - 1 \|_K < 1$ for all $n \geq N$ and such that $\sum_{n \geq N} \mathrm{Log}(f_n)$ converges normally to a function $w(z)$ on K. Now each function $\mathrm{Log}(f_n)$ is analytic on $\overset{\circ}{K}$, and $\sum_{n \geq N} \mathrm{Log}(f_n)$ converges uniformly on K, so w is analytic on $\overset{\circ}{K}$ and hence so is $f = \prod_{n=0}^{\infty} f_n = \exp(w) \cdot f_0 \dots f_{N-1}$. Since each point in R has a neighbourhood with compact closure, the result follows. □

For example, consider the function

$$S(z) = z \cdot \prod_{n=1}^{\infty} \left(1 - \frac{z^2}{n^2} \right).$$

The series $\sum_{n=1}^{\infty} z^2/n^2$ converges normally on all compact subsets of $\mathbb{C}$ (by comparison with the convergent series $\sum n^{-2}$), so by Theorem 3.8.4 the above product converges normally on all compact subsets, and hence $S(z)$ is analytic on $\mathbb{C}$ by Theorem 3.8.6.

Theorem 3.8.7. *Let f, f_n and R be as in Theorem 3.8.6, and let $z \in R$. Then $f(z) = 0$ if and only if $f_n(z) = 0$ for some $n \in \mathbb{N}$, in which case there are only finitely such n and the order of z as a zero of f is the sum of the orders of z as zeros of these functions f_n.*

Proof. If some $f_n(z) = 0$ then clearly $f = \prod f_n$ vanishes at z. Conversely it follows from the definition of infinite products of arbitrary complex numbers (given after Theorem 3.8.2) that if $f(z) = 0$ then some $f_n(z) = 0$. At most finitely many factors $f_n(z)$ can be zero (again by definition of convergence), and writing $f = (f_0 f_1 \dots f_{N-1}) \cdot \prod_{n \geq N} f_n$, with $f_n(z) \neq 0$ for $n \geq N$, we have $\prod_{n \geq N} f_n(z) \neq 0$ and so the final part of the theorem follows. □

For example, the function $S(z)$ defined above has simple zeros at each $n \in \mathbb{Z}$, and is non-zero on $\mathbb{C} \setminus \mathbb{Z}$.

If we consider the product $f = f_0 \dots f_m$ of a large number of analytic functions f_n, then the formula for the derivative of f, in terms of the functions f_n and their derivatives, is a little complicated. Rather simpler is the *logarithmic derivative* of f, which is defined to be

$$\frac{d}{dz}(\log f) = \frac{f'}{f},$$

and which can be expressed as

$$\frac{f'}{f} = \sum_{n=0}^{m} \frac{f'_n}{f_n}.$$

This formula can be considered as valid even at the zeros of f, where both sides have poles of the same order. We now consider the logarithmic derivative of an infinite product of analytic functions.

Theorem 3.8.8. *Let f, f_n and R be as in Theorem 3.8.6. Then $\sum_{n=0}^{\infty} f'_n/f_n$ converges uniformly to f'/f on all compact subsets of R.*

Proof. If K is any compact subset of R, then $\prod_{n=0}^{\infty} f_n$ converges normally on K, so there exists $N \in \mathbb{N}$ such that $n \geq N$ implies $\| f_n - 1 \|_K < 1$, and hence f_n has no zeros in K.

Let $g = f_0 f_1 f_1 \ldots f_{N-1}$ and $h = \prod_{n \geq N} f_n$, both analytic on $\mathring{K}$. We have $f = gh$, and so

$$\frac{f'}{f} = \frac{g'}{g} + \frac{h'}{h}$$

$$= \sum_{n=0}^{N-1} \frac{f'_n}{f_n} + \frac{h'}{h}.$$

Since $\prod_{n=0}^{\infty} f_n$ converges normally on K, $\sum_{n \geq N} \mathrm{Log}(f_n)$ converges normally to a function w on K, where

$$h = \prod_{n \geq N} f_n = \exp(w),$$

by Theorem 3.8.1. Since h is non-zero and analytic on $\mathring{K}$ (by Theorems 3.8.7 and 3.8.6 respectively), we have

$$\frac{h'}{h} = \frac{w' \exp(w)}{\exp(w)} = w'$$

on $\mathring{K}$. Now $w = \sum_{n \geq N} \mathrm{Log}(f_n)$ is a normally and hence uniformly convergent sequence of analytic functions on $\mathring{K}$, so we may differentiate term-by-term to give

$$w' = \sum_{n \geq N} \frac{f'_n}{f_n}.$$

Thus

$$\frac{f'}{f} = \sum_{n=0}^{N-1} \frac{f_n'}{f_n} + \sum_{n \geqslant N} \frac{f_n'}{f_n}$$

$$= \sum_{n=0}^{\infty} \frac{f_n'}{f_n},$$

this series converging uniformly on all compact subsets of $\overset{\circ}{K}$, by Corollary 3.7.2. This argument is valid for all compact $K \subseteq R$, so the result follows. $\square$

For example, the infinite product

$$S(z) = z \cdot \prod_{n=1}^{\infty} \left(1 - \frac{z^2}{n^2}\right) = z \cdot \prod_{n=1}^{\infty} \left(1 - \frac{z}{n}\right)\left(1 + \frac{z}{n}\right) \qquad (3.8.9)$$

converges normally on all compact subsets of $\mathbb{C}$, so by the previous theorem its logarithmic derivative $Z(z)$ is given by

$$Z(z) = \frac{S'(z)}{S(z)}$$

$$= \frac{1}{z} + \sum_{n=1}^{\infty} \frac{2z}{z^2 - n^2}$$

$$= \frac{1}{z} + \sum_{n=1}^{\infty} \left(\frac{1}{z-n} + \frac{1}{z+n}\right), \qquad (3.8.10)$$

this series converging uniformly on all compact subsets of $\mathbb{C}$. By Theorem 3.7.4 we can differentiate $Z(z)$ term-by-term to obtain a meromorphic function. Thus if $P(z) = -Z'(z)$ then

$$P(z) = \frac{1}{z^2} + \sum_{n=1}^{\infty} \left(\frac{1}{(z-n)^2} + \frac{1}{(z+n)^2}\right)$$

$$= \sum_{n=-\infty}^{\infty} \frac{1}{(z-n)^2}, \qquad (3.8.11)$$

this last step being valid since the series is absolutely convergent, as shown at the end of §3.7.

Clearly, $P(z)$ is a simply periodic meromorphic function, with $\mathbb{Z}$ as its group of periods. In the exercises we shall see that $P(z) = \pi^2 \operatorname{cosec}^2 \pi z$, and hence that $Z(z) = \pi \cot \pi z$ and $S(z) = \pi \sin \pi z$. In the next section we shall construct functions $\wp(z)$, $\zeta(z)$ and $\sigma(z)$ analogous to $P(z)$, $Z(z)$ and $S(z)$, in such a way that $\wp(z)$ is elliptic.

3.9 Weierstrass functions

Let $\Omega = \Omega(\omega_1, \omega_2)$ be a lattice with basis $\{\omega_1, \omega_2\}$ and let P be a fundamental parallelogram for Ω with no elements of Ω on ∂P. We need to construct non-constant functions f which are elliptic with respect to Ω. By Theorem 3.6.1 we know that such a function f cannot be analytic and so it must have poles in P. By Theorem 3.6.3 we know that f cannot have just one simple pole in P, so the simplest non-constant elliptic function has order 2, with either two simple poles or else a single pole of order 2 in P. In this section we shall introduce the Weierstrass function $\wp(z)$ which is elliptic of order 2 with respect to Ω and has a single pole of order 2 in P. This will be our basic elliptic function in the sense that every function which is elliptic with respect to Ω is a rational function of $\wp$ and its derivative $\wp'$ (see Theorem 3.11.1).

It is not difficult to construct elliptic functions of order $N \geqslant 3$ directly (see Theorem 3.9.3), but the method of construction does not apply so easily to the case $N = 2$. Instead of constructing $\wp(z)$ directly we shall derive it from the Weierstrass sigma-function $\sigma(z)$, which is related to $\wp(z)$ in much the same way as $S(z) = \pi \sin \pi z$ is related to $P(z) = \pi^2 \operatorname{cosec}^2 \pi z$ in the theory of simply periodic functions (see §3.8). Just as the convergence of the products and series defining $S(z), Z(z)$ and $P(z)$ depends upon the convergence of $\sum_{n=1}^{\infty} n^{-2}$, the convergence of the products and series defining the Weierstrass functions depends on similar sums indexed by the lattice Ω. To clarify the meaning of summation over Ω we must first describe a particular ordering of Ω.

The sets $\Pi_r = \{a\omega_1 + b\omega_2 \,|\, a, b \in \mathbb{R} \text{ and } \max(|a|, |b|) = r\}$, for integers $r \geqslant 1$, are similar parallelograms centred on 0, as shown in Fig. 3.15. Defining $\Omega_r = \Omega \cap \Pi_r$, we have

$$\Omega_r = \{m\omega_1 + n\omega_2 \,|\, m, n \in \mathbb{Z} \text{ and } \max(|m|, |n|) = r\}.$$

Now Ω is a disjoint union $\Omega = \{0\} \cup \Omega_1 \cup \Omega_2 \cup \cdots$, and for each $r \geqslant 1$ we have

$$|\Omega_r| = 8r. \tag{3.9.1}$$

Fig. 3.15

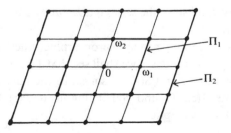

We can order the elements of Ω by starting at 0 and then listing the elements of $\Omega_1, \Omega_2, \ldots$ in turn, rotating around each Ω_r in the order $r\omega_1$, $r\omega_1 + \omega_2, \ldots, r\omega_1 - \omega_2$ so that the sequence spirals outwards from 0 (see Fig. 3.16).

Fig. 3.16

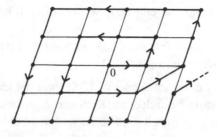

If we denote this ordering by $\omega^{(0)}, \omega^{(1)}, \omega^{(2)}, \ldots$ then $\omega^{(0)} = 0$, $\omega^{(1)} = \omega_1$, $\omega^{(2)} = \omega_1 + \omega_2$, $\omega^{(3)} = \omega_2, \ldots, \omega^{(8)} = \omega_1 - \omega_2$, $\omega^{(9)} = 2\omega_1$, $\omega^{(10)} = 2\omega_1 + \omega_2, \ldots$; clearly $|\omega^{(k)}| \to \infty$ as $k \to \infty$.

By $\sum_{\omega\in\Omega}$ and $\sum'_{\omega\in\Omega}$ we shall mean the sum over all (respectively all non-zero) lattice-points ω taken in the above order; thus $\sum_{\omega\in\Omega}h(\omega) = \sum_{k=0}^{\infty}h(\omega^{(k)})$ for any function h, and similarly $\sum'_{\omega\in\Omega}h(\omega) = \sum_{k=1}^{\infty}h(\omega^{(k)})$. By $\prod_{\omega\in\Omega}$ and $\prod'_{\omega\in\Omega}$ we shall mean the product over all (respectively all non-zero) lattice-points, again in the above order. For convenience we will often abbreviate the notation to $\sum, \sum'$, etc., the particular lattice Ω being understood. In practice, the particular ordering of Ω will not often be important, as the sums and products which concern us are usually absolutely convergent and hence invariant under rearrangements.

The convergence properties of the Weierstrass functions depend on the following result, which is a 2-dimensional analogue of the fact that the series $\sum_{r=1}^{\infty}r^{-s}$ defining the Riemann zeta-function converges if and only if $s > 1$.

Theorem 3.9.2. *If $s\in\mathbb{R}$, then $\sum'_{\omega\in\Omega}|\omega|^{-s}$ converges if and only if $s > 2$.*

Proof. If D and d are the greatest and least moduli of the elements of the parallelogram Π_1 containing Ω_1, then since

$$\Omega_r \subseteq \Pi_r = \{rz | z \in \Pi_1\}$$

we have $rD \geqslant |\omega| \geqslant rd$ for all $\omega\in\Omega_r$. Defining

$$\sigma_{r,s} = \sum_{\omega\in\Omega_r} |\omega|^{-s},$$

we see by (3.9.1) that $\sigma_{r,s}$ lies between $8r(rD)^{-s} = 8r^{1-s}D^{-s}$ and $8r(rd)^{-s} =$

$8r^{1-s}d^{-s}$. Thus $\sum_{r=1}^{\infty}\sigma_{r,s}$ converges if and only if $\sum_{r=1}^{\infty}r^{1-s}$ converges, that is, if and only if $s > 2$. Since the terms of $\sum'|\omega|^{-s}$ are positive and can be grouped together to give $\sum_{r=1}^{\infty}\sigma_{r,s}$, it follows that $\sum'|\omega|^{-s}$ converges if and only if $s > 2$. $\square$

It is now easy to construct elliptic functions of each order $N \geqslant 3$.

Theorem 3.9.3. *For each integer $N \geqslant 3$, the function $F_N(z) = \sum_{\omega\in\Omega}(z-\omega)^{-N}$ is elliptic of order N with respect to Ω.*

Proof. If K is any compact subset of $\mathbb{C}\backslash\Omega$ then the terms $(z-\omega)^{-N}$ are analytic and therefore bounded on K. Since K is bounded, for all but finitely many $\omega\in\Omega$ (say for all $\omega\in\Phi\subseteq\Omega$) we have $|\omega| \geqslant 2|z|$ for all $z\in K$. It follows that $|z-\omega| \geqslant |\omega| - |z| \geqslant \frac{1}{2}|\omega|$ for all $z\in K$, $\omega\in\Phi$, and hence $\|(z-\omega)^{-N}\|_K \leqslant 2^N|\omega|^{-N}$ for all $\omega\in\Phi$. If $N \geqslant 3$ then Theorem 3.9.2 and the comparison test imply that $\sum_{\omega\in\Omega}\|(z-\omega)^{-N}\|_K$ converges, so $\sum_{\omega\in\Omega}(z-\omega)^{-N}$ is normally convergent on K. Since each term $(z-\omega)^{-N}$ is analytic on K, Corollary 3.7.2 implies that $F_N(z)$ is analytic on $\mathbb{C}\backslash\Omega$. A similar argument, using Theorem 3.7.4, shows that $F_N(z)$ is meromorphic at each $\omega\in\Omega$, with a pole of order N corresponding to the term $(z-\omega)^{-N}$. Thus $F_N(z)$ is meromorphic on $\mathbb{C}$.

Since normal convergence implies absolute convergence, we may rearrange the series defining $F_N(z)$, and so if $\omega_0\in\Omega$ then we have

$$F_N(z+\omega_0) = \sum_{\omega\in\Omega}(z+\omega_0-\omega)^{-N}$$

$$= \sum_{\omega'\in\Omega}(z-\omega')^{-N}$$

$$= F_N(z),$$

where $\omega' = \omega - \omega_0$ ranges over Ω as ω does (though in a different order). Thus $F_N(z)$ is periodic with respect to Ω, and is therefore elliptic. Since $F_N(z)$ has a single class of poles of order N, $F_N(z)$ has order N. $\square$

Clearly, this method fails to produce an elliptic function $F_2(z)$ of order 2, since Theorem 3.9.2 cannot be used to prove convergence of $\sum(z-\omega)^{-2}$. In order to guarantee convergence, we make the terms of this series smaller, replacing $(z-\omega)^{-2}$ by $(z-\omega)^{-2} - \omega^{-2}$ for each $\omega \neq 0$. The resulting series

$$\wp(z) = \frac{1}{z^2} + \sum_{\omega\in\Omega}'\left(\frac{1}{(z-\omega)^2} - \frac{1}{\omega^2}\right) \tag{3.9.4}$$

represents an elliptic function of order 2, Weierstrass' *pe-function*. Since $\wp(z)$ does not have the form $\sum_{\omega\in\Omega}f(z-\omega)$, the periodicity of $\wp(z)$ is not obvious, and we shall prove it indirectly by integrating the derivative $\wp'(z)$, which is the elliptic function $-2F_3(z)$. It is straightforward and traditional to prove that $\wp(z)$ is meromorphic by imitating the proof of Theorem 3.9.3, comparing the series with $\sum'|\omega|^{-3}$ (see Exercise 3L), but we shall take a slightly different approach, and obtain $\wp(z)$ from the Weierstrass *sigma-function*

$$\left.\begin{array}{c} \sigma(z) = z \cdot \prod_{\omega \in \Omega}' g(\omega, z), \\[2mm] g(\omega, z) = \left(1 - \frac{z}{\omega}\right) \exp\left(\frac{z}{\omega} + \frac{1}{2}\left(\frac{z}{\omega}\right)^2\right). \end{array}\right\}$$

where $(3.9.5)$

(The factor $(1 - (z/\omega))$ is included in $g(\omega, z)$ to give $\sigma(z)$ a simple zero at each lattice-point ω, while the exponential factor is included to guarantee convergence of the infinite product.)

If K is any compact subset of $\mathbb{C}$, then since K is bounded and since $|\omega^{(k)}| \to \infty$ as $k \to \infty$, it follows that $g(\omega^{(k)}, z) \to 1$ uniformly on K as $k \to \infty$. Hence there exists an integer N_1 such that for all $k > N_1$, $\mathrm{Log}\,(g(\omega^{(k)}, z))$ is well defined for $z \in K$ and satisfies

$$\mathrm{Log}\,(g(\omega^{(k)}, z)) = \mathrm{Log}\left(1 - \frac{z}{\omega^{(k)}}\right) + \mathrm{Log}\left(\exp\left(\frac{z}{\omega^{(k)}} + \frac{1}{2}\left(\frac{z}{\omega^{(k)}}\right)^2\right)\right)$$

$$= \mathrm{Log}\left(1 - \frac{z}{\omega^{(k)}}\right) + \frac{z}{\omega^{(k)}} + \frac{1}{2}\left(\frac{z}{\omega^{(k)}}\right)^2.$$

Moreover, since K is bounded, there exists an integer N_2 such that $|\omega^{(k)}| > 2|z|$ for all $z \in K$, $k > N_2$. Thus for all $z \in K$ and $k > \max(N_1, N_2)$ we have

$$|\mathrm{Log}\,(g(\omega^{(k)}, z))| = \left|\mathrm{Log}\left(1 - \frac{z}{\omega^{(k)}}\right) + \frac{z}{\omega^{(k)}} + \frac{1}{2}\left(\frac{z}{\omega^{(k)}}\right)^2\right|$$

$$= \left|\frac{1}{3}\left(\frac{z}{\omega^{(k)}}\right)^3 + \frac{1}{4}\left(\frac{z}{\omega^{(k)}}\right)^4 + \ldots\right|$$

$$\leqslant \frac{1}{3}\left|\frac{z}{\omega^{(k)}}\right|^3 (1 + \tfrac{1}{2} + \tfrac{1}{4} + \ldots)$$

$$\leqslant \left|\frac{z}{\omega^{(k)}}\right|^3.$$

It follows from Theorem 3.9.2 that $\sum_k \mathrm{Log}\,(g(\omega^{(k)}, z))$ is normally convergent on K, and hence $z\prod_{\omega \in \Omega}' g(\omega, z)$ is normally convergent on K. By Theorem 3.8.6 this product converges to a function $\sigma(z)$ which is analytic on $\mathbb{C}$. Since $g(\omega, -z) = g(-\omega, z)$ it follows that $\sigma(-z) = -\sigma(z)$, that is $\sigma(z)$ is an odd function. The analogy between $\sigma(z)$ and the function $S(z)$ of §3.8

becomes clear if we write

$$S(z) = z . \prod_{\substack{n=-\infty \\ n \neq 0}}^{\infty} \left(1 - \frac{z}{n}\right) \exp\left(\frac{z}{n}\right).$$

By Theorem 3.8.8 the logarithmic derivative of $\sigma(z)$ will give an infinite series converging uniformly on compact subsets of $\mathbb{C}$ to a meromorphic function, which we denote by $\zeta(z)$. This is the Weierstrass *zeta-function* (not to be confused with the Riemann zeta-function $\zeta(s) = \sum_{r=1}^{\infty} r^{-s}$), given by

$$\zeta(z) = \sigma'(z)/\sigma(z)$$

$$= \frac{d}{dz}(\mathrm{Log}\, \sigma(z))$$

$$= \frac{1}{z} + \sum_{\omega \in \Omega}' \left(\frac{1}{z-\omega} + \frac{1}{\omega} + \frac{z}{\omega^2}\right). \tag{3.9.6}$$

Since $\sigma(z)$ is an odd function, $\zeta(z)$ is also odd. It has simple poles at the lattice-points, and is analytic on $\mathbb{C}\backslash\Omega$. As a series of meromorphic functions, $\zeta(z)$ converges uniformly on compact subsets of $\mathbb{C}$ (in the sense of the definition at the end of §3.7), so by Theorem 3.7.4 we may differentiate term-by-term to obtain a meromorphic function $\zeta'(z)$. Writing $\wp(z) = -\zeta'(z)$ we then have

$$\wp(z) = \frac{1}{z^2} + \sum_{\omega \in \Omega}' \left(\frac{1}{(z-\omega)^2} - \frac{1}{\omega^2}\right), \tag{3.9.7}$$

an even function which is analytic on $\mathbb{C}\backslash\Omega$ and has poles of order 2 at each $\omega \in \Omega$.

Theorem 3.9.8. $\wp(z)$ *is an elliptic function with* Ω *as its lattice* $\Omega_\wp$ *of periods.*

Proof. We have seen that $\wp(z)$ is meromorphic, so it is sufficient to prove that $\Omega = \Omega_\wp$. Since the series (3.9.7) defining $\wp(z)$ is uniformly convergent on compact subsets of $\mathbb{C}$, we can differentiate term-by-term, giving

$$\wp'(z) = -\frac{2}{z^3} - \sum_{\omega \in \Omega}' \frac{2}{(z-\omega)^3}$$

$$= -2 \sum_{\omega \in \Omega} (z-\omega)^{-3}$$

$$= -2F_3(z),$$

where $F_3(z)$ is the elliptic function of order 3 constructed in Theorem 3.9.3. It follows that for each $\omega \in \Omega$ the function $\wp'(z+\omega) - \wp'(z)$ is identically zero, and hence $\wp(z+\omega) - \wp(z)$ is constant, say $\wp(z+\omega) - \wp(z) = c_\omega$ for all $z \in \mathbb{C}$. Putting $z = -\omega/2$ we have $c_\omega = \wp(\omega/2) - \wp(-\omega/2) = 0$ since $\wp(z)$ is an even function. Thus $\wp(z + \omega) = \wp(z)$ for all $z \in \mathbb{C}$ and $\omega \in \Omega$, so $\Omega \subseteq \Omega_\wp$. Since 0 is a pole of $\wp(z)$, there is a pole at every point in $\Omega_\wp$; since $\wp(z)$ has no poles in $\mathbb{C} \setminus \Omega$ it follows that $\Omega_\wp \subseteq \Omega$, so $\Omega = \Omega_\wp$. $\square$

Theorem 3.9.9. $\wp(z)$ *has order* 2, *and* $\wp'(z)$ *has order* 3.

Proof. $\wp(z)$ has a single congruence class of poles (the lattice-points $\omega \in \Omega$), each of order 2, so $\wp(z)$ has order 2; similarly, $\wp'(z) = -2F_3(z)$ has a single class of poles of order 3, so $\wp'(z)$ has order 3. $\square$

It is important to note that the Weierstrass functions $\wp(z)$, $\zeta(z)$ and $\sigma(z)$ depend on the particular lattice Ω, so to be precise we should write $\wp(z, \Omega)$, etc. However, in most cases the lattice Ω is understood, and the abbreviated notation is unambiguous.

3.10 The differential equation for $\wp(z)$

In this section we derive an important equation connecting $\wp(z)$ and $\wp'(z)$, obtained from the Laurent series for $\wp(z)$ near $z = 0$. We start by finding the Laurent series for

$$\zeta(z) = \frac{1}{z} + {\sum_{\omega \in \Omega}}' \left(\frac{1}{z - \omega} + \frac{1}{\omega} + \frac{z}{\omega^2} \right). \tag{3.10.1}$$

Let $m = \min \{ |\omega| \, | \, \omega \in \Omega \setminus \{0\} \}$, and let $D = \{ z \in \mathbb{C} \, | \, |z| < m \}$, the largest open disc centred at 0 and containing no other lattice-points. Since

$$\frac{1}{z - \omega} + \frac{1}{\omega} + \frac{z}{\omega^2} = \frac{z^2}{\omega^2(z - \omega)},$$

we see, by comparison with $\sum' |\omega|^{-3}$, that $\sum'((z-w)^{-1} + (1/\omega) + (z/\omega^2))$ is absolutely convergent for each $z \in \mathbb{C} \setminus \Omega$. Moreover, for each $\omega \in \Omega \setminus \{0\}$ the binomial series

$$\frac{1}{z - \omega} = -\frac{1}{\omega} - \frac{z}{\omega^2} - \frac{z^2}{\omega^3} - \cdots$$

is absolutely convergent for $z \in D$, so we may substitute this in (3.10.1) and reverse the order of summation (see Apostol [1963], 13.9, for example), to obtain

$$\zeta(z) = \frac{1}{z} + {\sum}' \left(-\frac{z^2}{\omega^3} - \frac{z^3}{\omega^4} - \cdots \right)$$

$$= \frac{1}{z} - G_3 z^2 - G_4 z^3 - \cdots$$

for all $z \in D$, where

$$G_k = G_k(\Omega) = \sideset{}{'}\sum_{\omega \in \Omega} \omega^{-k},$$

(these series G_k are the *Eisenstein series* for Ω, absolutely convergent for $k \geqslant 3$ by Theorem 3.9.2). For odd k the terms ω^{-k} and $(-\omega)^{-k}$ cancel, giving $G_k = 0$, so the Laurent series for $\zeta(z)$ is

$$\zeta(z) = \frac{1}{z} - \sum_{n=2}^{\infty} G_{2n} z^{2n-1}, \qquad (3.10.2)$$

and hence

$$\wp(z) = -\zeta'(z) = \frac{1}{z^2} + \sum_{n=2}^{\infty} (2n-1) G_{2n} z^{2n-2}; \qquad (3.10.3)$$

this is the Laurent series for $\wp(z)$, valid for $z \in D$. Now a straightforward calculation gives

$$\wp'(z) = \frac{-2}{z^3} + 6 G_4 z + 20 G_6 z^3 + \dots,$$

and so

$$\wp'(z)^2 = \frac{4}{z^6} - \frac{24 G_4}{z^2} - 80 G_6 + z^2 \phi_1(z),$$

$$4 \wp(z)^3 = \frac{4}{z^6} + \frac{36 G_4}{z^2} + 60 G_6 + z^2 \phi_2(z),$$

$$60 G_4 \wp(z) = \frac{60 G_4}{z^2} + z^2 \phi_3(z),$$

where $\phi_1(z)$, $\phi_2(z)$ and $\phi_3(z)$ are power series convergent in D. These last three equations give

$$\wp'(z)^2 - 4 \wp(z)^3 + 60 G_4 \wp(z) + 140 G_6 = z^2 \phi(z),$$

where $\phi(z) = \phi_1(z) - \phi_2(z) + \phi_3(z)$ is a power series convergent in D. As $\wp$ and $\wp'$ are elliptic with respect to Ω, the function

$$f(z) = \wp'(z)^2 - 4 \wp(z)^3 + 60 G_4 \wp(z) + 140 G_6$$

is also elliptic. Since $f(z) = z^2 \phi(z)$ in D, with $\phi(z)$ analytic, f vanishes at 0 and hence at all $\omega \in \Omega$. However, by its construction f can have poles only where $\wp$ or $\wp'$ have poles, that is, at the lattice-points. Therefore f has no poles and so by Theorem 3.6.1 $f(z)$ is a constant, which must be zero since $f(0) = 0$. Thus we have proved

Theorem 3.10.4. $\wp'(z)^2 = 4 \wp(z)^3 - 60 G_4 \wp(z) - 140 G_6.$ $\square$

This is the differential equation for $\wp(z)$. It is customary to write

$$g_2 = 60\,G_4 = 60\sum{}'\omega^{-4},$$
$$g_3 = 140\,G_6 = 140\sum{}'\omega^{-6}, \qquad (3.10.5)$$

so that

$$\wp'(z)^2 = 4\,\wp(z)^3 - g_2\,\wp(z) - g_3. \qquad (3.10.6)$$

If we put $z = \wp(t)$ then we see that

$$\left(\frac{dz}{dt}\right)^2 = 4z^3 - g_2 z - g_3,$$

so that

$$\wp^{-1}(z) = t = \int \frac{dz}{\sqrt{(p(z))}},$$

where $p(z)$ is the cubic polynomial $4z^3 - g_2 z - g_3$. This shows how the inverses of elliptic functions appear as indefinite integrals, in much the same way as inverses of trigonometric functions do (see §3.6). This idea will be explored in greater detail in Chapter 6; for example, we will show that given any cubic polynomial

$$p(z) = 4z^3 - c_2 z - c_3$$

with distinct roots, there exists a lattice Ω such that $c_2 = g_2(\Omega)$ and $c_3 = g_3(\Omega)$. Here we will prove the converse, that the polynomial $4z^3 - g_2 z - g_3$ has distinct roots, by considering the zeros of $\wp'$.

Theorem 3.10.7. *Let Ω be a lattice with basis $\{\omega_1, \omega_2\}$, and let $\omega_3 = \omega_1 + \omega_2$. If P is a fundamental parallelogram with 0, $\tfrac{1}{2}\omega_1$, $\tfrac{1}{2}\omega_2$ and $\tfrac{1}{2}\omega_3$ in its interior, then $\tfrac{1}{2}\omega_1$, $\tfrac{1}{2}\omega_2$ and $\tfrac{1}{2}\omega_3$ are the zeros of $\wp'$ in P.*

Proof. By Theorem 3.9.9, $\wp'$ has order 3 and hence has three zeros in P. If $\omega \in \Omega$ then because $\tfrac{1}{2}\omega \sim -\tfrac{1}{2}\omega \bmod \Omega$ we have $\wp'(\tfrac{1}{2}\omega) = \wp'(-\tfrac{1}{2}\omega)$; since $\wp'$ is an odd function we have $\wp'(-\tfrac{1}{2}\omega) = -\wp'(\tfrac{1}{2}\omega)$ and hence $\wp'(\tfrac{1}{2}\omega) = 0$ or ∞. Since the only pole of $\wp'$ within P is the triple pole at 0 we have $\wp'(\tfrac{1}{2}\omega_j) = 0$ for $j = 1, 2, 3$. $\square$

We define $e_j = \wp(\tfrac{1}{2}\omega_j)$ for $j = 1, 2, 3$. Since $S = [\tfrac{1}{2}\omega_1] \cup [\tfrac{1}{2}\omega_2] \cup [\tfrac{1}{2}\omega_3]$ is the set of all zeros of $\wp'$ on $\mathbb{C}$, we see that $\{e_1, e_2, e_3\} = \wp(S)$ is independent of the particular basis $\{\omega_1, \omega_2\}$ chosen for Ω.

Corollary 3.10.8. *For each $c \in \Sigma \setminus \{e_1, e_2, e_3, \infty\}$ the equation $\wp(z) = c$ has* *two simple solutions; for $c = e_1, e_2, e_3$ or ∞ the equation has one double solution.*

Proof. Since $\wp$ is elliptic of order 2, it takes each value $c \in \Sigma$ twice by Theorem 3.6.4, giving either two simple solutions (at z and $-z$ since $\wp$ is even) or one double solution. If $c \in \mathbb{C}$ then $\wp(z) = c$ has a double solution if and only if $\wp'(z) = 0$, giving $z \sim \frac{1}{2}\omega_j (j = 1, 2, 3)$ and so $c = e_j$. The pole of order 2 at $z = 0$ shows that $\wp(z) = \infty$ has a double solution. $\square$

Theorem 3.10.9. *e_1, e_2 and e_3 are mutually distinct.*

Proof. Let $f_j(z) = \wp(z) - e_j$ for $j = 1, 2, 3$. As the poles of f_j are the same as those of $\wp$, f_j is an elliptic function of order 2 and therefore has two classes of zeros, counting multiplicities. As

$$f_j(\tfrac{1}{2}\omega_j) = f'_j(\tfrac{1}{2}\omega_j) = 0,$$

f_j has double zeros on $[\frac{1}{2}\omega_j]$ and hence has no other zeros. In particular, $f_j(\frac{1}{2}\omega_k) \neq 0$ for $j \neq k$. Since

$$f_j(\tfrac{1}{2}\omega_k) = \wp(\tfrac{1}{2}\omega_k) - e_j$$
$$= e_k - e_j,$$

it follows that $e_j \neq e_k$ for $j \neq k$. $\square$

By (3.10.6) the polynomial

$$p(z) = 4z^3 - g_2 z - g_3$$

has zeros at $z = \wp(t)$ where $\wp'(t) = 0$, so $p(z)$ has three distinct zeros $z = e_1$, e_2 and e_3.

3.11 The field of elliptic functions

In this section we consider a fixed lattice Ω; an *elliptic function* will mean a function which is elliptic with respect to Ω. If f and g are elliptic, then so are $f + g$, $f - g$ and fg, and if g is not identically zero then $1/g$ is elliptic. Thus the set of all elliptic functions is a field, which we shall denote by $E(\Omega)$. This field contains the subfield $E_1(\Omega)$ consisting of the even elliptic functions. The constant functions form a subfield of $E_1(\Omega)$ isomorphic to $\mathbb{C}$, so we may regard $E(\Omega)$ and $E_1(\Omega)$ as extension fields of $\mathbb{C}$. Since $E_1(\Omega)$ contains $\wp(z) = \wp(z, \Omega)$ it contains all rational functions of $\wp$ (with complex coefficients); these rational functions form a field $\mathbb{C}(\wp)$, the smallest field containing $\wp$ and the constant functions $\mathbb{C}$. Similarly, $E(\Omega)$ contains $\wp$

and $\wp'$, and hence contains the field $\mathbb{C}(\wp, \wp')$ of rational functions of $\wp$ and $\wp'$; this is the smallest field containing $\wp, \wp'$ and $\mathbb{C}$.

Theorem 3.11.1.

(i) *If f is an even elliptic function, then $f = R_1(\wp)$ for some rational function R_1; thus $E_1(\Omega) = \mathbb{C}(\wp)$.*

(ii) *If f is any elliptic function, then*

$$f = R_1(\wp) + \wp' R_2(\wp),$$

where R_1 and R_2 are rational functions; thus $E(\Omega) = \mathbb{C}(\wp, \wp')$.

Proof. (i) Let f be an even elliptic function. The result is obvious for constant functions, so suppose that f has order $N > 0$. If $k \in \mathbb{C}$, then $f(z) = k$ has multiple roots only where $f'(z) = 0$, and this occurs at only finitely many congruence classes of points z; thus $f(z) = k$ has all its roots simple for all but finitely many values of k. We can therefore choose two distinct complex numbers c and d so that the roots of $f(z) = c$ and of $f(z) = d$ are all simple, and so that none of these roots are congruent to 0 or $\frac{1}{2}\omega_j (j = 1, 2, 3)$. Since f is even, a complete set of roots of $f(z) = c$ will have the form $a_1, -a_1, a_2, -a_2, \ldots, a_n, -a_n$, these being simple and mutually non-congruent, and similarly for the roots $b_1, -b_1, \ldots, b_n, -b_n$ of $f(z) = d$. Hence the elliptic function

$$g(z) = \frac{f(z) - c}{f(z) - d}$$

has simple zeros at $a_1, -a_1, \ldots, a_n, -a_n$, and simple poles at $b_1, -b_1, \ldots, b_n, -b_n$.

Now Corollary 3.10.8 implies that the equations $\wp(z) = \wp(a_i)$ and $\wp(z) = \wp(b_i)$ have simple roots $z = \pm a_i$ and $z = \pm b_i$ respectively $(1 \leqslant i \leqslant n)$, so the elliptic function

$$h(z) = \frac{(\wp(z) - \wp(a_1))(\wp(z) - \wp(a_2))\ldots(\wp(z) - \wp(a_n))}{(\wp(z) - \wp(b_1))(\wp(z) - \wp(b_2))\ldots(\wp(z) - \wp(b_n))}$$

has the same zeros and poles as g, with the same multiplicities (all simple). Hence Theorem 3.6.6 implies that $g = \mu h$ for some constant $\mu \neq 0$. Solving

$$\frac{f(z) - c}{f(z) - d} = \mu \frac{(\wp(z) - \wp(a_1))\ldots(\wp(z) - \wp(a_n))}{(\wp(z) - \wp(b_1))\ldots(\wp(z) - \wp(b_n))}$$

for $f(z)$, we see that f is a rational function (with complex coefficients) $R_1(\wp)$ of $\wp$.

(ii) If f is odd, then $f / \wp'$ is even, so by (i) we have $f = \wp' R_2(\wp)$ for some rational function R_2. In general, if f is *any* elliptic function then

$$f(z) = \tfrac{1}{2}(f(z) + f(-z)) + \tfrac{1}{2}(f(z) - f(-z)),$$

where $\tfrac{1}{2}(f(z) + f(-z))$ is even and elliptic while $\tfrac{1}{2}(f(z) - f(-z))$ is odd and elliptic, so by the above arguments we have

$$f = R_1(\wp) + \wp' R_2(\wp),$$

where R_1 and R_2 are rational functions. $\quad\square$

Using the differential equation $(\wp')^2 = 4\wp^3 - g_2\wp - g_3$, we can reduce any rational function of $\wp$ and $\wp'$ to the form $R_1(\wp) + \wp' R_2(\wp)$ by eliminating powers of $\wp'$; for example

$$\frac{\wp\,\wp'}{\wp' + 1} = \frac{\wp\,\wp'(\wp' - 1)}{(\wp')^2 - 1} = \frac{(4\wp^4 - g_2\wp^2 - g_3\wp) - \wp'\wp}{4\wp^3 - g_2\wp - g_3 - 1}.$$

We can view Theorem 3.10.4 as an algebraic equation between the functions $\wp$ and $\wp'$. We now show that *any* two functions in $E(\Omega)$ are connected by an algebraic equation.

Theorem 3.11.2. *If f, $g \in E(\Omega)$ then there exists a non-zero irreducible polynomial $\Phi(x, y)$, with complex coefficients, such that $\Phi(f, g)$ is identically zero.*

Proof. If we choose any polynomial in two variables x, y, say

$$F(x, y) = \sum_{k=1}^{m} \sum_{l=1}^{n} \alpha_{kl} x^k y^l \quad (\alpha_{kl} \in \mathbb{C}),$$

then the function $h(z) = F(f(z), g(z))$ is an elliptic function with poles only at the poles of f or g. If f and g have M and N poles respectively, then h has at most $mM + nN$ poles (counting multiplicities in each case). Therefore, unless h is identically zero, it has at most $mM + nN$ zeros by Theorem 3.6.4. We now show that if m and n are large enough then we can choose the coefficients α_{kl} so that h has more than $mM + nN$ zeros and hence $h(z) \equiv 0$.

To do this, we let $z_1, \ldots, z_{mn-1}$ be $mn - 1$ non-congruent points distinct from the poles of f and of g. Now regard

$$h(z_j) = \sum_{k=1}^{m} \sum_{l=1}^{n} \alpha_{kl} f(z_j)^k g(z_j)^l = 0 \quad (j = 1, \ldots, mn - 1) \qquad (3.11.3)$$

as a set of $mn - 1$ homogeneous linear equations in the mn unknowns α_{kl}. As there are more unknowns than equations, this set of equations has a non-trivial solution, that is, there exist α_{kl}, not all zero, satisfying (3.11.3). Thus $F(x, y)$ is not identically zero, but $F(f(z), g(z)) = h(z) = 0$ at $z = z_1, \ldots, z_{mn-1}$. Now for m, n large enough,

$$mn - 1 > mM + nN,$$

and so by choosing the coefficients α_{kl} as above we must have $h(z) \equiv 0$, that is, $F(f, g) = 0$.

We can factorise $F(x, y)$ within the polynomial ring $\mathbb{C}[x, y]$ as a product

$$F(x, y) = F_1(x, y)F_2(x, y)\ldots F_r(x, y)$$

of irreducible polynomials. Thus $F_1(f, g)F_2(f, g)\ldots F_r(f, g) = 0$ within the field $E(\Omega)$, so some $F_i(f, g) = 0$, and we can take Φ to be F_i. □

3.12 Translation properties of $\zeta(z)$ and $\sigma(z)$

93-94

The functions $\zeta(z)$ and $\sigma(z)$ introduced in §3.9 are not elliptic, for, as we shall show, they are not invariant under the translations $z \mapsto z + \omega$ ($\omega \in \Omega$). However, an examination of their behaviour under translations will enable us to construct elliptic functions with prescribed properties in §§3.13 and 3.14.

Since $\zeta'(z) = -\wp(z)$, we have $\zeta'(z + \omega_j) = \zeta'(z)$ for $j = 1$, 2, so that integration with respect to z gives

$$\zeta(z + \omega_j) = \zeta(z) + \eta_j \quad (j = 1, 2),$$

where η_1, η_2 are constants, independent of z. If $\omega \in \Omega$ then $\omega = m\omega_1 + n\omega_2$, where m, $n \in \mathbb{Z}$, and hence

$$\zeta(z + \omega) = \zeta(z) + \eta, \tag{3.12.1}$$

where

$$\eta = m\eta_1 + n\eta_2. \tag{3.12.2}$$

Let P be a fundamental parallelogram for Ω, containing 0 within its interior, with vertices t, $t + \omega_1$, $t + \omega_1 + \omega_2$, $t + \omega_2$ and sides Γ_1, Γ_2, Γ_3, Γ_4 directed as shown in Fig. 3.17. Since $\zeta(z)$ is meromorphic and has a

Fig. 3.17

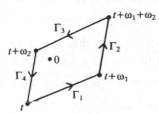

single pole in P (at 0) with residue equal to 1, we have

$$2\pi i = \int_{\partial P} \zeta(z)\, dz$$

$$= \sum_{j=1}^{4} \int_{\Gamma_j} \zeta(z)\, dz.$$

Now

$$\int_{\Gamma_3} \zeta(z)\, dz = -\int_{\Gamma_1} \zeta(z + \omega_2)\, dz$$

$$= -\int_{\Gamma_1} (\zeta(z) + \eta_2)\, dz,$$

so

$$\int_{\Gamma_1} \zeta(z)\, dz + \int_{\Gamma_3} \zeta(z)\, dz = -\int_{\Gamma_1} \eta_2\, dz$$

$$= -\eta_2 \omega_1.$$

Similarly,

$$\int_{\Gamma_2} \zeta(z)\, dz + \int_{\Gamma_4} \zeta(z)\, dz = \eta_1 \omega_2,$$

and hence

$$\eta_1 \omega_2 - \eta_2 \omega_1 = 2\pi i. \tag{3.12.3}$$

This equation is usually referred to as *Legendre's relation*. It implies that at least one of η_1, η_2 is non-zero, so $\zeta(z)$ is not elliptic.

To see how $\sigma(z)$ behaves under translation, we use

$$\frac{\sigma'(z)}{\sigma(z)} = \zeta(z).$$

From this and (3.12.1) we deduce

$$\frac{\sigma'(z + \omega)}{\sigma(z + \omega)} = \frac{\sigma'(z)}{\sigma(z)} + \eta,$$

where $\eta = m\eta_1 + n\eta_2$ for $\omega = m\omega_1 + n\omega_2$. Integrating this, we obtain

$$\log \sigma(z + \omega) = \log \sigma(z) + \eta z + c,$$

where c is a constant depending only on ω, and hence

$$\sigma(z + \omega) = \sigma(z) \exp(\eta z + c).$$

We now evaluate c. First suppose that $\frac{1}{2}\omega \notin \Omega$, so that $\sigma(\frac{1}{2}\omega) \neq 0$. Putting $z = -\frac{1}{2}\omega$ and using the fact that σ is an odd function, we obtain

$$\sigma(\tfrac{1}{2}\omega) = -\sigma(\tfrac{1}{2}\omega) \exp(-\tfrac{1}{2}\eta\omega + c),$$

and so cancelling $\sigma(\frac{1}{2}\omega)$ we get

$$\exp c = -\exp \tfrac{1}{2}\eta\omega.$$

Thus

$$\sigma(z + \omega) = -\sigma(z)\exp\eta(z + \tfrac{1}{2}\omega), \qquad (3.12.4)$$

provided $\tfrac{1}{2}\omega\notin\Omega$. Repeating this, we have

$$\sigma(z + 2\omega) = \sigma(z)\exp\eta(z + \tfrac{1}{2}\omega)\exp\eta\left(z + \frac{3\omega}{2}\right)$$

$$= \sigma(z)\exp 2\eta(z + \omega),$$

so if $\omega' = 2\omega$ satisfies $\tfrac{1}{2}\omega'\in\Omega$ then

$$\sigma(z + \omega') = \sigma(z)\exp\eta'(z + \tfrac{1}{2}\omega'), \qquad (3.12.5)$$

where $\eta' = 2\eta$. Combining (3.12.4) and (3.12.5) we have

$$\sigma(z + \omega) = \varepsilon\sigma(z)\exp\eta(z + \tfrac{1}{2}\omega), \qquad (3.12.6)$$

where $\eta = m\eta_1 + n\eta_2$ for $\omega = n\omega_1 + n\omega_2$, and where

$$\varepsilon = \begin{cases} +1 & \text{if } \tfrac{1}{2}\omega\in\Omega, \\ -1 & \text{otherwise.} \end{cases}$$

Putting $\omega = m\omega_1 + n\omega_2$, we see that $\tfrac{1}{2}\omega\in\Omega$ if and only if both m and n are even, so $\varepsilon = (-1)^{mn + m + n}$.

3.13 The construction of elliptic functions with given zeros and poles

We now return to the problem, posed in §3.6, of finding elliptic functions $f\in E(\Omega)$ with given sets of zeros and poles. At the end of §3.6 we showed that if f has $[a_1],\ldots,[a_r]$ and $[b_1],\ldots,[b_s]$ as its congruence classes of zeros and poles in $\mathbb{C}$, with multiplicities $k_1,\ldots,k_r$ and $l_1,\ldots,l_s$ respectively, then

(i) $\sum_{j=1}^r k_j = \sum_{j=1}^s l_j$,
(ii) the sets $[a_1]\cup\ldots\cup[a_r]$ and $[b_1]\cup\ldots\cup[b_s]$ are disjoint,
(iii) $\sum_{j=1}^r k_j a_j \sim \sum_{j=1}^s l_j b_j$ mod Ω.

As promised in §3.6, we now show that these conditions are not only necessary but also sufficient for the existence of f; this is in contrast with the situation for rational functions on the sphere, where conditions corresponding to (i) and (ii) are necessary and sufficient.

Theorem 3.13.1. *Let $[a_1],\ldots,[a_r]$ and $[b_1],\ldots,[b_s]$ be elements of $\mathbb{C}/\Omega$ for a lattice Ω, and let $k_1,\ldots,k_r, l_1,\ldots,l_s$ be positive integers. If conditions (i), (ii), and (iii) hold, then there exists an elliptic function $f\in E(\Omega)$ with zeros of multiplicity k_j at each $[a_j]$, poles of multiplicity l_j at each $[b_j]$, and no other zeros or poles.*

Proof. Let $u_1, \ldots, u_n$ be the elements $a_1, \ldots, a_r$, each a_j being listed k_j times, so that $n = \sum_{j=1}^r k_j$; similarly let $v_1, \ldots, v_n$ be the elements $b_1, \ldots, b_s$, counting multiplicities l_j. Then condition (iii) takes the form

$$\sum_{j=1}^n u_j - \sum_{j=1}^n v_j = \omega \in \Omega,$$

and there is no loss in replacing u_1 by $u_1 + \omega$ so that (iii) now gives

$$\sum_{j=1}^n u_j = \sum_{j=1}^n v_j. \tag{3.13.2}$$

Now consider the function

$$f(z) = \frac{\sigma(z - u_1) \ldots \sigma(z - u_n)}{\sigma(z - v_1) \ldots \sigma(z - v_n)}.$$

Since $\sigma(z)$ is an analytic function, $f(z)$ is meromorphic. By (3.12.6) we have

$$\sigma(z - u_j + \omega_i) = -\sigma(z - u_j)\exp(\eta_i(z - u_j + \tfrac{1}{2}\omega_i)) \quad (j = 1, \ldots, n; i = 1, 2),$$

and similarly for $\sigma(z - v_j + \omega_i)$, so for $i = 1, 2$ we have

$$f(z + \omega_i) = \frac{(-1)^n \exp\left(\sum_{j=1}^n \eta_i(z - u_j + \tfrac{1}{2}\omega_i)\right)}{(-1)^n \exp\left(\sum_{j=1}^n \eta_i(z - v_j + \tfrac{1}{2}\omega_i)\right)} \cdot f(z)$$

$$= \exp\left(\eta_i \sum_{j=1}^n (v_j - u_j)\right) \cdot f(z)$$

$$= f(z),$$

using (3.13.2). Thus $f(z)$ is doubly periodic with respect to Ω, and hence $f(z)$ is elliptic. Applying Theorem 3.8.7 to the infinite product (3.9.5) for $\sigma(z)$, we see that $\sigma(z)$ has simple zeros at the lattice-points $z \in \Omega$ and that $\sigma(z) \neq 0$ for $z \notin \Omega$. Hence the zeros and poles of $f(z)$ are at $[a_1], \ldots, [a_r]$ and $[b_1], \ldots, [b_s]$ with multiplicities $k_1, \ldots, k_r$ and $l_1, \ldots, l_s$ respectively. $\square$

If g is any other elliptic function with the same zeros and poles as f, then by Theorem 3.6.6, $g(z) = cf(z)$ for some constant $c \neq 0$.

3.14 The construction of elliptic functions with given principal parts

It is possible to construct a rational function on Σ with any given finite set of poles and with any given principal parts at those poles. That this is not possible for elliptic functions on $\mathbb{C}/\Omega$ is shown by Theorem 3.6.2 which imposes a constraint on the residues at these poles. Specifically, if an elliptic function f has s distinct classes of poles $[b_1], \ldots, [b_s]$ in $\mathbb{C}$, with principal part

$$\sum_{k=1}^{l_j} \frac{a_{k,j}}{(z-b_j)^k} \tag{3.14.1}$$

at each b_j $(1 \leqslant j \leqslant s)$, then

$$\sum_{j=1}^{s} a_{1,j} = 0. \tag{3.14.2}$$

The main result of this section is that (3.14.2) is not only necessary but also sufficient for the existence of an elliptic function $f(z)$ with principal parts given by (3.14.1). First we need the following lemma.

Lemma 3.14.3. *Let* $c_1, \ldots, c_s$ *be complex numbers. Then* $g(z) = \sum_{j=1}^{s} c_j \zeta(z - b_j)$ *is elliptic if and only if* $\sum_{j=1}^{s} c_j = 0$.

Proof. Since $\zeta(z)$ is meromorphic, so is $g(z)$. If $\omega \in \Omega$ then by (3.12.1)

$$g(z + \omega) = \sum_{j=1}^{s} c_j \zeta(z + \omega - b_j)$$

$$= \sum_{j=1}^{s} c_j \zeta(z - b_j) + \sum_{j=1}^{s} c_j \eta$$

$$= g(z) + \eta \sum_{j=1}^{s} c_j,$$

where $\eta = m\eta_1 + n\eta_2$ for $\omega = m\omega_1 + n\omega_2$. By Legendre's relation (3.12.3) at least one of η_1, η_2 is non-zero, so $\eta \neq 0$ for some $\omega \neq 0$, and hence $g(z)$ is elliptic with respect to Ω if and only if $\sum_{j=1}^{s} c_j = 0$. $\square$

Theorem 3.14.4. *Let* $b_1, \ldots, b_s$ *be complex numbers, mutually non-congruent with respect to the lattice* Ω, *and let* $l_1, \ldots, l_s$ *be positive integers. If* $a_{k,j}$ *are complex numbers* $(1 \leqslant k \leqslant l_j, 1 \leqslant j \leqslant s)$ *such that* (3.14.2) *holds and* $a_{l_j,j} \neq 0$ *for each* j, *then there exists an elliptic function* $f \in E(\Omega)$ *with poles at* $[b_1], \ldots, [b_s]$, *the principal part at* b_j *being as in* (3.14.1).

Proof. We define

$$f(z) = \sum_{j=1}^{s} \sum_{k=1}^{l_j} a_{k,j} F_k(z - b_j). \tag{3.14.5}$$

where $F_1 = \zeta$, $F_2 = \wp$, and for $k \geqslant 3$, $F_k(z)$ is the elliptic function $\sum_{\omega}(z - \omega)^{-k}$ constructed in Theorem 3.9.3. Now $\sum_{j=1}^{s} a_{1,j} F_1(z - b_j)$ is elliptic by (3.14.2) and Lemma 3.14.3, and since F_k is elliptic for each $k \geqslant 2$

it follows that f is elliptic. Each F_k has a single class of poles of order k, at the lattice-points $\omega \in \Omega$, the principal part at 0 being z^{-k}, so it follows that the poles of f are at $[b_1], \ldots, [b_j]$, the principal part at b_j being given by (3.14.1), using the fact that $a_{l_j, j} \neq 0$. $\square$

If g is another elliptic function with the same poles and principal parts as f, then Theorem 3.6.5 implies that g differs from f by an additive constant.

Now let $V = V(l_1, b_1; l_2, b_2; \ldots; l_s, b_s)$ be the set of all elliptic functions (with respect to Ω) which are analytic on $\mathbb{C} \backslash ([b_1] \cup \ldots \cup [b_j])$ and which are analytic or have poles of order at most l_j on each class $[b_j]$, $1 \leqslant j \leqslant s$. If $f, g \in V$ then $f + g \in V$ and $cf \in V$ for all constants $c \in \mathbb{C}$, so V is a vector space over $\mathbb{C}$.

Theorem 3.14.6.   *V has dimension $l_1 + \ldots + l_s$ over $\mathbb{C}$.*

Proof.   By the proof of Theorem 3.14.4, and the remark following it, the most general form for an element g of V is given by $g = f + c$, where $c \in \mathbb{C}$ and f is as in (3.14.5), the constants $a_{k,j}$ being arbitrary apart from the relation (3.14.2) (we allow $a_{l_j, j} = 0$, so that g may have a pole of order less than l_j at $[b_j]$). It follows that V is spanned by the $l_1 + \ldots + l_s - s$ functions $F_k(z - b_j)$ ($2 \leqslant k \leqslant l_j$, $1 \leqslant j \leqslant s$), the $s - 1$ functions $F_1(z - b_j) - F_1(z - b_1)$ ($2 \leqslant j \leqslant s$), and the constant function 1, so that V has dimension at most $l_1 + \ldots + l_s$. By considering their principal parts it is easily seen that these functions are linearly independent over $\mathbb{C}$, so they form a basis for V and hence V has the required dimension. $\square$

For example, a basis for $V = V(2, 0)$ is given by $\{\wp, 1\}$; thus dim $V = 2$, and the most general elliptic function with poles of order at most 2 on Ω (and analytic on $\mathbb{C} \backslash \Omega$) has the form $a \wp(z) + c$, with a, c arbitrary elements of $\mathbb{C}$.

We can consider Theorem 3.14.6 as a statement about spaces of meromorphic functions on the torus $\mathbb{C}/\Omega$. As such, it is closely related to the important Riemann–Roch theorem, which gives information about the dimensions of spaces of meromorphic functions defined on compact Riemann surfaces (see, for example, Springer [1957], Chapter 10).

3.15 Topological properties of elliptic functions

We saw in §1.5 that a rational function $f : \Sigma \rightarrow \Sigma$ of degree $d > 0$ may be regarded as a d-sheeted branched covering of the Riemann sphere Σ

by itself. Now suppose that a function $f:\mathbb{C}\to\Sigma$ is elliptic with respect to a lattice Ω and that f has order $N>0$. Then f induces a function $\hat{f}:\mathbb{C}/\Omega\to\Sigma$ by $\hat{f}([z])=f(z)$ for each $[z]\in\mathbb{C}/\Omega$, and we shall see that $\hat{f}$ is an N-sheeted branched covering of the sphere Σ by the torus $\mathbb{C}/\Omega$.

Suppose that $a\in\mathbb{C}$, and $f(a)=c\in\Sigma$ with multiplicity k. If U is any neighbourhood of a, sufficiently small that no two points of U are congruent mod Ω, then as shown in Fig. 3.18 the projection $p:\mathbb{C}\to\mathbb{C}/\Omega$, $z\mapsto[z]$, maps U homeomorphically onto a neighbourhood $\hat{U}$ of $[a]$ in $\mathbb{C}/\Omega$ (such a neighbourhood U exists since Ω is discrete). The arguments used in §1.5, applied to U, show that f is an open mapping and is locally k-to-one at a; using the homeomorphism $U\to\hat{U}$ we see that $\hat{f}$ is an open mapping and is locally k-to-one at $[a]$.

Fig. 3.18

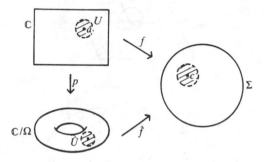

The branch-points of $\hat{f}$ are the points of multiplicity $k>1$, that is, the zeros of $\hat{f}'$ and the multiple poles (if any) of $\hat{f}$. Since f and f' are meromorphic, these sets are discrete and therefore finite (by compactness of $\mathbb{C}/\Omega$). The arguments of §1.5 show that away from the branch-points, $\hat{f}$ is a covering map, the number of sheets being $|\hat{f}^{-1}([a])|=N$ by Theorem 3.6.4. If we include the branch-points then we see that $\hat{f}$ is an N-sheeted branched covering of the sphere Σ by the torus $\mathbb{C}/\Omega$.

For example, let f be Weierstrass' elliptic function $\wp$. Then $N=2$, so $\hat{\wp}$ is a 2-sheeted branched covering of Σ by $\mathbb{C}/\Omega$. By Theorem 3.10.7, $\wp'$ has three congruence classes of zeros, namely $[\tfrac{1}{2}\omega_1]$, $[\tfrac{1}{2}\omega_2]$ and $[\tfrac{1}{2}\omega_3]$, where $\{\omega_1,\omega_2\}$ is a basis for Ω and $\omega_3=\omega_1+\omega_2$. The multiplicity k of $\wp$ at each of these zeros must satisfy $1<k\leqslant N=2$, so $k=2$. There is a single congruence class of poles of $\wp$, namely $[0]=\Omega$, these poles having multiplicity $k=2$. Hence $\hat{\wp}:\mathbb{C}/\Omega\to\Sigma$ has four branch-points of order $k-1=1$, these being the classes $[\tfrac{1}{2}\omega]$ for $\omega\in\Omega$.

We can visualise $\hat{\wp}$ as follows. Let P be the fundamental parallelogram for Ω with vertices $\tfrac{1}{2}(\pm\omega_1\pm\omega_2)$, so that 0 is the centre of P. Now $\wp$ is even and of order $N = 2$, so if $z_1, z_2 \in \overset{\circ}{P}$ then $\wp(z_1) = \wp(z_2)$ if and only if $z_2 = \pm z_1$, the corresponding condition for points $z_1, z_2 \in \partial P$ being $z_2 \sim \pm z_1$. Hence the image $\hat{\wp}(\mathbb{C}/\Omega) = \wp(\mathbb{C}) = \wp(P)$ may be obtained from P by identifying each point $z \in \overset{\circ}{P}$ with $-z$, and each point $z \in \partial P$ with the points in $[\pm z] \cap \partial P$. Now the transformation $\rho : z \mapsto -z$ is a rotation of P about 0 by an angle π (see Fig. 3.19), and since ρ maps congruent pairs of points in ∂P to congruent pairs, ρ induces a map $\hat{\rho}$ from the torus $\mathbb{C}/\Omega$ to itself, given by $\hat{\rho}([z]) = [\rho(z)] = [-z]$. The fixed-points of $\hat{\rho}$ are the classes $[z] = [-z]$, that is the classes satisfying $[2z] = [0] = \Omega$, so these are the four branch-points $[\frac{1}{2}\omega]$ of $\hat{\wp}$. We can visualise $\mathbb{C}/\Omega$ as a torus

Fig. 3.19

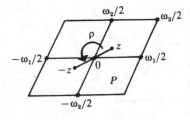

Fig. 3.20

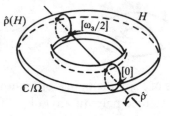

Fig. 3.21

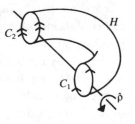

Fig. 3.22

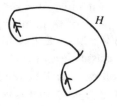

in $\mathbb{R}^3$, with $\hat{\rho}$ represented by a rotation through an angle π about an axis meeting $\mathbb{C}/\Omega$ in these four branch-points (see Fig. 3.20). As shown in Figs. 3.20 and 3.21, we may divide $\mathbb{C}/\Omega$ into two halves H and $\hat{\rho}(H)$, which are interchanged by $\hat{\rho}$ and which meet across the two circular boundary components C_1 and C_2 of H. Each point in $\mathbb{C}/\Omega$ is equivalent under $\hat{\rho}$ to a unique point in H, except that for each $i = 1, 2$, points on C_i are equivalent in pairs. We therefore obtain the quotient space $\hat{\wp}(\mathbb{C}/\Omega)$ from H by identifying pairs of points on $C_1 \cup C_2$ as indicated by the arrows in Fig. 3.21. The resulting space (Fig. 3.22) is homeomorphic to a sphere, as we must expect since $\hat{\wp}(\mathbb{C}/\Omega)$ is the Riemann sphere Σ. To sum up, the effect of $\hat{\wp}$ on $\mathbb{C}/\Omega$ is to identify each class $[z]$ with $[-z]$; since the rotation $\hat{\rho}$ has the same effect, the image $\hat{\wp}(\mathbb{C}/\Omega)$ is just the quotient space under the action of $\hat{\rho}$, and this is homeomorphic to a sphere. We shall return to this branched covering map in §4.9(iv), where we show that if $p(z)$ is a cubic polynomial with distinct roots, then the Riemann surface of the equation $w^2 = p(z)$ is a torus, a two-sheeted covering of Σ with four branch-points.

3.16 Real elliptic curves

We have seen that $\wp$ satisfies a differential equation $(\wp')^2 = p(\wp)$, where $p(x)$ is the cubic polynomial $4x^3 - g_2 x - g_3$, so every point $t \in \mathbb{C}/\Omega$ determines a point $(\wp(t), \wp'(t))$ on the *elliptic curve*

$$E = \{(x, y) \in \Sigma \times \Sigma \mid y^2 = p(x)\}.$$

We can think of E as the graph of the equation $y^2 = p(x)$, for $x, y \in \Sigma$. As a subset of $\Sigma \times \Sigma$, E has a natural topology; the next result shows that E is homeomorphic to a torus.

Lemma 3.16.1. *The map* $\theta : \mathbb{C}/\Omega \to E$, $t \mapsto (\wp(t), \wp'(t))$, *is a homeomorphism.*

Proof. The points $t = [0]$, $[\frac{1}{2}\omega_j]$ (and no others) are mapped by θ to (∞, ∞) and $(e_j, 0)$ respectively, for $j = 1, 2, 3$. For each remaining point $(x, y) \in E$, we have $x \neq \infty$, e_j and $y \neq \infty$, 0; since $\wp$ is even and of order 2, with a simple point at each $t \neq [0], [\frac{1}{2}\omega_j]$, it follows that for each $x \neq \infty$, e_j there are two distinct solutions $t = \pm t_1$ of $\wp(t) = x$. Now $\wp'(t_1) = -\wp'(-t_1) \neq \wp'(-t_1)$, so $\wp'(t_1)$ and $\wp'(-t_1)$ take the two values of $\sqrt{p(\wp(t))} = \sqrt{p(x)}$, and one of these values is y. Thus there is a unique $t (= t_1$ or $-t_1)$ in $\mathbb{C}/\Omega$ satisfying $\theta(t) = (x, y)$, so θ is a bijection.

Being meromorphic and not constant, $\wp$ and $\wp'$ are continuous open functions, and hence so is θ, so θ is a homeomorphism. $\square$

Fig. 3.23

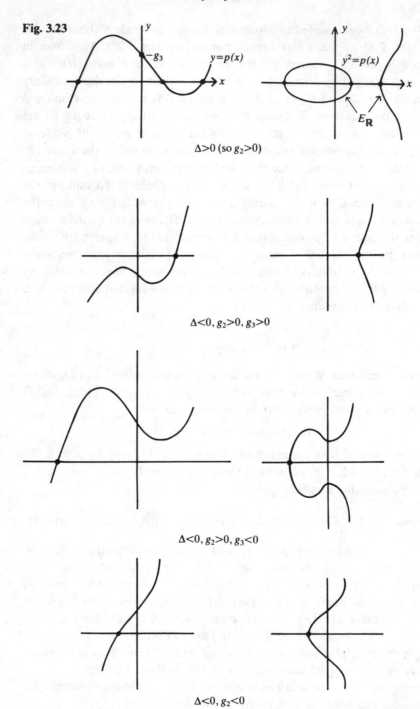

$\Delta>0$ (so $g_2>0$)

$\Delta<0, g_2>0, g_3>0$

$\Delta<0, g_2>0, g_3<0$

$\Delta<0, g_2<0$

This embedding of a torus $\mathbb{C}/\Omega$, as a subset E of a product $\Sigma \times \Sigma$ of two spheres, is rather hard to visualise, and we shall return to it in §4.9(iv), when we construct the Riemann surface of the equation $y^2 = p(x)$. For the time being, we shall concentrate on the 'real points' of E, those for which $x, y \in \mathbb{R}$, under the extra assumption that *the coefficients g_2, g_3 of $p(x)$ are real.* Assuming this, we define the *real elliptic curve* $E_\mathbb{R}$ to be $\{(x, y) \in \mathbb{R}^2 | y^2 = p(x)\}$, the graph of $y^2 = p(x)$ as an equation between real variables. Clearly $E_\mathbb{R}$ is symmetric about the x-axis of $\mathbb{R}^2$, and other properties of $E_\mathbb{R}$ are easily found by sketching the graph of $p(x)$ and then taking square roots whenever $p(x) \geqslant 0$. For example, being a real cubic polynomial with no repeated roots, $p(x)$ has one or three real roots, as the discriminant $\Delta = g_2^3 - 27g_3^2$ is negative or positive (this can easily be seen by considering the stationary points of $p(x)$); the curve $E_\mathbb{R}$ then has one or two components respectively as shown in Fig. 3.23, the curves on the left representing $y = p(x)$, those on the right $y^2 = p(x)$.

First we examine the conditions under which the coefficients g_2, g_3 are real. We define a meromorphic function $f : \mathbb{C} \to \Sigma$ to be *real* if $f(\bar{z}) = \overline{f(z)}$ for all $z \in \mathbb{C}$ (where $\bar{\infty}$ is interpreted as ∞). We define a lattice Ω to be *real* if $\bar{\Omega} = \Omega$ (where $\bar{\Omega}$ denotes $\{\bar{\omega} | \omega \in \Omega\}$).

Theorem 3.16.2. *The following conditions are equivalent:*

(i) $g_2, g_3 \in \mathbb{R}$;
(ii) $G_k \in \mathbb{R}$ for all $k \geqslant 3$;
(iii) $\wp$ is a real function;
(iv) Ω is a real lattice.

Proof. (i) $\Rightarrow$ (ii). Differentiating $(\wp')^2 = 4\wp^3 - g_2\wp - g_3$, and then dividing by $2\wp'$ (which is not identically zero), we have

$$\wp'' = 6\wp^2 - \frac{g_2}{2}. \tag{3.16.3}$$

Now by (3.10.3), $\wp(z)$ has a Laurent expansion

$$\wp(z) = z^{-2} + \sum_{n=1}^{\infty} a_n z^{2n},$$

valid near $z = 0$, where

$$a_n = (2n + 1)G_{2n+2} = (2n + 1)\sum_\omega{}' \omega^{-2n-2}.$$

The coefficient of z^{2n} in the expansion of $\wp''(z)$ is therefore $(2n + 2)(2n + 1)a_{n+1}$, while the coefficient of z^{2n} in $\wp(z)^2$ is $2a_{n+1} + \sum_{r+s=n} a_r a_s$.

Equating coefficients in (3.16.3) we therefore have, for each $n \geqslant 1$,

$$(2n + 2)(2n + 1)a_{n+1} = 12a_{n+1} + 6 \sum_{r+s=n} a_r a_s,$$

and hence

$$(2n + 5)(n - 1)a_{n+1} = 3 \sum_{r+s=n} a_r a_s.$$

For $n \geqslant 2$ we therefore have

$$a_{n+1} = \frac{3}{(2n + 5)(n - 1)} \sum_{r+s=n} a_r a_s,$$

expressing a_{n+1} in terms of $a_1, \ldots, a_n$. Thus

$$a_3 = \tfrac{1}{3}a_1^2,$$

$$a_4 = \tfrac{3}{11}a_1 a_2,$$

$$a_5 = \tfrac{1}{13}(2a_1 a_3 + a_2^2) = \tfrac{1}{39}(2a_1^2 + 3a_2^2), \text{ etc.}$$

By induction on n, we see that each coefficient a_n is a polynomial in a_1 and a_2, with rational coefficients. Using $a_n = (2n + 1)G_{2n+2}$, $g_2 = 60G_4$, and $g_3 = 140G_6$, we see that each G_k (k even, $k \geqslant 4$) is a polynomial in g_2 and g_3 with rational coefficients, so if g_2 and g_3 are real then so is G_k; since $G_k = 0$ for all odd k, (ii) is proved.

(ii) $\Rightarrow$ (iii). If $G_k \in \mathbb{R}$ for all $k \geqslant 3$, then the coefficients of the Laurent series for $\wp(z)$ are real, so $\overline{\wp(\bar{z})} = \wp(z)$ near $z = 0$. Now $\wp(z)$ and $\overline{\wp(\bar{z})}$ are meromorphic functions, identically equal on a neighbourhood of 0, so they are identically equal on $\mathbb{C}$ by Theorem A.8. Thus $\wp$ is a real function.

(iii) $\Rightarrow$ (iv). Let $\omega \in \Omega$. Then $\wp(z + \bar{\omega}) = \overline{\wp(\bar{z} + \omega)} = \overline{\wp(\bar{z})} = \wp(z)$ since $\wp$ is real and has ω as a period. Thus $\bar{\omega} \in \Omega$, so $\bar{\Omega} \subseteq \Omega$. Taking complex conjugates, we have $\Omega = \bar{\bar{\Omega}} \subseteq \bar{\Omega}$, so $\Omega = \bar{\Omega}$ and Ω is real.

(iv) $\Rightarrow$ (i). This follows immediately from $g_2 = 60 \sum'_\omega \omega^{-4}$ and $g_3 = 140 \sum'_\omega \omega^{-6}$.   □

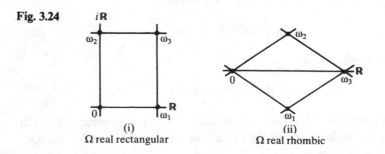

Fig. 3.24

(i) Ω real rectangular

(ii) Ω real rhombic

This raises the problem of determining the real lattices. We say that Ω is *real rectangular* if $\Omega = \Omega(\omega_1, \omega_2)$ where ω_1 is real and ω_2 is purely imaginary, and that Ω is *real rhombic* if $\Omega = \Omega(\omega_1, \omega_2)$ where $\bar{\omega}_1 = \omega_2$. (The fundamental parallelogram with vertices $0, \omega_1, \omega_2$ and $\omega_3 = \omega_1 + \omega_2$ is rectangular or rhombic respectively, hence the names; see Fig. 3.24.)

Theorem 3.16.4. *A lattice Ω is real if and only if it is real rectangular or real rhombic.*

Proof. If $\Omega = \Omega(\omega_1, \omega_2)$ is real rectangular, with $\omega_1 \in \mathbb{R}$ and $\omega_2 \in i\mathbb{R}$, then $\bar{\Omega} = \Omega(\bar{\omega}_1, \bar{\omega}_2) = \Omega(\omega_1, -\omega_2) = \Omega(\omega_1, \omega_2) = \Omega$, so Ω is real. A similar argument applies to real rhombic lattices.

Conversely, suppose that Ω is real. If $\omega \in \Omega$ then $\omega + \bar{\omega}$, $\omega - \bar{\omega} \in \Omega$, so Ω contains both real and purely imaginary elements, and these form discrete subgroups $\Omega \cap \mathbb{R} = \lambda \mathbb{Z}$ and $\Omega \cap i\mathbb{R} = \mu i \mathbb{Z}$ for certain $\lambda, \mu \in \mathbb{R}$, $\lambda, \mu > 0$. Clearly $\Omega \supseteq \lambda \mathbb{Z} + \mu i \mathbb{Z}$, and if we have equality then Ω is real rectangular since $\{\lambda, \mu\}$ is a basis. Hence suppose that there exists $\omega \in \Omega \setminus (\lambda \mathbb{Z} + \mu i \mathbb{Z})$. By adding a suitable element of $\lambda \mathbb{Z} + \mu i \mathbb{Z}$ to ω we may assume that $0 \leqslant \mathrm{Re}(\omega) < \lambda$ and $0 \leqslant \mathrm{Im}(\omega) < \mu$. Now

$$2\omega = (\omega + \bar{\omega}) + (\omega - \bar{\omega}),$$

with $\omega + \bar{\omega} \in \Omega \cap \mathbb{R} = \lambda \mathbb{Z}$ and $\omega - \bar{\omega} \in \Omega \cap i\mathbb{R} = \mu i \mathbb{Z}$, so we have

$$2\omega = m\lambda + n\mu i$$

for integers m, n, and the conditions on $\mathrm{Re}(\omega)$ and $\mathrm{Im}(\omega)$ force m and n to take the values 0 or 1. Since ω is neither real nor purely imaginary, we must have $m = n = 1$ and so $\omega = \frac{1}{2}(\lambda + \mu i)$. Thus every element of $\Omega \setminus (\lambda \mathbb{Z} + \mu i \mathbb{Z})$ has the form

$$\frac{1}{2}(\lambda + \mu i) + a\lambda + b\mu i = (a + b + 1)\left(\frac{\lambda + \mu i}{2}\right) + (a - b)\left(\frac{\lambda - \mu i}{2}\right),$$

for integers a, b, while every element of $\lambda \mathbb{Z} + \mu i \mathbb{Z}$ has the form

$$a\lambda + b\mu i = (a + b)\left(\frac{\lambda + \mu i}{2}\right) + (a - b)\left(\frac{\lambda - \mu i}{2}\right).$$

Thus $\Omega = \Omega(\frac{1}{2}(\lambda + \mu i), \frac{1}{2}(\lambda - \mu i))$, which is real rhombic. $\square$

Theorem 3.16.5. *Let Ω be a real lattice. Then the real elliptic curve $E_{\mathbb{R}}$ has one or two components as Ω is real rhombic or real rectangular respectively.*

Proof. Since $E_{\mathbb{R}}$ is the graph of $y^2 = p(x)$, we can count its components by counting the roots of $p(x)$, that is, the points $(x, y) \in \mathbb{R}^2$ for which $y =$

$\wp'(z) = 0$ and $x = \wp(z) \in \mathbb{R}$ for some $z \in \mathbb{C}$. Since $p(x)$ is a cubic polynomial with distinct roots, there are one or three such points, and $E_{\mathbb{R}}$ has one or two components respectively. The only solutions of $\wp'(z) = 0$ are of the form $z \sim \frac{1}{2}\omega_j (j = 1, 2, 3)$, so it is sufficient to determine which of $e_j = \wp(\frac{1}{2}\omega_j)$ are real.

Suppose that $\Omega = \Omega(\omega_1, \omega_2)$ is real rectangular, with $\omega_1 \in \mathbb{R}$ and $\omega_2 \in i\mathbb{R}$ (see Fig. 3.25). Since $\wp$ is real, $\wp(\mathbb{R}) \subseteq \mathbb{R} \cup \{\infty\}$, so $e_1 = \wp(\frac{1}{2}\omega_1) \in \mathbb{R}$. If $z \in \frac{1}{2}\omega_2 + \mathbb{R}$ then $\bar{z} = z - \omega_2$ and hence $\wp(z) = \wp(z - \omega_2) = \wp(\bar{z}) = \overline{\wp(z)}$, so $\wp(z) \in \mathbb{R}$; in particular, putting $z = \frac{1}{2}\omega_2, \frac{1}{2}\omega_3$ we get $e_2, e_3 \in \mathbb{R}$. Thus $p(x)$ has three real roots, so $E_{\mathbb{R}}$ meets the x-axis in three points e_1, e_2, e_3 and hence has two components, one unbounded, containing $(e_1, 0)$, corresponding to $z \in \mathbb{R}$, the other bounded, containing $(e_2, 0)$ and $(e_3, 0)$, corresponding to $z \in \frac{1}{2}\omega_2 + \mathbb{R}$ (see Fig. 3.26).

Fig. 3.25

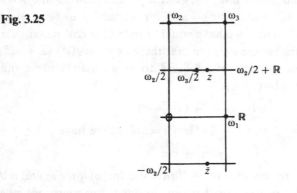

Fig. 3.26

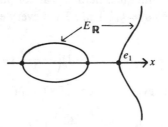

Fig. 3.27

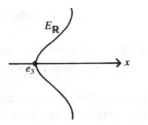

If $\Omega = \Omega(\omega_1, \omega_2)$ is real rhombic, with $\bar{\omega}_1 = \omega_2$, then again we have $\wp(\mathbb{R}) \subseteq \mathbb{R} \cup \{\infty\}$ and so $e_3 \in \mathbb{R}$. Now $\frac{1}{2}\bar{\omega}_1 = \frac{1}{2}\omega_2$, so $e_2 = \wp(\frac{1}{2}\omega_2) = \wp(\frac{1}{2}\bar{\omega}_1)$ $= \overline{\wp(\frac{1}{2}\omega_1)} = \bar{e}_1$; since e_1 and e_2 are distinct, neither of them can be real, so $p(x)$ has a unique real root, and $E_{\mathbb{R}}$ has just one component, meeting the x-axis at $(e_3, 0)$, and corresponding to $z \in \mathbb{R}$ (see Fig. 3.27) (Also see Du Val [1973] and Alling [1981]).

3.17 The addition theorem

A function f is said to possess an *addition theorem* if there is an identity between $f(z_1)$, $f(z_2)$ and $f(z_1 + z_2)$ of the form

$$R(f(z_1), f(z_2), f(z_1 + z_2)) = 0 \tag{3.17.1}$$

for all z_1, z_2, where R is a non-zero rational function (with complex coefficients) of its three variables; if we multiply (3.17.1) by the denominator of R then we may assume that R is a polynomial.

The first important examples of addition theorems which we meet are those involving the exponential and trigonometric functions, for instance

$$\exp(z_1 + z_2) = \exp(z_1).\exp(z_2) \tag{3.17.2}$$

and

$$\tan(z_1 + z_2) = \frac{\tan z_1 + \tan z_2}{1 - \tan z_1.\tan z_2},$$

both of which are equivalent to identities of the form (3.17.1). We obtain an addition theorem for the sine function from

$$\sin(z_1 + z_2) = \sin z_1.\cos z_2 + \sin z_2.\cos z_1$$

by putting $\cos z_j = \sqrt{(1 - \sin^2 z_j)}$ for $j = 1, 2$ and then squaring twice to eliminate square roots; similar techniques may be applied to the other trigonometric functions.

The rational functions also possess addition theorems (though in this case (3.17.1) is generally neither obvious nor particularly useful). If $f = p/q$ is a rational function of z, with p and q polynomials, then

$$g(u) = p(u) - f(z_1)q(u),$$
$$h(v) = p(v) - f(z_2)q(v),$$

and

$$k(u, v) = p(u + v) - f(z_1 + z_2)q(u + v)$$

are polynomials in u, v, and in u and v respectively, with coefficients of the form $a + bf(z_1)$, $a + bf(z_2)$ and $a + bf(z_1 + z_2)$ for various constants

a, b obtained from the coefficients of p and q. Moreover, the equations

$$\left.\begin{aligned} g(u) &= 0, \\ h(v) &= 0, \\ k(u, v) &= 0, \end{aligned}\right\} \tag{3.17.3}$$

have a common solution at $u = z_1$, $v = z_2$. Now a necessary and sufficient condition for two polynomial equations $s(u) = 0$ and $t(u) = 0$ to have a common solution u is the vanishing of a polynomial in the coefficients of s and t called their *resultant* (see Appendix 3). Thus we may eliminate u between $g(u) = 0$ and $k(u, v) = 0$ to obtain the resultant polynomial $l(v)$, and then we may eliminate v between $l(v) = 0$ and $h(v) = 0$ to obtain an addition theorem of the form (3.17.1). For example, if $f(z) = z/(z + 1)$ then $g(u) = u - f(z_1)(u + 1)$, etc., so equations (3.17.3) take the form

$$u(1 - f(z_1)) - f(z_1) = 0,$$
$$v(1 - f(z_2)) - f(z_2) = 0,$$
$$(u + v)(1 - f(z_1 + z_2)) - f(z_1 + z_2) = 0,$$

and eliminating u and v we have the addition theorem

$$\frac{f(z_1)}{1 - f(z_1)} + \frac{f(z_2)}{1 - f(z_2)} = \frac{f(z_1 + z_2)}{1 - f(z_1 + z_2)}.$$

(The elimination of u and v is usually much harder to perform when f has degree greater than 1.)

A similar argument, also using the addition theorem (3.17.2) for the exponential function, shows that if $\omega \in \mathbb{C} \setminus \{0\}$ then any rational function of $\exp(2\pi i z / \omega)$ has an addition theorem; taking $\omega = 2\pi$ we get the addition theorems for the trigonometric functions.

We shall now show that elliptic functions have addition theorems; in fact the development of the theory of elliptic functions can be traced back to the investigations of C.G. Fagnano and L. Euler in the eighteenth century into the addition theorems of certain elliptic integrals. Conversely, a deep result of K. Weierstrass shows that a meromorphic function with an addition theorem must be a rational function, a rational function of $\exp(2\pi i z / \omega)$ for some $\omega \neq 0$, or an elliptic function with respect to some lattice (see Forsyth [1918], Chapter XIII); this illustrates the strong links between the rational, simply periodic and doubly periodic functions we have considered so far in this book.

We first show that Weierstrass' function $\wp$ has an addition theorem; it is convenient to do this by considering the group structure on the elliptic curve E.

We have seen in Lemma 3.16.1 that there is a bijection $\theta : t \mapsto (\wp(t), \wp'(t))$

from the torus $\mathbb{C}/\Omega$ to the elliptic curve $E = \{(x, y) \in \Sigma \times \Sigma \,|\, y^2 = p(x)\}$. Since $\mathbb{C}/\Omega$ is a group we can use θ to impose a group structure on E: if $P_j \in E$ for $j = 1, 2$ then we define $P_1 + P_2$ to be $\theta(t_1 + t_2)$ where $P_j = \theta(t_j)$; thus $\theta : \mathbb{C}/\Omega \to E$ is an isomorphism. For example, the zero element of E is $\theta([0]) = (\wp(0), \wp'(0)) = (\infty, \infty)$, and if $P = (x, y) = (\wp(t), \wp'(t)) \in E$ then the inverse of P is $-P = (\wp(-t), \wp'(-t)) = (\wp(t), -\wp'(t)) = (x, -y)$, so taking inverses corresponds to 'reflection in the x-axis'.

It is a little harder to describe addition in E in terms of the coordinates x, y. Let $P_j = (x_j, y_j) = \theta(t_j) \in E$ for $j = 1, 2$, so $P_1 + P_2 = \theta(t_1 + t_2)$. To avoid trivial cases, we assume that P_1, P_2 and $P_1 + P_2$ are non-zero in E, that is, $t_1, t_2, t_1 + t_2 \neq [0]$. Now suppose that we had an elliptic function g of order 3 with a triple pole at $[0]$ and simple zeros at t_1 and t_2 (or a double zero at t_1 if $t_1 = t_2$); then g must have a third zero at $t_3 \in \mathbb{C}/\Omega$ where $t_1 + t_2 + t_3 = [0]$ by Theorem 3.6.7, so putting $P_3 = (x_3, y_3) = \theta(t_3)$ we get $P_1 + P_2 + P_3 = 0$ and hence $P_1 + P_2 = -P_3 = (x_3, -y_3)$. (Notice that our hypotheses about t_1 and t_2 guarantee that none of the zeros t_j coincides with the pole $[0]$, and that $t_3 \neq t_1, t_2$.) We now show that such a function g exists.

Consider the function

$$g(t) = \wp'(t) - \alpha \wp(t) - \beta \tag{3.17.4}$$

for suitably chosen constants $\alpha, \beta \in \mathbb{C}$. Certainly, any function of this form has order 3, with a triple pole at $[0]$. Suppose first that $t_1 \neq t_2$. Then g will have the required zeros provided that

$$\wp'(t_j) = \alpha \wp(t_j) + \beta,$$

that is,

$$y_j = \alpha x_j + \beta \tag{3.17.5}$$

for $j = 1, 2$; since $t_1 \neq \pm t_2$ and $t_1, t_2 \neq [0]$, we have x_1 and x_2 distinct and finite, so there exist $\alpha, \beta \in \mathbb{C}$ satisfying (3.17.5). Since $g(t_3) = 0$, (3.17.5) holds for $j = 3$ also, and thus the points P_1, P_2 and P_3 are collinear in $\mathbb{C} \times \mathbb{C}$. Putting $P_3 = (\wp(t_1 + t_2), -\wp'(t_1 + t_2))$ we deduce from this that

$$\begin{vmatrix} \wp(t_1) & \wp'(t_1) & 1 \\ \wp(t_2) & \wp'(t_2) & 1 \\ \wp(t_1 + t_2) & -\wp'(t_1 + t_2) & 1 \end{vmatrix} = 0,$$

which is valid also when $t_1 = t_2$. Expanding this determinant, we have a form of addition theorem involving $\wp'$ as well as $\wp$.

We now show that $\wp$ has an addition theorem of the form

$$R(\wp(t_1), \wp(t_2), \wp(t_1 + t_2)) = 0 \tag{3.17.6}$$

for some rational function R; we do this by calculating the coefficient α

in (3.17.4) in two different ways. Firstly, if $t_1 \neq t_2$ then (3.17.5) implies that

$$\alpha = \frac{y_1 - y_2}{x_1 - x_2}. \tag{3.17.7}$$

Secondly, using $y_j^2 = p(x_j)$ and $y_j = \alpha x_j + \beta$ we obtain

$$p(x_j) - (\alpha x_j + \beta)^2 = 0$$

for $j = 1, 2, 3$, so that x_1, x_2 and x_3 are the roots of the cubic polynomial

$$p(x) - (\alpha x + \beta)^2 = 4x^3 - \alpha^2 x^2 - (g_2 + 2\alpha\beta)x - (g_3 + \beta^2).$$

The formula for the sum of the roots of a polynomial now gives

$$x_1 + x_2 + x_3 = \frac{\alpha^2}{4}, \tag{3.17.8}$$

so that using (3.17.7) and $\wp(t_1 + t_2) = \wp(-t_3) = \wp(t_3) = x_3$, we obtain

$$\wp(t_1 + t_2) = \frac{1}{4}\left(\frac{\wp'(t_1) - \wp'(t_2)}{\wp(t_1) - \wp(t_2)}\right)^2 - \wp(t_1) - \wp(t_2), \tag{3.17.9}$$

valid for $t_1, t_2, t_1 \pm t_2 \neq [0]$. This is usually referred to as the *addition theorem* for $\wp$, though strictly speaking we should now use $(\wp')^2 = p(\wp)$ to eliminate the derivatives and get an equation of the form (3.17.6). For each fixed $t_2 \neq [0]$, equation (3.17.6) is valid for all $t_1 \neq [0]$, $\pm t_2$; we can regard $R(\wp(t_1), \wp(t_2), \wp(t_1 + t_2))$ as an elliptic function of t_1, continuous as a function $\mathbb{C}/\Omega \to \Sigma$, so by taking limits as t_1 approaches each of these exceptional values we see that (3.17.6) is true for all t_1 and for all $t_2 \neq [0]$; now fixing t_1 and letting $t_2 \to [0]$ we see that (3.17.6) is true for all t_1, $t_2 \in \mathbb{C}/\Omega$. As an example of this limiting process, if we fix $t_2(=t$, say) and let $t_1 \to t_2$ then we get the *duplication theorem*

$$\wp(2t) = \frac{1}{4}\left(\frac{\wp''(t)}{\wp'(t)}\right)^2 - 2\wp(t) \tag{3.17.10}$$

(see Exercise 3T). Alternatively, we can modify our proof of the addition theorem to prove (3.17.10): when $t_1 = t_2(=t)$, we need the function $g = \wp' - \alpha\wp - \beta$ to have a *double* root at t, so that $\wp''(t) - \alpha\wp'(t) = 0$, that is, $\alpha = \wp''(t)/\wp'(t)$, and now (3.17.10) follows from (3.17.8).

Returning to the group-structure of E, we see that $P_1 + P_2$ is the point $(x_3, -y_3) \in E$ where

$$x_3 = \frac{\alpha^2}{4} - x_1 - x_2$$

$$= \frac{1}{4}\left(\frac{y_1 - y_2}{x_1 - x_2}\right)^2 - x_1 - x_2. \tag{3.17.11}$$

We have

$$y_3 = \alpha x_3 + \beta$$

with

$$\beta = \frac{x_1 y_2 - x_2 y_1}{x_1 - x_2},$$

and so

$$y_3 = \frac{(x_3 - x_2)y_1 - (x_3 - x_1)y_2}{x_1 - x_2}. \tag{3.17.12}$$

This shows that the coordinates of $P_1 + P_2$ may be expressed as rational functions (with rational coefficients) of the coordinates of P_1 and P_2; similarly inversion $(x, y) \mapsto (x, -y)$ is expressed rationally, so E is an example of an *algebraic group* (see Borel [1956]).

Now suppose that Ω is a real lattice (see §3.16). If we adjoin the zero element (∞, ∞) of E to the real elliptic curve $E_\mathbb{R} = \{(x, y) \in E \mid x, y \in \mathbb{R}\}$, then

$$\hat{E}_\mathbb{R} = E_\mathbb{R} \cup \{(\infty, \infty)\}$$

is a subgroup of E, since by the previous paragraph it is closed under the group operations. Given $P_1 \neq P_2$ in $\hat{E}_\mathbb{R}$, we can find $P_1 + P_2$ by using the collinearity of P_1, P_2 and P_3: we take the straight line $y = \alpha x + \beta$ through P_1 and P_2, find its third intersection P_3 with $\hat{E}_\mathbb{R}$, and reflect this in the x-axis to get $P_1 + P_2 = -P_3$ (see Fig. 3.28). If $P_1 = P_2$ then we take the tangent to $\hat{E}_\mathbb{R}$ at P_1 and proceed as before.

A similar argument shows that if g_2, $g_3 \in \mathbb{Q}$ then the 'rational elliptic curve'

$$\hat{E}_\mathbb{Q} = \{(x, y) \in E \mid x, y \in \mathbb{Q}\} \cup \{(\infty, \infty)\}$$

is a subgroup of E; this is of great importance for the number-theoretic

Fig. 3.28

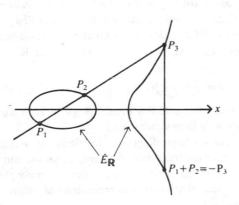

problem of finding solutions x, $y \in \mathbb{Z}$ of $y^2 = p(x)$ (see Mordell [1969], and Lang [1978]).

Finally, we fulfil our promise to show that *every* elliptic function f has an addition theorem. By Theorem 3.11.2 there is a polynomial Φ of two variables such that $\Phi(\wp, f) = 0$. If we eliminate $\wp(z_1)$, $\wp(z_2)$ and $\wp(z_1 + z_2)$ from the equations

$$\Phi(\wp(z_1), f(z_1)) = 0,$$
$$\Phi(\wp(z_2), f(z_2)) = 0,$$
$$\Phi(\wp(z_1 + z_2), f(z_1 + z_2)) = 0,$$
$$R(\wp(z_1), \wp(z_2), \wp(z_1 + z_2)) = 0,$$

then we get a polynomial identity between $f(z_1)$, $f(z_2)$ and $f(z_1 + z_2)$, that is, an addition theorem for f.

EXERCISES

3A. (i) Show that $\mathbb{Z}[i] = \{m + ni \,|\, m, n \in \mathbb{Z}\}$ and $\mathbb{Z}[\rho] = \{m + n\rho \,|\, m, n \in \mathbb{Z}\}$, where $\rho = \frac{1}{2}(-1 + \sqrt{-3})$ is a cube root of 1, both form discrete subgroups of $\mathbb{C}$.

     (ii) Show that $\mathbb{Z}[\sqrt{2}] = \{m + n\sqrt{2} \,|\, m, n \in \mathbb{Z}\}$ is a subgroup of $\mathbb{C}$ but that it is not discrete.

3B. Use Theorem 3.2.2 to show that if Ω is a discrete subgroup of $\mathbb{R}^n$ then there exists $\omega_1 \in \Omega \setminus \{0\}$ with least positive value of $|\omega_1|$, where $|\omega|$ denotes the Euclidean distance of ω from 0. (This can be used to shorten the proof of Theorem 3.1.3.)

3C. Show that a non-constant rational function cannot be periodic.

3D. Using the ideas presented in §3.3 find the Fourier series expansion for cosec $2\pi z$ valid in the region $\{z \,|\, \text{Im}\, z > 0\}$ and also the Fourier series expansion valid in the region $\{z \,|\, \text{Im}\, z < 0\}$.

3E. Show that $a\omega_2 + b\omega_1 \in \Omega(\omega_1, \omega_2)$ can be an element of a basis for $\Omega(\omega_1, \omega_2)$ if and only if a and b are co-prime, or $a = \pm 1$ and $b = 0$, or $a = 0$ and $b = \pm 1$.

3F. Find the Dirichlet regions for the lattices $\mathbb{Z}[i]$ and $\mathbb{Z}[\rho]$ of Exercise 3A.

3G. An *automorphism* of a lattice Ω is a bijection $\alpha : \Omega \to \Omega$ such that $\alpha(u + v) = \alpha(u) + \alpha(v)$, for all $u, v \in \Omega$. Prove that if $\omega \mapsto \lambda\omega (\omega \in \Omega)$ is an automorphism of Ω then

$$\lambda = \pm 1, \pm i, \tfrac{1}{2}(1 \pm \sqrt{-3}) \quad \text{or} \quad \tfrac{1}{2}(-1 \pm \sqrt{-3}).$$

(Hint: Show that λ has modulus 1 and, using §3.4, is an eigenvalue of a 2×2 integer matrix with determinant ± 1.)

3H. Use 3G to obtain the 'crystallographic restriction' (Coxeter [1969], §4.5): the only possibly finite orders, greater than one, for an automorphism of a lattice are 2, 3, 4 or 6. Show that every lattice admits an automorphism of order 2 and describe the lattices that admit automorphisms of orders 3, 4, or 6.

3I. Verify the claim, made at the end of §3.7, that $\sum_{n=-\infty}^{\infty}(z-n)^{-2}$ converges normally (and hence uniformly and absolutely) on all compact subsets of $\mathbb{C}\setminus\mathbb{Z}$ and so represents a meromorphic function with period 1.

3J. Let $h(z) = \pi^2 \operatorname{cosec}^2 \pi z - \sum_{n=-\infty}^{\infty}(z-n)^{-2}$.
Prove the following:
  (i) $h(z)$ has period 1;
  (ii) $h(z)$ has a finite limit at $z = 0$ and hence, by periodicity, at all the integers.

Thus $h(z)$ is continuous and $|h(z)| \leqslant M$ for all $z \in [0,1]$ and hence, by periodicity, $|h(z)| \leqslant M$ for all real z. Show that

$$h\left(\frac{z}{2}\right) + h\left(\frac{z+1}{2}\right) = 4h(z)$$

and deduce that $M = 0$ and hence

$$\pi^2 \operatorname{cosec}^2 \pi z = \sum_{n=-\infty}^{\infty} (z-n)^{-2}.$$

3K. Use 3J and the ideas of §3.8 to verify that

$$\frac{1}{z} + \sum_{n=1}^{\infty}\left(\frac{1}{z-n} + \frac{1}{z+n}\right) = \pi \cot \pi z$$

and

$$z \prod_{n=1}^{\infty}\left(1 - \frac{z^2}{n^2}\right) = \pi \sin \pi z.$$

3L. Give a direct proof (using the ideas of the proof of Theorem 3.9.3) that $\wp(z)$, defined by the series (3.9.4), is an elliptic function.

3M. If f is an elliptic function of order $m > 0$ show that f' is elliptic and that its order n satisfies $m + 1 \leqslant n \leqslant 2m$. Give examples to show that both bounds are attained.

3N. Let f be elliptic with respect to a lattice Ω, and suppose that f has order m. Let A be the set of points $c \in \mathbb{C} \cup \{\infty\}$ for which $f(u) = c$ has less than m distinct solutions. Prove that A is a non-empty finite set. Interpret this result in terms of the projection map $\hat{f}:\mathbb{C}/\Omega \to \Sigma$ introduced in §3.15.

3P. Prove that

$$\wp'(z) = \frac{2\sigma(z - \tfrac{1}{2}\omega_1)\sigma(z - \tfrac{1}{2}\omega_2)\sigma(z - \tfrac{1}{2}\omega_3)}{\sigma(\tfrac{1}{2}\omega_1)\sigma(\tfrac{1}{2}\omega_2)\sigma(\tfrac{1}{2}\omega_3)\sigma^3(z)}$$

where $\omega_3 = -\omega_1 - \omega_2$. (Hint: Compare the zeros and poles of both sides and also their behaviour near $z = 0$.)

3Q. If $\wp(z_1) = \wp(z_2)$ show that $z_1 \sim \pm z_2 \bmod \Omega$.

3R. Let g be elliptic with respect to $\Omega(\omega_1, \omega_2)$. Suppose that all points of the form $\frac{1}{2}(p\omega_1 + q\omega_2)$ where p and q are integers, $p + q$ odd, are zeros of order one of g, and that g has no other zeros. Suppose that all points of the form $\frac{1}{2}(r\omega_1 + s\omega_2)$, where r and s are integers, $r + s$ even, are poles of order one of g, and that g has no other poles. Prove that $g(z)$ is a constant multiple of

$$\frac{\wp'(z)}{\wp(z) - e_3}.$$

3S.   Let b_1, b_2 be two complex numbers which are not congruent mod Ω. Write down a function which is elliptic with respect to Ω and has poles at b_1, b_2 with principal parts

$$\frac{1}{z - b_1} + \frac{2}{(z - b_1)^2} \quad \text{and} \quad \frac{-1}{z - b_2}$$

respectively.

3T.   Show in detail how the duplication theorem (3.17.10) follows from (3.17.9).

3U.   Prove that

$$\wp(u) - \wp(v) = \frac{\sigma(v - u)\sigma(v + u)}{\sigma^2(u)\sigma^2(v)}.$$

By taking logarithmic derivatives with respect to u deduce that

$$\frac{\wp'(u)}{\wp(u) - \wp(v)} = \zeta(u - v) + \zeta(u + v) - 2\zeta(u).$$

Hence show that

$$\frac{\wp'(u) - \wp'(v)}{\wp(u) - \wp(v)} = 2\zeta(u + v) - 2\zeta(u) - 2\zeta(v)$$

and deduce the following 'addition theorem' for $\wp$:

$$\wp(u + v) = \wp(u) - \frac{1}{2}\frac{\partial}{\partial u}\left(\frac{\wp'(u) - \wp'(v)}{\wp(u) - \wp(v)}\right).$$

4

Meromorphic continuation and Riemann surfaces

So far in this book, we have taken a particular surface S (usually the Riemann sphere Σ or a torus $\mathbb{C}/\Omega$) and have considered the functions which are meromorphic on S (in these particular cases, the rational and elliptic functions). In this chapter we will reverse the process: we take a function f, and consider what is the most natural surface to take as the domain of definition of f. Two particular problems arise:

(i) If f is meromorphic (or analytic) on some region $D \subseteq \Sigma$, can we extend f to a function which is meromorphic (or analytic) on some larger region $E \supset D$?

(ii) If f is a so-called 'many-valued function' (such as $\sqrt{z}$ or $\log(z)$), can we represent f by a *single*-valued function on some suitable domain?

To solve problem (i) we introduce the concepts of meromorphic and analytic continuation, and we then show how this leads to the construction of Riemann surfaces, the 'suitable domains' in problem (ii). After examining in detail some of the surfaces which arise in this way, we then show how Riemann surfaces may be defined abstractly (as objects in their own right, independent of many-valued functions), and finally we investigate some of the topological properties of these surfaces.

4.1 Meromorphic and analytic continuation

Recall that a *region* is a non-empty open subset of Σ which is connected (or, equivalently, path-connected). We define a *function element* to be a pair (D, f) where D is a region and $f: D \to \Sigma$ is a (single-valued) meromorphic function on D. In some books, where the emphasis is on analytic functions, one requires that f should be analytic on D; under these circumstances we shall call (D, f) an *analytic function element*. For example, if f is any rational function then (Σ, f) is a function element; if $\mathscr{D}$ denotes the open unit disc $\{z \in \mathbb{C} \mid |z| < 1\}$ (a convention we will retain throughout this chapter), and if $f(z) = \sum_{n=0}^{\infty} z^n$, then $(\mathscr{D}, f)$ is an analytic function element. In both of these cases, the domain D was chosen to be the largest on which f is meromorphic or analytic, but this need not always be the case – it is often convenient to restrict attention to some smaller region, such as $|z| < \frac{1}{2}$ for $\sum_{n=0}^{\infty} z^n$ for instance.

Lemma 4.1.1. *Let (D, f) and (D, g) be function elements defined on the same region D. If $f \equiv g$ on some non-empty open subset U of D, then $f \equiv g$ on D.*

Proof. The function $h = f - g$ is meromorphic on D, and its poles form an isolated set $h^{-1}(\infty)$, so the set $D' = D \backslash h^{-1}(\infty)$ is non-empty, open and path-connected, that is, D' is a region on which h is analytic. We have $h \equiv 0$ on the non-empty open subset $U' = U \cap D'$ of D', so by choosing a sequence of points $z_n \in U'$ converging to a limit $z^* \in U'$, and by applying Theorem 1.3.1 to h, we see that $h \equiv 0$ on D'. Each pole of h is a limit of points in D', so by continuity (since h is meromorphic) we have $h \equiv 0$ on D, and hence $f \equiv g$ on D. $\square$

Thus a function element (D, f) is determined by the behaviour of f near any given point $a \in D$, since we can take U to be a neighbourhood of a in D.

Corollary 4.1.2. *If (D_1, f_1) is a function element and D_2 is a region with $D_1 \cap D_2 \neq \varnothing$, then there is at most one meromorphic function f_2 on D_2 such that $f_1 \equiv f_2$ on $D_1 \cap D_2$ (as illustrated in Fig. 4.1).*

Fig. 4.1

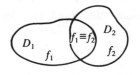

Proof. Suppose that f_2 and g_2 are meromorphic on D_2 and are identically equal to f_1 on $U = D_1 \cap D_2$. Then $f_2 \equiv g_2$ on U, and U is open since D_1 and D_2 are, so $f_2 \equiv g_2$ on D_2 by Lemma 4.1.1. $\square$

When such a function f_2 exists, we call the function element (D_2, f_2) a *direct meromorphic continuation* of (D_1, f_1), or a *direct analytic continuation* if f_2 is analytic; in either case, we write $(D_1, f_1) \sim (D_2, f_2)$, meaning that $D_1 \cap D_2 \neq \varnothing$ and $f_1 \equiv f_2$ on $D_1 \cap D_2$. We then have a function element (D, f) where $D = D_1 \cup D_2$ and $f(z) = f_j(z)$ for $z \in D_j$ (this is unambiguous for $z \in D_1 \cap D_2$). For example, consider $D_1 = \mathcal{D}$ and $f_1(z) = \sum_{n=0}^{\infty} z^n$; the function $f_2(z) = (1 - z)^{-1}$ is meromorphic on $D_2 = \Sigma$, and satisfies $f_1 \equiv f_2$ on $D_1 \cap D_2 = \mathcal{D}$, so $(D_1, f_1) \sim (D_2, f_2)$, giving a direct meromorphic continuation onto Σ. By Corollary 4.1.2, this continuation is unique.

The relation $\sim$ between function elements is reflexive and symmetric, but not transitive. For example, if $D_1 \cap D_2 \neq \emptyset \neq D_2 \cap D_3$, then it does not follow that $D_1 \cap D_3 \neq \emptyset$, and even if $D_1 \cap D_3 \neq \emptyset$ then it does not follow from $(D_1, f_1) \sim (D_2, f_2) \sim (D_3, f_3)$ that $f_1 \equiv f_3$ on $D_1 \cap D_3$. The most familiar example of this behaviour involves the logarithm function (which we will examine in more detail later in this section): if D_1, D_2 and D_3 are regions in $\mathbb{C} \backslash \{0\}$ encircling the origin as illustrated in Fig. 4.2, and if f_1 is a single-valued analytic branch of the many-valued function $\log(z)$ on D_1, then we have direct analytic continuations $(D_1, f_1) \sim (D_2, f_2) \sim (D_3, f_3)$ but we find that $f_3 \equiv f_1 + 2\pi i$ on $D_1 \cap D_3$, so that (D_3, f_3) is not a direct analytic continuation of (D_1, f_1).

Fig. 4.2

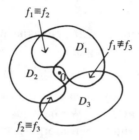

Thus one function element (such as (D_1, f_1)) can sometimes give rise to two distinct functions (here f_1 and f_3) on the same region; this illustrates how repeated continuation of a *single*-valued function element can produce a so-called *many*-valued function. Nevertheless, it is often possible to construct a sequence of direct meromorphic continuations

$$(D_1, f_1) \sim (D_2, f_3) \sim (D_3, f_3) \sim \dots.$$

in such a way that they give a single-valued meromorphic function on $\bigcup_n D_n$; we call this process *meromorphic continuation* (or *analytic continuation* if each f_n is analytic).

For example, we can construct the gamma-function $\Gamma(z)$ in this way. Consider the improper integral

$$f_1(z) = \int_0^{\infty} t^{z-1} e^{-t} dt$$

$$= \lim_{a \to 0+} \int_a^1 t^{z-1} e^{-t} dt + \lim_{b \to +\infty} \int_1^b t^{z-1} e^{-t} dt,$$

where $z \in \mathbb{C}$ and the path of integration is along the positive part of the real line. This integral converges as $b \to +\infty$ for all $z \in \mathbb{C}$, and it converges as $a \to 0+$ provided $\text{Re}(z) > 0$ (by comparison with $\int t^{x-1} dx$, for example, where $x = \text{Re}(z)$). Thus $f_1(z)$ is defined on the region

$$D_1 = \{z \in \mathbb{C} \mid \text{Re}(z) > 0\}.$$

One can show that $\int_0^\infty (\partial/\partial z)(t^{z-1} e^{-t}) dt$ is uniformly convergent with respect to z, so we may differentiate $f_1(z)$ with respect to z under the integral sign. This shows that f_1 is analytic on D_1, so (D_1, f_1) is an analytic function element.

Integrating by parts, we have

$$f_1(z+1) = \int_0^\infty t^z e^{-t} dt$$

$$= [-t^z e^{-t}]_0^\infty + z \int_0^\infty t^{z-1} e^{-t} dt$$

$$= z f_1(z)$$

so that $f_1(z) = f_1(z+1)/z$ for all $z \in D_1$. Now the function $\dot{f}_2(z) = f_1(z+1)/z$ is meromorphic on the larger region

$$D_2 = \{z \in \mathbb{C} \mid \text{Re}(z) > -1\},$$

since $f_1(z+1)$ and z are analytic on D_2. By direct integration, $f_1(1) = 1$, so f_2 has a simple pole at $z = 0$. We have just seen that $f_2 \equiv f_1$ on the region $D_1 \subset D_2$, so f_2 is a direct meromorphic continuation of f_1 onto D_2.

We now iterate this process. For each integer $n \geq 1$, let

$$D_n = \{z \in \mathbb{C} \mid \text{Re}(z) > 1 - n\},$$

and suppose that for some $n \geq 1$ there is a function element (D_n, f_n) satisfying

$$f_n(z) = \frac{f_n(z+1)}{z}$$

for all $z \in D_n$. (We have seen that this is so for $n = 1$.) Then the function

$$f_{n+1}(z) = \frac{f_n(z+1)}{z}$$

is meromorphic on D_{n+1} (since $f_n(z+1)$ and z are), and $f_{n+1} \equiv f_n$ on the region $D_n \subset D_{n+1}$, so f_{n+1} is a direct meromorphic continuation of f_n onto D_{n+1}, satisfying

$$f_{n+1}(z) = \frac{f_{n+1}(z+1)}{z}.$$

for all $z \in D_{n+1}$ (both sides of this equation being equal to $f_n(z+1)/z$).

We may therefore proceed by induction on n to obtain a sequence of direct meromorphic continuations $(D_1, f_1) \sim (D_2, f_2) \sim \cdots$, with each f_n satisfying

$$f_n(z) = \frac{f_n(z+1)}{z}$$

for all $z \in D_n$. We define the *gamma-function*

$$\Gamma(z) = f_n(z)$$

for $z \in D_n$, this definition being unambiguous since if $z \in D_m \cap D_n$ then $f_m(z) = f_n(z)$. Thus $\Gamma(z)$ is meromorphic on $\mathbb{C} = \bigcup_{n=1}^{\infty} D_n$, and satisfies

$$\Gamma(z) = \frac{\Gamma(z+1)}{z}$$

for all $z \in \mathbb{C}$. Since f_1 has a simple pole at $z = 0$, so does Γ, and hence it follows from the above functional equation that Γ has simple poles at $z = -1, -2, \ldots$. Since these poles converge to ∞, Γ is not meromorphic on Σ.

Having dealt with this fairly straightforward example, we now return to the logarithm function for a more detailed examination of some of the difficulties involved in meromorphic continuation.

For each $z \in \mathbb{C} \setminus \{0\}$, the values w of $\log(z)$ are the solutions of $\exp(w) = z$. Putting $z = re^{i\theta}$, with $r, \theta \in \mathbb{R}$ and $r > 0$, we see that $w = \ln(r) + i\theta$, where

$$\ln(r) = \int_1^r t^{-1}\, dt$$

is the real-valued logarithm of r obtained by integration along $\mathbb{R}$ from 1 to r; there are infinitely many choices for θ, all differing by multiples of 2π, and we must arrange our function elements so that θ is single-valued on the chosen regions. The difficulty is that if the region $D \subseteq \mathbb{C} \setminus \{0\}$ encircles the origin, then we cannot make a single-valued continuous choice of $\theta = \arg(z)$ on D. For example, suppose that D contains the unit circle $C = \{z \in \mathbb{C} \mid |z| = 1\}$, and that $\theta(z)$ is a continuous choice of $\arg(z)$ on C; then the function $\phi(z) = \theta(z) + \theta(\bar{z})$ is continuous on C, and since $\theta(\bar{z}) = -\theta(z) + 2n\pi (n \in \mathbb{Z})$ we see that $\phi(C) \subseteq 2\pi\mathbb{Z}$. Now ϕ is continuous and C is connected, so $\phi(C)$ is a connected subset of $2\pi\mathbb{Z}$ and is therefore a single point $2\pi N$, that is, ϕ is constant on C. Then

$$\theta(1) = \tfrac{1}{2}\phi(1) = \pi N = \tfrac{1}{2}\phi(-1) = \theta(-1),$$

which is clearly false since $\theta(1) \in 2\pi\mathbb{Z}$ and $\theta(-1) \in 2\pi\mathbb{Z} + \pi$.

To avoid this difficulty, we will use a particular class of regions which do not encircle 0. Let J be any open interval

$$J = (\alpha, \beta) = \{\theta \in \mathbb{R} \mid \alpha < \theta < \beta\},$$

where $\alpha < \beta \leqslant \alpha + 2\pi$, let D_J be the region

$$D_J = \{z = re^{i\theta} \mid r > 0 \quad \text{and} \quad \theta \in J\},$$

and for each such $z \in D_J$ let

$$f_J(z) = \ln(r) + i\theta.$$

Since J has length at most 2π, each $z \in D_J$ determines a unique $\theta = \arg(z) \in J$, so f_J is single-valued on D_J. Writing $z = x + iy$ one easily verifies the Cauchy–Riemann equations for f_J:

$$\frac{\partial}{\partial x} \ln(r) = \frac{\partial \theta}{\partial y} \quad \text{and} \quad \frac{\partial}{\partial y} \ln(r) = -\frac{\partial \theta}{\partial x}.$$

Thus f_J is analytic on D_J, so we have an analytic function element $L_J = (D_J, f_J)$, representing a branch of $\log(z)$ on D_J.

For example, making J as large as possible we could take $J = (-\pi, \pi)$ so that $D_J = \mathbb{C} \setminus \{z \in \mathbb{R} \mid z \leqslant 0\}$. However, this is inconvenient for our purposes since there is no *direct* analytic continuation of L_J beyond D_J: for instance, if (D, f) is a direct analytic continuation with $-1 \in D$ then every neighbourhood U of -1 in D contains elements z', z'' with $\mathrm{Im}(z') > 0 > \mathrm{Im}(z'')$, as shown in Fig. 4.3; we have $f(z') = f_J(z') \to i\pi$ as $z' \to -1$, and $f(z'') = f_J(z'') \to -i\pi$ as $z'' \to -1$, so f is not analytic (or even meromorphic) at -1. Similar problems arise at all points $z < 0$. Nevertheless, we *can* continue f_J outside D_J by first restricting f_J to a smaller subregion of D_J. Let K be the interval $(0, \pi)$, so that $K \subset J$ and hence $D_K \subset D_J$ and $f_K \equiv f_J$ on D_K, giving a direct analytic continuation $L_J \sim L_K$. Since D_K is the upper half-plane, there are no points $z'' \in D_K$ close to the line $z \leqslant 0$ satisfying $\mathrm{Im}(z'') < 0$, and there is no difficulty in continuing L_K across the negative real line.

Fig. 4.3

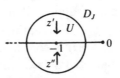

This suggests that it is wisest to use intervals J of length strictly less than 2π. Let D_A, D_B and D_C be the open half-planes corresponding to the intervals

$$A = \left(-\frac{\pi}{2}, \frac{\pi}{2}\right),$$

$$B = \left(\frac{\pi}{6}, \frac{7\pi}{6}\right),$$

$$C = \left(\frac{5\pi}{6}, \frac{11\pi}{6}\right),$$

Fig. 4.4

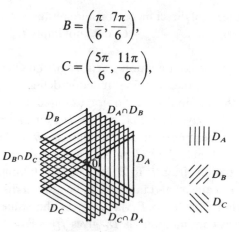

as shown in Fig. 4.4, and let L_A, L_B and L_C be the corresponding branches of $\log(z)$. Now $D_A \cap D_B \neq \varnothing$, and for $z = re^{i\theta} \in D_A \cap D_B$ we have $\theta \in A \cap B = (\pi/6, \pi/2)$, so $f_A(z) = \ln(r) + i\theta = f_B(z)$ and hence $L_A \sim L_B$. A similar argument shows that $L_B \sim L_C$. However, although $D_A \cap D_C \neq \varnothing$ we do not have $L_A \sim L_C$: the elements of $D_A \cap D_C$ are of the form $z = re^{i\theta}$ with $\theta \in (-\pi/2, -\pi/6) \subset A$, giving $f_A(z) = \ln(r) + i\theta$, and also of the form $z = e^{i(\theta + 2\pi)}$ with $\theta + 2\pi \in (3\pi/2, 11\pi/6) \subset C$, so $f_C(z) = \ln(r) + i(\theta + 2\pi) = f_A(z) + 2\pi i$. Thus analytic continuation of L_C onto D_A produces the function element $(D_A, f_A + 2\pi i) = L_{(A + 2\pi)}$, where $A + 2\pi$ is the interval $\{\theta + 2\pi \mid \theta \in A\} = (3\pi/2, 5\pi/2)$. If we iterate this process n times, using the sequence of intervals

$$A, B, C, A + 2\pi, B + 2\pi, \ldots, C + 2(n-1)\pi, A + 2n\pi,$$

then we obtain the function element $L_{(A + 2n\pi)}$, giving the value $f_{(A + 2n\pi)}(z) = f_A(z) + 2n\pi i$; this is done by continuing around the origin n times in the positive direction, and if we continue n times in the *negative* direction, using the sequence of intervals

$$A, C - 2\pi, B - 2\pi, A - 2\pi, C - 4\pi, \ldots, B - 2n\pi, A - 2n\pi,$$

then we obtain the function element $L_{(A - 2n\pi)}$, giving the value $f_{(A - 2n\pi)}(z) = f_A(z) - 2n\pi i$. In this way, starting at any $z \neq 0$ and continuing around the origin an appropriate number of times, in the positive or negative direction, we can return to z and obtain any of the infinitely many values of $\log(z)$.

This example illustrates several of the difficulties associated with meromorphic continuation:

(i) if an unsuitable region D is chosen (such as $D = D_J$ with $J = (-\pi, \pi)$) then a function element (D, f) may admit no further direct meromorphic continuation beyond D;

(ii) the relation $\sim$ of direct meromorphic continuation is not transitive, since we have seen that $L_A \sim L_B \sim L_C$ does not imply $L_A \sim L_C$, even though $D_A \cap D_C \neq \varnothing$;

(iii) if meromorphic continuation starts from, and eventually returns to a particular region D, then the final function defined on D may be quite different from the original function; for example f_A may be continued meromorphically to give $f_A + 2n\pi i$ on $D = D_A$, for any $n \in \mathbb{Z}$;

(iv) if (E, g) is a meromorphic continuation of (D, f), then the values of the function g on E may depend on the chosen sequence of direct meromorphic continuations from D to E; for example, taking $(D, f) = L_A$ with $A = (-\pi/2, \pi/2)$ as above, and taking E to be a small disc containing -1, then continuation from D to E via D_B gives the value $f_B(-1) = i\pi$ for $\log(-1)$, whereas continuation via D_C gives $f_C(-1) = -i\pi$, so we obtain two distinct meromorphic continuations g on E;

(v) if (D, f) and (E, g) are function elements and $f \equiv g$ on a non-empty open subset U of $D \cap E$, it does not follow that $f \equiv g$ on $D \cap E$, so we need not have $(D, f) \sim (E, g)$; the point is that $D \cap E$, although open, need not be path-connected and hence need not be a region, so Lemma 4.1.1 does not apply. For example, take $(D, f) = L_{(A \cup B)}$ and $(E, g) = L_C$ as above, where $A \cup B = (-\pi/2, 7\pi/6)$ and $C = (5\pi/6, 11\pi/6)$, so $D \cap E$ is the disjoint union of the two regions $D_B \cap D_C$ and $D_C \cap D_A$ shown in Fig. 4.5. On $D_B \cap D_C$ we have $g \equiv f_C \equiv f_B \equiv f_{(A \cup B)} \equiv f$, whereas on $D_C \cap D_A$ we have $g \equiv f_C \equiv f_A + 2\pi i \equiv f_{(A \cup B)} + 2\pi i \equiv f + 2\pi i$.

Fig. 4.5

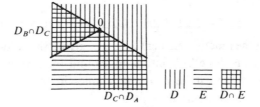

We can avoid this last difficulty by using only *convex* regions, such as discs (in the above example D is not convex): the intersection of two convex sets is convex and therefore path-connected. This will be useful in the next section where the function elements considered will be power series convergent on open discs.

4.2 Analytic continuation using power series

An effective method of constructing analytic continuations is to use the fact that a function f is analytic at some point $a \in \mathbb{C}$ if and only if it is represented by a power series $\sum_{n=0}^{\infty} a_n(z-a)^n$ near a (we can also deal with $a = \infty$ by considering the behaviour of $f \circ J$ near 0).

Suppose that (D, f) is an analytic function element with $D \subseteq \mathbb{C}$, and that $a \in D$. Let D_1 be the largest open disc centred at a and contained in D, as shown in Fig. 4.6 (we will regard $\mathbb{C}$ as an open disc centred at a, to allow for the case $D = \mathbb{C}$). Then f is analytic on D_1, so if we define $a_n = f^{(n)}(a)/n!$ for each integer $n \geqslant 0$, then by Taylor's theorem the power series

$$f_1(z) = \sum_{n=0}^{\infty} a_n(z-a)^n$$

converges absolutely to $f(z)$ on D_1, giving a direct analytic continuation $(D, f) \sim (D_1, f_1)$. (It is important that $D_1 \subseteq D$: if we take D_1 to be the largest disc, centred at a, on which $\sum_{n=0}^{\infty} a_n(z-a)^n$ converges, then $D \cap D_1$ need not be a region and we need not have $(D, f) \sim (D_1, f_1)$, as shown by Fig. 4.7, with f a branch of $\log(z)$.)

Now let $b \in D_1$ and let U be the largest open disc centred at b and contained in D_1, as in Fig. 4.8. Since f_1 is analytic on U, Taylor's theorem gives

$$f_1(z) = \sum_{n=0}^{\infty} b_n(z-b)^n$$

Fig. 4.6

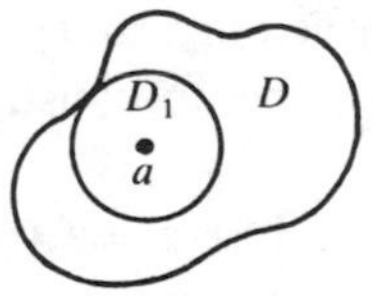

Fig. 4.7

Fig. 4.8

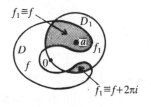

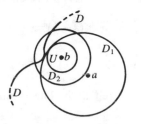

for all $z \in U$, where

$$b_n = \frac{1}{n!} f_1^{(n)}(b)$$

$$= \frac{1}{n!} \frac{d^n}{dz^n} \left(\sum_{m=0}^{\infty} a_m(z-a)^m \right)_{z=b}$$

$$= \sum_{m=n}^{\infty} \binom{m}{n} a_m(b-a)^{m-n},$$

using absolute convergence to differentiate term by term. Now the largest region on which this power series $\sum_{n=0}^{\infty} b_n(z-b)^n$ converges is an open disc D_2, centred at b and containing U; on D_2 the series represents an analytic function f_2, so we have an analytic function element (D_2, f_2). Since D_1 and D_2 are discs, $D_1 \cap D_2$ is a region, and since $f_1 \equiv f_2$ on the non-empty open set $U \subseteq D_1 \cap D_2$, Lemma 4.1.1 implies that $f_1 \equiv f_2$ on $D_1 \cap D_2$ and hence $(D_1, f_1) \sim (D_2, f_2)$. If, as is often the case, D_2 strictly contains U, then $D_2 \nsubseteq D_1$ and so we have a non-trivial analytic continuation of (D_1, f_1).

We can now iterate this process, expanding $f_2(z)$ as a power series $f_3(z) = \sum_{n=0}^{\infty} c_n(z-c)^n$ convergent on an open disc D_3 centred at some point $c \in D_2$; since $D_2 \cap D_3$ is a region, we have $(D_2, f_2) \sim (D_3, f_3)$, and so on. Since we are using discs, we can (and generally will) take each D_j to be as large as possible, subject only to f_j being analytic on D_j. In this way, it may eventually be possible to construct an analytic continuation of f outside D.

For example, let $a = 0$ and $f(z) = f_1(z) = \sum_{n=0}^{\infty} z^n$, with region of convergence $D = D_1 = \mathscr{D}$, the open unit disc. If $b \in D_1$, then since $a_m = 1$ for all $m \geq 0$ we have

$$b_n = \sum_{m=n}^{\infty} \binom{m}{n} b^{m-n},$$

and this converges to $(1-b)^{-1-n}$ since $|b| < 1$. (This can be seen either by differentiating the series $(1-b)^{-1} = \sum_{m=0}^{\infty} b^m$ n times with respect to b, or by applying the binomial theorem directly to $(1-b)^{-1-n}$ and comparing coefficients.) We therefore have an analytic function

$$f_2(z) = \sum_{n=0}^{\infty} b_n(z-b)^n$$

$$= \sum_{n=0}^{\infty} (1-b)^{-1-n}(z-b)^n$$

on some open disc D_2 centred at b. The radius of convergence of this power series is

$$\lim_{n \to \infty} (|1 - b|^{1+n})^{1/n} = |1 - b|,$$

so D_2 is $\{z \in \mathbb{C} \mid |z - b| < |1 - b|\}$, the largest open disc centred at b and avoiding 1 (where f_1 has a pole). We therefore have $D_2 \nsubseteq D_1$ except when $b \in \mathbb{R}$ and $0 \leqslant b < 1$, so by avoiding such a choice of b we can continue f_1 analytically outside D_1.

4.3 Regular and singular points

In the examples we have considered so far, there have been certain points $c \in \Sigma$ at which meromorphic continuation is impossible: for instance $c = \infty$ for $\Gamma(z)$ (which cannot be continued to a meromorphic function at ∞ since it has a sequence of poles converging to ∞), and $c = 0$ or ∞ for $\log(z)$ (which has no *single*-valued meromorphic branch near either of these points). If (D, f) is a function element then a point c on the boundary ∂D of D is called a *regular point* for (D, f) if there is a direct meromorphic continuation $(E, g) \sim (D, f)$ with $c \in E$; there is no loss of generality in taking E to be an open disc centred at c, as in Fig. 4.9. If there is no such direct meromorphic continuation then c is called a *singular point* for (D, f); thus singular points represent obstacles to meromorphic continuation. If all points $c \in \partial D$ are singular, then ∂D is called the *natural boundary* of (D, f).

Fig. 4.9

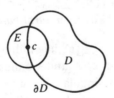

(It should be pointed out that in some books, where the emphasis is on analytic rather than meromorphic functions, c is defined to be a singular point if there is no direct *analytic* continuation at c: under this convention, though not under ours, $c = 1$ would be a singular point for $f(z) = \sum_{n=0}^{\infty} z^n$ on the open unit disc $\mathscr{D}$.)

Lemma 4.3.1. *Let (D, f) be a function element with $c \in \partial D$. If c is regular for (D, f) then $f(z)$ approaches some limit in Σ as $z \to c$ with $z \in D$.*

Proof. By hypothesis, there is a direct meromorphic continuation $(E, g) \sim (D, f)$ with $c \in E$, so for $z \in D \cap E$ we have $f(z) = g(z) \to g(c) \in \Sigma$ as $z \to c$. $\square$

The contrapositive is equally useful: if $f(z)$ has no limit as $z \to c$ then c is singular. A similar proof shows that if there is a direct analytic continuation at c, then $f(z)$ has a finite limit as $z \to c$. For example, for each integer $k \geqslant 1$ the power series $f_k(z) = \sum_{n=0}^{\infty} z^{kn}$ converges to $(1 - z^k)^{-1}$ on the open unit disc $\mathcal{D}$, so if c is a kth root of unity then there is no direct analytic continuation of $(\mathcal{D}, f_k)$ at c (though, of course, meromorphic continuation is possible, so c is regular).

In the above series $f_k(z)$, the 'gap' between successive non-zero coefficients is equal to k, and the larger this gap is, the more obstacles there are to analytic continuation. More generally, Hadamard's gap theorem (Rudin [1974], 16.6) states that if a power series

$$f(z) = \sum_{n=0}^{\infty} a_n z^n$$

has radius of convergence 1, and if the gaps between the suffixes n of successive non-zero coefficients a_n increase sufficiently rapidly as $n \to \infty$, then every point on the unit circle C is singular, that is, C is the natural boundary of $(\mathcal{D}, f)$. (As normally stated, the theorem states that *analytic* continuation across C is impossible, but the extension to meromorphic continuation is trivial.) We shall not investigate this subject in great detail, since such functions do not often arise naturally; instead, we shall consider a simple example (see also Exercise 4A).

Theorem 4.3.2. *The series $f(z) = \sum_{n=1}^{\infty} z^{n!} = z + z^2 + z^6 + z^{24} + \ldots$ has the unit circle $C = \{z \in \mathbb{C} \mid |z| = 1\}$ as its natural boundary.*

Proof. By the root test, $\sum_{n=1}^{\infty} z^{n!}$ has radius of convergence 1, so f is analytic on the open unit disc $\mathcal{D}$, and we must show that every point c on the unit circle $C = \partial\mathcal{D}$ is singular. If c were regular, then there would be a direct meromorphic continuation $(E, g) \sim (\mathcal{D}, f)$ with $c \in E$; since g is meromorphic we could take E sufficiently small that g is analytic on $E \backslash \{c\}$, and so for each point $d \neq c$ in $C \cap E$ we would have $f(z) = g(z) \to g(d) \in \mathbb{C}$ as $z \to d$, $z \in \mathcal{D}$. However, we shall show that no matter how small E is, there exists some $d \neq c$ in $C \cap E$ such that $f(z) \to \infty$ as $z \to d$ along a radius from 0 to d (see Fig. 4.10).

Fig. 4.10

We first show that if $r \in \mathbb{R}$ and $0 \leqslant r < 1$, then $f(r) \to \infty$ as $r \to 1$. Given any $M \in \mathbb{R}$, choose a positive integer $m > M$; since $0 \leqslant r < 1$ we have

$$f(r) = \sum_{n=1}^{\infty} r^{n!}$$

$$\geqslant \sum_{n=1}^{m} r^{n!}$$

$$\geqslant mr^{m!},$$

and as $r \to 1$ we have $mr^{m!} \to m > M$, so provided r is sufficiently close to 1 we have $mr^{m!} \geqslant M$ and hence $f(r) \geqslant M$, as required.

Now take any point $c \in C$, and put $c = e^{2\pi i s}$ with $s \in \mathbb{R}$. If E is any open neighbourhood of c in Σ then we can choose a rational number $t \neq s$ sufficiently close to s that the point $d = e^{2\pi i t}$ lies in $(C \cap E) \setminus \{c\}$. Since $t \in \mathbb{Q}$ we have $t = p/q$ with $p, q \in \mathbb{Z}$ and $q > 0$. Let $z = rd$ where $r \in \mathbb{R}$, $0 \leqslant r < 1$, so that $z \in \mathscr{D}$ and

$$z^{n!} = r^{n!} d^{n!} = r^{n!} e^{2\pi i p n! / q}.$$

If $n \geqslant q$ then q divides $n!$, so $z^{n!} = r^{n!}$ and hence

$$f(z) = \sum_{n=1}^{q-1} z^{n!} + \sum_{n=q}^{\infty} r^{n!}$$

$$= P(z) + f(r) - P(r),$$

where $P(z)$ is the polynomial $\sum_{n=1}^{q-1} z^{n!}$. As $r \to 1$ we have $z \to d$, so $P(z)$ and $P(r)$ have finite limits $P(d)$ and $P(1)$, whereas we have seen that $f(r) \to \infty$. Thus $f(z) \to \infty$, giving the required contradiction. $\square$

At the other extreme, there exist power series with radius of convergence 1 such that every point $c \in C$ is regular: $\sum_{n=0}^{\infty} z^n$ has this property, for example, though there is a point ($c = 1$) where *analytic* continuation is impossible. The following result shows that this is typical.

Theorem 4.3.3. *If a power series $f(z) = \sum_{n=0}^{\infty} a_n (z-a)^n$ has radius of convergence $\rho \neq 0, \infty$, then there is at least one point c on the circle $|z - a| = \rho$ at which direct analytic continuation is impossible.*

Proof. Writing $f(z) = \sum_{n=0}^{\infty} \rho^n a_n \zeta^n$, with $\zeta = (z-a)/\rho$, we may assume without loss of generality that $a = 0$ and $\rho = 1$, so f is analytic on $\mathscr{D} = \{z \in \mathbb{C} \mid |z| < 1\}$, which has as its boundary the unit circle $C = \{z \in \mathbb{C} \mid |z| = 1\}$.

Suppose that for each point $c \in C$ there is a direct analytic continuation $(E_c, g_c) \sim (\mathscr{D}, f)$ with $c \in E_c$. Putting $c = e^{i\phi}$ with $\phi \in \mathbb{R}$, we can take the region

E_c to be of the form

$$E_c = \{re^{i\theta} \mid 1 - r_c < r < 1 + r_c,\ \phi - \alpha_c < \theta < \phi + \alpha_c\},$$

where $0 < r_c < 1$ and $0 < \alpha_c < \pi$, as pictured in Fig. 4.11. This condition on α_c guarantees that for any $c, d \in C$, if $E_c \cap E_d$ is non-empty, as in Fig. 4.12, then it is a region (being path-connected); in this case $g_c \equiv f \equiv g_d$ on the non-empty open subset $\mathcal{D} \cap E_c \cap E_d$ of $E_c \cap E_d$, so $(E_c, g_c) \sim (E_d, g_d)$.

Fig. 4.11

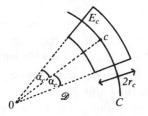

Fig. 4.12

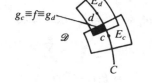

Being compact, C is covered by finitely many such regions $E_{c_1}, \ldots, E_{c_k}$. Let

$$\rho' = 1 + \min_{1 \leqslant i \leqslant k} r_{c_i},$$

so that $\rho' > 1$, and let $E = \{z \in \mathbb{C} \mid |z| < \rho'\}$, so that

$$E \subseteq \mathcal{D} \cup E_{c_1} \cup \ldots \cup E_{c_k}.$$

We define a function g on E by

$$g(z) = \begin{cases} f(z) & \text{if } z \in \mathcal{D}, \\ g_{c_i}(z) & \text{if } z \in E_{c_i}. \end{cases}$$

This definition is consistent since if $z \in \mathcal{D} \cap E_{c_i}$ or if $z \in E_{c_i} \cap E_{c_j}$ then $f(z) = g_{c_i}(z)$ or $g_{c_i}(z) = g_{c_j}(z)$ respectively. Near any point in E, g is identically equal to f or g_{c_i} for some i, so g is analytic on E. By Taylor's theorem the series

$$\sum_{n=0}^{\infty} \frac{g^{(n)}(0)}{n!} z^n$$

must converge to $g(z)$ on the largest open disc centred at 0 and contained in

E, and clearly this disc is E itself. Since $g \equiv f$ near 0, we have

$$\frac{g^{(n)}(0)}{n!} = \frac{f^{(n)}(0)}{n!} = a_n$$

for all $n \geqslant 0$, so we have shown that $\sum_{n=0}^{\infty} a_n z^n$ converges on E and hence has radius of convergence at least $\rho' > 1$, against our assumption. This contradiction shows that there must be some point $c \in C$ at which direct analytic continuation of $(\mathcal{D}, f)$ is impossible. $\quad\square$

Thus, for a power series with radius of convergence $\rho \neq 0$, ∞, the 'best possible' behaviour is that exhibited by $\sum_{n=0}^{\infty} z^n$, where there is just one point $(c = 1)$ where we need meromorphic rather than analytic continuation.

Finally, a brief warning. It is important to note that whether or not a point c is regular for (D, f) depends on D as well as on f: for example, even though *direct* analytic or meromorphic continuation may be impossible at c, we may nevertheless be able to continue f across ∂D at c *indirectly*, using a sequence of direct continuations, as we saw in §4.1 in connection with $\log(z)$. If we take $J = (-\pi, \pi)$ then the function element L_J has domain $D = D_J = \{z = re^{i\theta} \mid r > 0 \text{ and } -\pi < \theta < \pi\}$, with boundary ∂D consisting of the negative real line together with its end-points 0 and ∞; as shown in §4.1, each $c \in \partial D$ is a singular point for L_J, but nevertheless by first restricting to the smaller region D_K with $K = (0, \pi)$ it *is* possible to continue $\log(z)$ across ∂D at any point $c \neq 0$, ∞, using a sequence of two direct continuations. Examples like this suggest that the concept of the natural boundary of a *function* (rather than a function element) is rather hard to define, so we will regard this line of thought as having reached its natural boundary, and continue it no further!

4.4 Meromorphic continuation along a path

To deal with the problems of non-uniqueness of meromorphic continuation, such as those we met in §4.1 in connection with $\log(z)$, we introduce the concept of meromorphic continuation along a path. We shall show that meromorphic continuation along a given path is unique, and then we shall consider the relationship between meromorphic continuations along different paths between two given points.

A *path* γ is a continuous function $\gamma: I \to \Sigma$, where I is the closed unit interval $[0, 1] = \{s \in \mathbb{R} \mid 0 \leqslant s \leqslant 1\}$. Since γ is continuous and I is compact and connected, the image $\gamma(I)$ is compact and connected. By abuse of language, we often refer to $\gamma(I)$ as 'the path γ'. If $a = \gamma(0)$ and $b = \gamma(1)$ then we say that γ is a path 'from a to b'; γ is a *closed* path if $a = b$, and γ is *simple* if $\gamma(s) = \gamma(s')$ implies either $s = s'$ or else $s = 0$ and $s' = 1$ (that is, γ has no self-intersections except possibly where $a = b$). It if often useful to regard $\gamma(s)$ as a point moving continuously with respect to time s for $0 \leqslant s \leqslant 1$.

Let (D, f) be a function element, let $a \in D$ and let γ be a path in Σ from a to some point $b \in \Sigma$. Then a *meromorphic continuation of (D, f) along γ* is a finite sequence of direct meromorphic continuations $(D, f) \sim (D_1, f_1) \sim (D_2, f_2) \sim \ldots \sim (D_m, f_m)$ such that:

(i) each region D_i is an open disc in Σ (as defined in §2.8) with $a \in D_1 \subseteq D$;
(ii) there is a subdivision $0 = s_0 < s_1 < \ldots < s_m = 1$ of I such that $\gamma([s_{i-1}, s_i]) \subseteq D_i$ for $i = 1, 2, \ldots, m$ (and hence $b \in D_m$) (see Fig. 4.13).

If all the function elements (D_i, f_i) are analytic, we call this an *analytic continuation along γ*.

Fig. 4.13

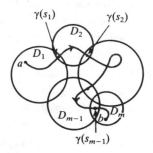

The reader may wonder why we require the regions D_i to be discs, and why we allow only finitely many of them. There is no loss of generality in using discs, since any region in Σ is a union of open discs, and indeed, if we construct the continuations by means of power series as in §4.2, then we will be forced to use discs. Moreover, it is easily seen that if two open discs on the sphere S^2 have non-empty intersection, then that intersection is path-connected and is therefore a region; using the homeomorphism $\pi : S^2 \to \Sigma$ we see that discs in Σ also have this useful property. Since γ is compact, we can reduce any covering of γ by regions to a finite subcovering, so there is no loss of generality in restricting attention to finite sequences of direct continuations.

As an example of analytic continuation along a path, let γ be the unit circle parametrised by $\gamma(s) = e^{2\pi i s}$ $(s \in I)$, so $a = b = 1$. We can continue $\log(z)$ analytically around γ using the sequence of direct analytic continuations

$L_A \sim L_B \sim L_C \sim L_{(A+2\pi)}$ as defined in §4.1 (recall that according to the definition given in §2.8, each open half-plane in $\mathbb{C}$, such as D_A, D_B, D_C, is an example of an open disc in Σ). This is illustrated in Fig. 4.14. In this example, we can take the subdivision of I in condition (ii) to be $0 < \frac{1}{6} < \frac{1}{2} < \frac{5}{6} < 1$. Thus we start with $f_1(a) = f_A(1) = 0$ at $a = 1$, and end with $f_4(b) = f_{(A+2\pi)}(1) = 2\pi i$ at $b = 1$. Similarly, if $n \in \mathbb{Z}$ and we take $\gamma''(s) = e^{2\pi n i s}$, then the path γ'' winds n times around 0, and we continue along γ'' from $f_A(1) = 0$ to $f_{(A+2n\pi)}(1) = 2n\pi i$, as we saw in §4.1.

Fig. 4.14

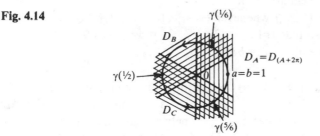

The aim of meromorphic continuation along a path γ is to start with a meromorphic function $f = f_1$ at $a = \gamma(0)$ and to allow this function to vary meromorphically along γ until we reach a meromorphic function f_m at $b = \gamma(1)$. The following result shows that the resulting value $f_m(b)$ depends only on the initial function f and the path γ from a to b, and not on the particular choice of discs D_i covering γ, so we may denote this value by $f_\gamma(b)$.

Theorem 4.4.1. *Let* $(D, f) \sim (D_1, f_1) \sim \ldots \sim (D_m, f_m)$ *and* $(D, f) \sim (E_1, g_1) \sim \ldots \sim (E_n, g_n)$ *be meromorphic continuations of (D, f) along a path γ from a to b, and let $0 = s_0 < s_1 < \ldots < s_m = 1$ and $0 = t_0 < t_1 < \ldots < t_n = 1$ be the corresponding subdivisions of I. Then $(D_i, f_i) \sim (E_j, g_j)$ whenever $[s_{i-1}, s_i] \cap [t_{j-1}, t_j] \neq \emptyset$, and in particular $f_m(b) = g_n(b)$.*

Proof. Suppose that $[s_{i-1}, s_i] \cap [t_{j-1}, t_j] \neq \emptyset$, so that $D_i \cap E_j \neq \emptyset$, and suppose (for an eventual contradiction) that $f_i \not\equiv g_j$ on $D_i \cap E_j$. We may assume that i and j are chosen so that $i + j$ is minimal with respect to this

Fig. 4.15

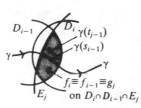

property, and (interchanging the two meromorphic continuations if necessary) that $t_{j-1} \leqslant s_{i-1}$, as in Fig. 4.15. Since $(D_1, f_1) \sim (E_1, g_1)$ it follows that $i > 1$. Since $[s_{i-1}, s_i]$ meets $[t_{j-1}, t_j]$ we have $s_{i-1} \in [t_{j-1}, t_j]$, so $\gamma(s_{i-1}) \in E_j$. Clearly, $\gamma(s_{i-1}) \in D_i \cap D_{i-1}$, so $D_i \cap D_{i-1} \cap E_j$ is non-empty.

Now $s_{i-1} \in [s_{i-2}, s_{i-1}] \cap [t_{j-1}, t_j]$, so by the minimality of $i + j$ we have $f_{i-1} \equiv g_j$ on $D_{i-1} \cap E_j$. Since $(D_{i-1}, f_{i-1}) \sim (D_i, f_i)$ we have $f_{i-1} \equiv f_i$ on $D_{i-1} \cap D_i$, and hence $f_i \equiv g_j$ on $D_i \cap D_{i-1} \cap E_j$ which is a non-empty open subset of the region $D_i \cap E_j$. Thus $f_i \equiv g_j$ on $D_i \cap E_j$ by Lemma 4.1.1, contradicting our choice of i and j.

Thus $(D_i, f_i) \sim (E_j, g_j)$ whenever $[s_{i-1}, s_i] \cap [t_{j-1}, t_j] \neq \varnothing$, and taking $i = m$ and $j = n$ (so that $1 \in [s_{i-1}, s_i] \cap [t_{j-1}, t_j]$) we have $(D_m, f_m) \sim (E_n, g_n)$, and hence $f_m(b) = g_n(b)$ since $b \in D_m \cap D_n$. $\square$

As a further example of analytic continuation along paths, we consider the many-valued function $z^{1/q}$, where q is an integer, $q \geqslant 2$. For any $z \in \mathbb{C} \setminus \{0\}$ there are q values of $z^{1/q}$, namely the solutions w of $w^q = z$; we can express these as $w = \exp(q^{-1} \log(z))$, different values of $z^{1/q}$ corresponding to different values of $\log(z)$. More precisely, if (D, f) is a function element representing a branch of $\log(z)$, so that $\exp(f(z)) \equiv z$ on D, then the function $g(z) = \exp(q^{-1} f(z))$ is analytic and satisfies $g(z)^q \equiv z$ on D, so the function element (D, g) represents a branch of $z^{1/q}$ on D. If γ is any path in $\mathbb{C} \setminus \{0\}$ then we can find the analytic continuation of $z^{1/q}$ along γ by considering the continuation of $\log(z)$. For example, if $(D, f) = L_A = (D_A, f_A)$ as above, giving the branch $g_A = \exp(q^{-1} f_A)$ of $z^{1/q}$ on D_A satisfying $g_A(1) = \exp(0) = 1$, then by continuing from 1 to 1 along the path $\gamma^n(s) = e^{2n\pi i s}$ we get

$$g_{(A+2n\pi)} \equiv \exp\left(\frac{1}{q} f_{(A+2n\pi)}\right) \equiv \exp\left(\frac{1}{q}(f_A + 2n\pi i)\right) \equiv \varepsilon^n g_A,$$

where

$$\varepsilon = \exp\left(\frac{2\pi i}{q}\right);$$

this gives a branch of $z^{1/q}$ taking the value ε^n at 1. Since ε has order q under multiplication we have $g_{(A+2n\pi)} \equiv g_{(A+2m\pi)}$ if and only if $n \equiv m \bmod (q)$; in particular, $g_{(A+2q\pi)} \equiv g_A$. Thus the q branches of $z^{1/q}$ on D_A are

$$g_A, g_{(A+2\pi)} = \varepsilon g_A, \ldots, g_{(A+2(q-1)\pi)} = \varepsilon^{q-1} g_A,$$

taking the values $1, \varepsilon, \ldots, \varepsilon^{q-1}$ at 1. Similarly, if we start with a particular value w of $z^{1/q}$ at *any* point $z \in \mathbb{C} \setminus \{0\}$, then by continuing around 0 repeatedly we obtain the values $w, \varepsilon w, \ldots, \varepsilon^{q-1} w$ of $z^{1/q}$ in succession. This shows that we cannot find an analytic (or even meromorphic) continuation of $z^{1/q}$ at 0: if we had a function element (D, g) representing a branch of $z^{1/q}$ on a region D containing 0, then g would be single-valued on D; however, by continuing g analytically along a closed path which winds once around 0 within D, we replace g by $\varepsilon g \neq g$, giving a contradiction. A similar argument shows that there is no continuation at ∞, so 0 and ∞ are singular points for $z^{1/q}$.

4.5 The monodromy theorem

If a function element (D, f) can be continued meromorphically along a path γ from a to b, then by Theorem 4.4.1 the value $f_\gamma(b)$ of the resulting continuation at b is independent of the method of continuation along γ. However, different paths γ_0 and γ_1 from a to b may give different values at b, as we saw in §4.4 for $\log(z)$; we now consider sufficient conditions for $f_{\gamma_0}(b) = f_{\gamma_1}(b)$.

If γ_0 and γ_1 are paths from a to b in a topological space X, then a *homotopy* from γ_0 to γ_1 in X is a continuous function $\Gamma : I^2 \to X$ such that $\Gamma(s, 0) = \gamma_0(s)$, $\Gamma(s, 1) = \gamma_1(s)$, $\Gamma(0, t) = a$, and $\Gamma(1, t) = b$ for all $s, t \in I$. Thus, for each $t \in I$ we have a path γ_t from a to b in X given by $\gamma_t(s) = \Gamma(s, t)$; as t increases from 0 to 1, γ_t is continuously deformed (within X) from γ_0 to γ_1, keeping the end-points fixed at a and b, as depicted in Fig. 4.16. We say that γ_0 and γ_1 are *homotopic* in X, written $\gamma_0 \simeq \gamma_1$, if there is a homotopy from γ_0 to γ_1 in X; $\simeq$ is an equivalence relation, and the equivalence classes are called *homotopy classes*.

Fig. 4.16

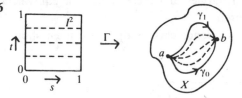

For example, if γ_0 and γ_1 are the paths from 1 to -1 in $\mathbb{C}$ defined by $\gamma_0(s) = e^{\pi i s}$ and $\gamma_1(s) = e^{-\pi i s}$ then $\Gamma(s, t) = e^{\pi i s} - 2ti \sin(\pi s)$ gives a homotopy from γ_0 to γ_1 in $\mathbb{C}$. However, it is intuitively clear (and we shall shortly prove) that γ_0 and γ_1 are *not* homotopic in $\mathbb{C} \setminus \{0\}$: some intermediate path γ_t would have to pass through 0 or ∞, impossible since $\Gamma(I^2) \subseteq \mathbb{C} \setminus \{0\}$ (see Fig. 4.17). As a second example, we shall show in §4.6 that each closed path δ from 1 to 1 in $\mathbb{C} \setminus \{0\}$ is homotopic to the path

$\gamma^n(s) = e^{2n\pi is}$ for some unique $n \in \mathbb{Z}$, called the *winding number* of δ since it represents the number of times δ winds around 0.

Fig. 4.17

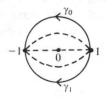

The following result shows that homotopic paths give rise to the same meromorphic continuation.

Theorem 4.5.1 *Let $X \subseteq \Sigma$, let a, $b \in X$, and let Γ be a homotopy in X between two paths γ_0 and γ_1 from a to b in X; if (D, f) is a function element which can be continued meromorphically along each path $\gamma_t : s \mapsto \Gamma(s, t)(s, t \in I)$, then the meromorphic functions at b, resulting from the continuations along γ_0 and γ_1, are identically equal in some neighbourhood of b, and in particular, $f_{\gamma_0}(b) = f_{\gamma_1}(b)$.*

Proof. For all $t, t' \in I$, we will write $t \sim t'$ if the meromorphic continuations of (D, f) along γ_t and $\gamma_{t'}$ produce functions which are identically equal in some neighbourhood of b; by Theorem 4.4.1, this is independent of the particular continuations along these paths. Clearly $\sim$ is an equivalence relation on I, and we have to prove that $0 \sim 1$. To do this, it is sufficient to prove that each equivalence class is open in I, for then each class is also closed (its complement, being a union of equivalence classes, is open), so the connectedness of I implies that there is just one equivalence class and hence $0 \sim 1$ as required.

To show that each equivalence class is open, we must show that for

Fig. 4.18

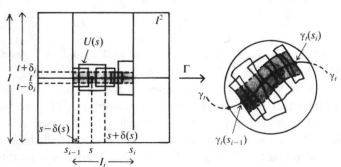

each $t \in I$ there exists some $\delta > 0$ such that $t' \sim t$ for all $t' \in I$ satisfying $|t' - t| < \delta$. By hypothesis, there is a sequence of direct meromorphic continuations $(D, f) \sim (D_1, f_1) \sim \ldots \sim (D_m, f_m)$ and a subdivision $0 = s_0 < s_1 < \ldots < s_m = 1$ of I such that $\gamma_t(I_i) \subseteq D_i$ for $1 \leqslant i \leqslant m$, where I_i denotes the closed subinterval $[s_{i-1}, s_i]$ of I. Since Γ is continuous and each D_i is open, for every $s \in I_i$ there exists an open neighbourhood

$$U(s) = \{(s', t') \in I_i \times I \mid |s' - s| < \delta(s), |t' - t| < \delta(s)\}$$

of (s, t) in $I_i \times I$, with $\delta(s) > 0$, such that $(s', t') \in U(s)$ implies $\Gamma(s', t') \in D_i$, that is $\gamma_{t'}(s') \in D_i$ (see Fig. 4.18). Being compact, I_i is covered by *finitely* many such intervals $(s - \delta(s), s + \delta(s))$, so let δ_i be the least of the finitely many corresponding values of $\delta(s)$. Then we have $\delta_i > 0$, and if $t' \in I$ then $|t' - t| < \delta_i$ implies $\gamma_{t'}(I_i) \subseteq D_i$. Defining δ to be the least δ_i for $1 \leqslant i \leqslant m$, we have $\delta > 0$, and if $t' \in I$ then $|t' - t| < \delta$ implies $\gamma_{t'}(I_i) \subseteq D_i$ for all $i = 1, 2, \ldots, m$. For such values of t' we therefore have a meromorphic continuation of (D, f) along $\gamma_{t'}$ given by $(D, f) \sim (D_1, f_1) \sim \ldots \sim (D_m, f_m)$, with the same subdivision $0 = s_0 < s_1 < \ldots < s_m = 1$ and the same function elements (D_i, f_i) as were used for continuation along γ_t, as in Fig. 4.19. Thus the continuations along γ_t and $\gamma_{t'}$ both give rise to the meromorphic function f_m near b, so $t' \sim t$, as required. $\square$

Fig. 4.19

We can now prove our earlier claim that the paths $\gamma_0(s) = e^{\pi i s}$ and $\gamma_1(s) = e^{-\pi i s}$ from 1 to -1 cannot be homotopic in $\mathbb{C} \setminus \{0\}$. If we take (D, f) to be the function element L_A, the branch of $\log(z)$ defined in §4.1, then it is not hard to show that (D, f) can be continued analytically along any path from 1 to -1 in $\mathbb{C} \setminus \{0\}$ (using suitable function elements L_J, for example); since $f_{\gamma_0}(-1) = \pi i$ and $f_{\gamma_1}(-1) = -\pi i$, it follows from Theorem 4.5.1 that γ_0 and γ_1 are not homotopic in $\mathbb{C} \setminus \{0\}$. Similarly, if γ^n is the closed path $\gamma^n(s) = e^{2\pi n i s} (n \in \mathbb{Z})$, then by continuing $\log(z)$ analytically around γ^n as in §4.4 we see that γ^m and γ^n are homotopic in $\mathbb{C} \setminus \{0\}$ if and only if $m = n$.

In order to apply Theorem 4.5.1, we need a condition on X which guarantees that any two paths from a to b in X are homotopic. For each $a \in X$ let $\gamma_{(a)}$ denote the constant path $\gamma_{(a)}(s) = a$ for all $s \in I$; then a closed path γ from a to a in X is said to be *null-homotopic* if it is homotopic in X to $\gamma_{(a)}$, and X is said to be *simply connected* if it is path-connected and all closed paths in X are null-homotopic. For example, $\mathbb{C}$ is simply connected since if γ is any closed path from a to a in $\mathbb{C}$ then there is a homotopy $\Gamma:\gamma \simeq \gamma_{(a)}$ given by $\Gamma(s,t) = \gamma(s) + t(a - \gamma(s))$; on the other hand $\mathbb{C}\backslash\{0\}$ is not simply connected since (as we have just seen) the unit circle $\gamma(s) = e^{2\pi i s}$ is not null-homotopic in $\mathbb{C}\backslash\{0\}$.

Theorem 4.5.2. *A topological space X is simply connected if and only if for each pair of points a, $b \in X$ there is a single homotopy class of paths from a to b in X.*

Proof. Suppose that for each a, $b \in X$ there is a single homotopy class of paths from a to b in X. Then X is path-connected, and taking $a = b$ we see that any closed path from a to a in X is homotopic to $\gamma_{(a)}$, that is, X is simply connected.

Conversely, suppose that X is simply connected, so (by definition) X is path-connected. Let γ_0 and γ_1 be paths from a to b in X, and let δ be the closed path from a to a given by

$$\delta(s) = \begin{cases} \gamma_0(2s) & 0 \leqslant s \leqslant \tfrac{1}{2}, \\ \gamma_1(2 - 2s) & \tfrac{1}{2} \leqslant s \leqslant 1. \end{cases}$$

Thus as s increases from 0 to $\tfrac{1}{2}$, $\delta(s)$ travels forwards along γ_0 from a to b, while as s increases from $\tfrac{1}{2}$ to 1, $\delta(s)$ travels backwards along γ_1 from b to a. Since X is simply connected there is a homotopy Δ from δ to $\gamma_{(a)}$, and we shall use Δ to construct a homotopy Γ from γ_0 to γ_1. Before defining Γ explicitly, we shall give an informal description of the intermediate paths $\gamma_t (t \in I)$, as illustrated in Fig. 4.20.

For $0 \leqslant t \leqslant \tfrac{1}{2}$ we form γ_t by adjoining to the end of γ_0 a loop which travels backwards along γ_1 from b to $\gamma_1(1 - 2t)$ and then travels forwards along γ_1 from $\gamma_1(1 - 2t)$ to b. Thus γ_t is a path in X from a to b, and as

Fig. 4.20

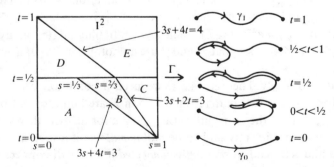

t increases from 0 to $\frac{1}{2}$, $1 - 2t$ decreases from 1 to 0, so the loop gradually stretches outwards from b until (when $t = \frac{1}{2}$) it travels twice along the entire length of γ_1, once in each direction. Thus when $t = \frac{1}{2}$, γ_t follows γ_0 (forwards), then γ_1 (backwards), and finally γ_1 (forwards), or equivalently, γ_t follows δ and then γ_1 (both forwards). For $\frac{1}{2} \leqslant t \leqslant 1$ we use the homotopy $\Delta:\delta \simeq \gamma_{(a)}$ to pull δ in towards a, leaving the final section of γ_t along γ_1 unchanged, until when $t = 1$ we have $\gamma_t = \gamma_1$, as required.

To be more specific, we divide I^2 into five subsets A, B, C, D, E as shown, and define $\Gamma(s, t)$ on these subsets by

$$
\Gamma(s,t) = \begin{cases}
\gamma_0\left(\dfrac{3s}{3 - 4t}\right) & 0 \leqslant t \leqslant \frac{1}{2}, 0 \leqslant s \leqslant 1 - \dfrac{4t}{3} & \text{(A)}; \\[2mm]
\gamma_1(4 - 3s - 4t) & 0 \leqslant t \leqslant \frac{1}{2}, 1 - \dfrac{4t}{3} \leqslant s \leqslant 1 - \dfrac{2t}{3} & \text{(B)}; \\[2mm]
\gamma_1(3s - 2) & 0 \leqslant t \leqslant \frac{1}{2}, 1 - \dfrac{2t}{3} \leqslant s \leqslant 1 & \text{(C)}; \\[2mm]
\Delta\left(\dfrac{3s}{4 - 4t}, 2t - 1\right) & \frac{1}{2} \leqslant t \leqslant 1, 0 \leqslant s \leqslant \frac{4}{3}(1 - t) & \text{(D)}; \\[2mm]
\gamma_1\left(\dfrac{4 - 3s - 4t}{1 - 4t}\right) & \frac{1}{2} \leqslant t \leqslant 1, \frac{4}{3}(1 - t) \leqslant s \leqslant 1 & \text{(E)}.
\end{cases}
$$

It is straightforward to check that $\Gamma(s, 0) = \gamma_0(s)$, $\Gamma(s, 1) = \gamma_1(s)$, $\Gamma(0, t) = a$, and $\Gamma(1, t) = b$ for all s, $t \in I$. Continuity of Γ follows from the continuity of γ_0, γ_1 and Δ and the fact that the various definitions of $\Gamma(s, t)$ agree when (s, t) is in the intersection of two or more subsets $A, \ldots, E$. The only case which needs any comment is when $s = 0$ and $t = 1$: as $s \to 0$ and $t \to 1$ with $0 \leqslant s \leqslant \frac{4}{3}(1 - t)$ (so that $(s, t) \in D$) we have $\Gamma(s, t) = \Delta(\sigma, \tau)$ with $\tau = 2t - 1 \to 1$ but with $\sigma = 3s/(4 - 4t)$ having no limit; however, $\lim_{\tau \to 1}\Delta(\sigma, \tau) = a$ for *all* $\sigma \in I$ (since Δ is a homotopy from δ to $\gamma_{(a)}$), so $\Gamma(s, t) \to a$ as required. Thus Γ is a homotopy from γ_0 to γ_1 in X, so X has a single homotopy class of paths from a to b. $\square$

The simplest way of seeing that the above definition of Γ agrees with the informal description preceding it is to consider the intermediate paths $\gamma_t(s) = \Gamma(s, t)$ for various values of $t \in I$, in each case considering how the point $\gamma_t(s)$ moves around X as s increases from 0 to 1. Five typical paths γ_t are illustrated above, corresponding to the cases $t = 0$, $0 < t < \frac{1}{2}$, $t = \frac{1}{2}$, $\frac{1}{2} < t < 1$, and $t = 1$; the reader is encouraged to check that the illustration does represent γ_t accurately in each case.

Combining Theorems 4.5.1 and 4.5.2 we see that continuation of a function element onto a simply connected region E is independent of the path of continuation, so we have a single-valued meromorphic function on E. The *monodromy theorem* states this more precisely:

Theorem 4.5.3. *Let E be a simply connected region in Σ, and let (D, f) be a function element with $D \subseteq E$. If (D, f) can be continued meromorphically along all paths in E starting at some point $a \in D$, then there is a direct meromorphic continuation $(E, g) \sim (D, f)$.*

Proof. Since E is simply connected, E is path-connected, so for each $b \in E$ there is a path γ in E from a to b. By hypothesis, we may continue (D, f) along γ; we denote the value of this continuation at b by $f_\gamma(b)$. Since E is simply connected, any two paths in E from a to b are homotopic, by Theorem 4.5.2, so by Theorem 4.5.1 they induce the same value of $f_\gamma(b)$. Thus $f_\gamma(b)$ is independent of the path γ from a to b, so we have a single-valued function $g: E \to \mathbb{C}$ given by $g(b) = f_\gamma(b)$.

Now suppose that the value $g(b)$ is produced by a meromorphic continuation $(D, f) \sim (D_1, f_1) \sim \ldots \sim (D_m, f_m)$ along γ, so that $g(b) = f_m(b)$; there is no loss of generality in assuming that the disc D_m is contained in E. For each $c \in D_m$ there is a path δ in D_m from b to c, and hence there is a path ε in E from a to c, following γ and then δ, given by

$$\varepsilon(s) = \begin{cases} \gamma(2s) & 0 \leqslant s \leqslant \tfrac{1}{2}, \\ \delta(2s - 1) & \tfrac{1}{2} \leqslant s \leqslant 1 \end{cases}$$

(see Fig. 4.21). Since $\delta \subseteq D_m$, the sequence $(D, f) \sim (D_1, f_1) \sim \ldots \sim (D_m, f_m)$ is a meromorphic continuation along ε, so $g(c) = f_\varepsilon(c) = f_m(c)$. Thus $g \equiv f_m$ on D_m, so g is meromorphic at b, since f_m is, and hence (E, g) is a function element. Since meromorphic continuation of (D, f) along paths within D must give $g(b) = f(b)$ for $b \in D$, we have $(E, g) \sim (D, f)$ as required. $\square$

Fig. 4.21

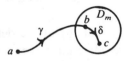

Thus, as shown in §4.1, we can continue a branch of $\log(z)$ from a neighbourhood D of 1 onto a region $E = D_J$, where J is an open interval $(\alpha, \beta) \subset \mathbb{R}$ with $\beta - \alpha \leqslant 2\pi$, since such a region is simply connected. Taking $J = ((2n - 1)\pi, (2n + 1)\pi)$, with $n \in \mathbb{Z}$, we 'cut' Σ along the negative real line from 0 to ∞, as in Fig. 4.22, and the remaining region $E = D_J =$

$\mathbb{C}\backslash\{z\in\mathbb{R}|z\leqslant 0\}$ is simply connected, so we have a branch f_J of $\log(z)$, defined on E. This branch, which we shall denote by f_n, is determined by the value $f_n(1) = 2n\pi i$; the cut from 0 to ∞ prevents any path in E from winding around 0 (or equivalently, around ∞), so analytic continuation along paths in E is single-valued. When we consider other many-valued functions we shall use a similar technique, cutting Σ along suitable lines to produce a simply connected region on which we can define a single-valued meromorphic branch of the function.

Fig. 4.22

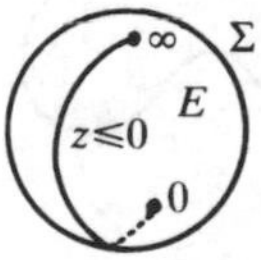

4.6 The fundamental group

Using homotopy, we can associate to each path-connected space X a group $\pi_1(X)$, the fundamental group of X, an important concept in algebraic topology (see, for example, Massey [1967]).

If γ and δ are paths in a topological space X with $\gamma(1) = \delta(0)$, then the *product* $\gamma\delta$ of γ and δ is the path

$$(\gamma\delta)(s) = \begin{cases} \gamma(2s) & 0 \leqslant s \leqslant \tfrac{1}{2}, \\ \delta(2s-1) & \tfrac{1}{2} \leqslant s \leqslant 1, \end{cases}$$

shown in Fig. 4.23. (Intuitively, the point $(\gamma\delta)(s)$ travels along γ and then along δ, at twice the usual speed in each case in order to complete the journey in unit time – see the proof of Theorem 4.5.3, where $\varepsilon = \gamma\delta$.) The *inverse* path of γ is

$$\gamma^{-1}(s) = \gamma(1-s);$$

thus $\gamma^{-1}(s)$ follows the path γ in the reverse direction.

Fig. 4.23

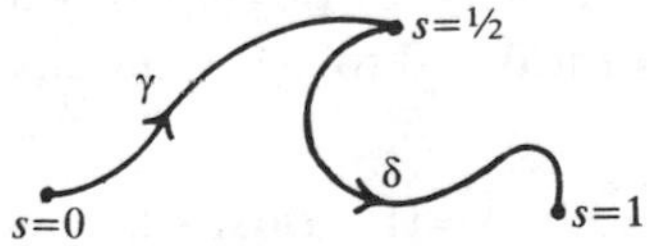

The homotopy classes $[\gamma]$ of closed paths γ from a given point $a\in X$ to itself form a group $\pi_1(X, a)$: the product $[\gamma][\delta]$ of two classes is the class $[\gamma\delta]$, the identity element is the class containing the constant path $\gamma_{(a)}(s) = a$ for all $s\in I$, and the inverse of $[\gamma]$ is $[\gamma^{-1}]$. It is straightforward to verify that these definitions of group operations are independent of the choice of representatives γ, δ of $[\gamma]$, $[\delta]$, and that the group axioms are satisfied.

Fig. 4.24

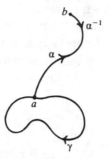

If X is path-connected, then there is a path α from any point $a \in X$ to any point $b \in X$, as in Fig. 4.24, and the map $[\gamma] \to [\alpha^{-1}\gamma\alpha]$ is an isomorphism $\pi_1(X, a) \cong \pi_1(X, b)$; thus as an abstract group, $\pi_1(X, a)$ is independent of the choice of $a \in X$, so we denote this group by $\pi_1(X)$, the *fundamental group* of X. For example, X is (by definition) simply connected if and only if $\pi_1(X)$ is the trivial group. For our purposes, the most important example of a space which is not simply connected is the punctured plane:

Theorem 4.6.1. $\pi_1(\mathbb{C}\backslash\{0\})$ *is an infinite cyclic group, generated by* $[\gamma]$, *where* γ *is the unit circle parametrised by* $\gamma(s) = e^{2\pi i s}$, $s \in I$.

Proof. Since $\mathbb{C}\backslash\{0\}$ is path-connected, it is sufficient to show that every closed path δ from 1 to 1 in $\mathbb{C}\backslash\{0\}$ is homotopic to $\gamma^n(s) = e^{2n\pi i s}$ for some unique $n \in \mathbb{Z}$.

Putting $\delta(s) = re^{i\theta}$, with $r = r(s) > 0$ and $\theta = \theta(s) \in \mathbb{R}$, we see that $\theta(s) = \arg(\delta(s)) = \operatorname{Im}(\log(\delta(s)))$ is a many-valued function of s for $s \in I$. If we continue $\log(z)$ analytically along δ, starting with $\log(z) = 0$ at $z = \delta(0) = 1$, we eventually obtain $\log(z) = 2n\pi i$ at $z = \delta(1) = 1$, where $n \in \mathbb{Z}$, and the corresponding values of $\theta(s) = \operatorname{Im}(\log(\delta(s)))$ vary continuously with respect to s from $\theta(0) = \operatorname{Im}(0) = 0$ to $\theta(1) = \operatorname{Im}(2n\pi i) = 2n\pi$.

If we define

$$R = R(s, t) = (1 - t)r(s) + t,$$
$$\Theta = \Theta(s, t) = (1 - t)\theta(s) + 2n\pi st,$$

then

$$\Gamma(s, t) = Re^{i\Theta} \quad (s, t \in I)$$

is a homotopy within $\mathbb{C}\backslash\{0\}$ from δ (at $t = 0$) to γ^n (at $t = 1$): the effect of

$R(s, t)$ is to deform δ radially onto the unit circle, while $\Theta(s, t)$ deforms δ until θ increases linearly from 0 to $2n\pi$ (see Fig. 4.25).

Fig. 4.25

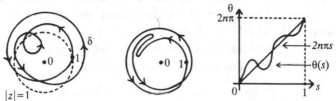

Thus $[\delta] = [\gamma''] = [\gamma]''$, so $\pi_1(\mathbb{C}\setminus\{0\})$ is a cyclic group, generated by $[\gamma]$. By applying Theorem 4.5.1 to the analytic continuations of $\log(z)$ along γ^m and γ^n, we see that $\gamma^m \simeq \gamma^n$ if and only if $m = n$, so δ determines n (its winding number) uniquely, and $\pi_1(\mathbb{C}\setminus\{0\})$ is an infinite cyclic group. $\square$

More generally, if $a \in \mathbb{C}$ then $\pi_1(\mathbb{C}\setminus\{a\})$ is infinite cyclic, generated by $[\gamma + a]$ where $\gamma + a$ is the closed path $(\gamma + a)(s) = e^{2\pi i s} + a$. We define the *winding number* $n_a(\delta)$ of a closed path δ around a to be the unique integer n such that $\delta \simeq (\gamma + a)^n$ in $\mathbb{C}\setminus\{a\}$.

4.7 The Riemann surface of $\log(z)$

We have seen, in connection with $\log(z)$ and $z^{1/q}$, how meromorphic continuation of a function element can give rise to a many-valued function (or more precisely to many different functions on the same region). This is unsatisfactory, since it means that we have to use great care in using such apparently harmless statements as $\log(ab) = \log(a) + \log(b)$ (as we saw in §3.8). The solution adopted in §3.8 was to restrict attention to one particular branch of $\log(z)$, denoted there by $\mathrm{Log}(z)$, but this is not completely satisfactory, since $\mathrm{Log}(z)$ is not continuous when $z < 0$, and we still do not have $\mathrm{Log}(ab) = \mathrm{Log}(a) + \mathrm{Log}(b)$ for all $a, b \neq 0$. A better solution, which applies to many-valued functions in general, is due to Riemann: instead of restricting the values of the function, we extend its domain. Specifically, we construct a surface S, a covering map $\psi : S \to \mathbb{C}\setminus\{0\}$

Fig. 4.26

and a function $\phi: S \to \mathbb{C}$ such that for each $z \in \mathbb{C} \setminus \{0\}$ the elements of $\psi^{-1}(z)$ are mapped bijectively by ϕ onto the different values of $\log(z)$, so that $\exp \circ \phi = \psi: S \to \mathbb{C} \setminus \{0\}$ (see Fig. 4.26). Thus each sheet of S (regarded as a covering-space of $\mathbb{C} \setminus \{0\}$) corresponds to a particular branch of $\log(z)$ represented by the restriction of ϕ to that sheet, and we may regard ϕ as composed of all the different branches of $\log(z)$.

The surface S, known as the *Riemann surface* of $\log(z)$, can be constructed from the sequence of direct analytic continuations

$$\ldots \sim L_{(C-2\pi)} \sim L_A \sim L_B \sim L_C \sim L_{(A+2\pi)} \sim \ldots,$$

considered in §4.1. The general idea is to regard the underlying regions of these function elements as being disjoint, and then to glue them together wherever they carry identically equal values of $\log(z)$; the resulting

Fig. 4.27

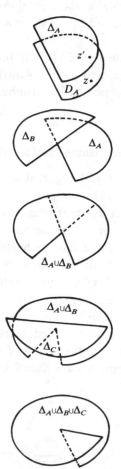

surface S is the domain of a single-valued function ϕ, which is locally equal to a branch of $\log(z)$.

Let Δ_A be a horizontal surface, lying above the region D_A of the plane $\mathbb{C}$; each $z' \in \Delta_A$ lies above a unique $z \in D_A$, and we define a function $\phi: \Delta_A \to \mathbb{C}$ by $\phi(z') = f_A(z)$.

Now let Δ_B be a horizontal surface lying above D_B, at a different level from Δ_A; we define $\phi(z'') = f_B(z)$ whenever $z'' \in \Delta_B$ lies above $z \in D_B$. If $z' \in \Delta_A$ and $z'' \in \Delta_B$ lie above the same point $z \in D_A \cap D_B$, then since $L_A \sim L_B$ we have $\phi(z') = f_A(z) = f_B(z) = \phi(z'')$; we therefore identify all such pairs z', z'' of overlapping points with each other, and we have a well-defined function ϕ on the resulting surface $\Delta_A \cup \Delta_B$.

We now introduce a third surface Δ_C lying above D_C and disjoint from $\Delta_A \cup \Delta_B$, and we define $\phi(z''') = f_C(z)$ whenever $z''' \in \Delta_C$ lies above $z \in D_C$. Since $L_B \sim L_C$ we have $\phi(z'') = \phi(z''')$ whenever z'' and z''' lie above some point $z \in D_B \cap D_C$, so if we identify z'' with z''' then ϕ is a well-defined function on $\Delta_A \cup \Delta_B \cup \Delta_C$. This process is illustrated in Fig. 4.27. Notice that if $z' \in \Delta_A$ and $z''' \in \Delta_C$ lie above some $z \in D_A \cap D_C$, then $\phi(z''') = f_C(z) = f_A(z) + 2\pi i = \phi(z') + 2\pi i \neq \phi(z')$, so instead of identifying z' with z''' we allow Δ_A and Δ_C to overlap at different levels above $\mathbb{C}$.

We continue this process, following the sequence

$$\ldots \sim L_{(C-2\pi)} \sim L_A \sim L_B \sim L_C \sim L_{(A+2\pi)} \sim \ldots$$

in both directions. For example, when we introduce $\Delta_{(A+2\pi)}$ lying above $D_{(A+2\pi)} = D_A$, we identify overlapping points of Δ_C and $\Delta_{(A+2\pi)}$ since $L_C \sim L_{(A+2\pi)}$, but we keep $\Delta_{(A+2\pi)}$ separate from Δ_A, letting them lie one over the other at different levels, since they carry different branches of $\log(z)$ (see Fig. 4.28). Each point s of the resulting surface $S = \ldots \cup \Delta_{(C-2\pi)} \cup \Delta_A \cup \Delta_B \cup \Delta_C \cup \Delta_{(A+2\pi)} \cup \ldots$ lies above a unique point $z = \psi(s)$ in $\mathbb{C} \setminus \{0\}$, and the function $\psi: S \to \mathbb{C} \setminus \{0\}$ is easily seen to be a covering map (as defined in §1.5), each point $z \in \mathbb{C} \setminus \{0\}$ being covered by infinitely many points $s \in S$. As illustrated in Fig. 4.29, we can visualise S as a spiral with infinitely many turns extending upwards and downwards, with ψ corresponding to projection onto the horizontal surface $\mathbb{C} \setminus \{0\}$. We have

Fig. 4.28

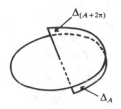

a function $\phi: S \to \mathbb{C}$, the value of $\phi(s)$ for each $s \in S$ being one of the values of $\log(\psi(s))$, depending on which sheet of S contains s; conversely, for each $z \in \mathbb{C} \setminus \{0\}$ and for each value w of $\log(z)$ there is a point $s \in \psi^{-1}(z) \subset S$ with $\phi(s) = w$.

The comments at the end of §4.5 give us a second, equivalent way of constructing S which can be adapted to apply to other functions. If we cut Σ along the line $z \leqslant 0$ from 0 to ∞, then the remaining region $E = \mathbb{C} \setminus \{z \in \mathbb{R} \mid z \leqslant 0\}$ is simply connected and is the domain of branches f_n of $\log(z)$ satisfying $f_n(1) = 2n\pi i$, for each $n \in \mathbb{Z}$. We take disjoint copies E_n of E, each the domain of f_n; we can think of these surfaces E_n as lying one above the other over $E \subset \mathbb{C}$, like pages of a book, as in Fig. 4.30, or as being wrapped around the sphere Σ like layers of an onion, as in Fig. 4.31.

Fig. 4.29

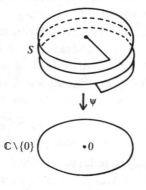

Fig. 4.30

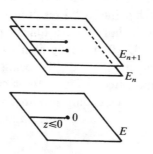

Fig. 4.31

Now let (D, f) be a function element representing a branch of $\log(z)$ on some disc D containing a point $a \in \mathbb{R}$, $a < 0$, see Fig. 4.32. On the region $D_+ = \{z \in D | \operatorname{Im}(z) > 0\}$ we have $f \equiv f_n$ for some $n \in \mathbb{Z}$, and on the region $D_- = \{z \in D | \operatorname{Im}(z) < 0\}$ we have $f \equiv f_m$ for some $m \in \mathbb{Z}$. As $z \to a$ with $z \in D_+$ we have

$$f(z) = f_n(z) \to \ln(|a|) + (2n + 1)\pi i,$$

while as $z \to a$ with $z \in D_-$ we have

$$f(z) = f_m(z) \to \ln(|a|) + (2m - 1)\pi i.$$

Since f is analytic at a, $f(z)$ has a unique limit as $z \to a$, so $m = n + 1$.

Fig. 4.32

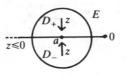

Thus as we cross the cut $z \leqslant 0$ at any point $a < 0$, passing from D_+ to D_-, analytic continuation takes us from f_n on D_+ to $f_m = f_{n+1}$ on D_-. We therefore join the edge $\operatorname{Im}(z) > 0$ of E_n to the edge $\operatorname{Im}(z) < 0$ of E_{n+1} along a line L_n from 0 to ∞ (but not including 0 or ∞); performing this operation on successive pairs E_n and E_{n+1} is rather like buttoning one's shirt and jacket together, and then one's jacket and overcoat, etc. We now define the surface S to be the union of all these sheets E_n and lines L_n, for $n \in \mathbb{Z}$, as illustrated in Fig. 4.33. We define $\phi : S \to \mathbb{C}$ by $\phi(s) = f_n(s)$ if $s \in E_n$, and $\phi(s) = \ln(|a|) + (2n + 1)\pi i$ if $s \in L_n$ lies above $a < 0$. For each $s \in S$, we define $\psi(s)$ to be the unique element of $\mathbb{C} \backslash \{0\}$ above which s lies, giving a covering map $\psi : S \to \mathbb{C} \backslash \{0\}$. It is easily seen that S, ψ and ϕ, as constructed here, are essentially the same as in the original construction.

Fig. 4.33

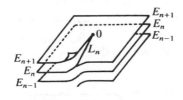

This method of construction shows that S is homeomorphic to $\mathbb{C}$. Since $f_n(re^{i\theta}) = \ln(|r|) + i\theta$, where $(2n - 1)\pi < \theta < (2n + 1)\pi$, ϕ maps E_n homeomorphically onto the strip given by $(2n - 1)\pi < \operatorname{Im}(w) < (2n + 1)\pi$; similarly, L_n is mapped homeomorphically onto the line $\operatorname{Im}(w) = (2n + 1)\pi$. Now these sets $\phi(E_n)$ and $\phi(L_n)$ form a partition of $\mathbb{C}$, so ϕ is a bijection from S to $\mathbb{C}$, and it is easily seen from our construction that both ϕ and ϕ^{-1} are continuous, so that $\phi : S \to \mathbb{C}$ is a homeomorphism. We can regard $\mathbb{C}$, partitioned in this way, as representing the infinite spiral S 'straightened out' by ϕ, each parallel strip on $\mathbb{C}$ corresponding to a sheet of S (see Fig. 4.34). If we let a point $z \in \mathbb{C} \setminus \{0\}$ follow a path winding around 0, then any point $s \in \psi^{-1}(z)$ will move continuously in a spiral above z, passing successively through sheets $\ldots, E_{-1}, E_0, E_1, \ldots$ of S, while the corresponding point $\phi(s) = \log(z)$ passes through the strips $\ldots, \phi(E_{-1})$, $\phi(E_0)$, $\phi(E_1), \ldots$ associated with the various branches $\ldots, f_{-1}, f_0, f_1, \ldots$ of $\log(z)$.

Fig. 4.34

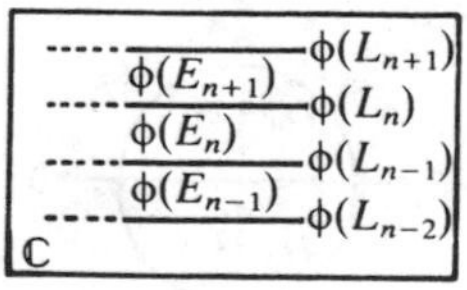

$$
\begin{array}{l}
\text{-----} \overline{} \phi(L_{n+1}) \\
\text{-----} \overline{\phi(E_{n+1})} \phi(L_n) \\
\overline{\phi(E_n)} \phi(L_{n-1}) \\
\text{-----} \overline{\phi(E_{n-1})} \phi(L_{n-2}) \\
\mathbb{C}
\end{array}
$$

4.8 The Riemann surface of $z^{1/q}$

In §4.4 we were able to derive the analytic continuation of $z^{1/q}$ from that of $\log(z)$; similarly we can now adapt the method of construction of the Riemann surface of $\log(z)$ to obtain the Riemann surface of $z^{1/q}$.

As in §4.7, let $E = \mathbb{C} \setminus \{z \in \mathbb{R} \,|\, z \leqslant 0\}$. This simply connected region is the domain of branches $g_n = \exp(q^{-1} f_n)$ of $z^{1/q}$ satisfying $g_n(1) = \varepsilon^n$, where $\varepsilon = \exp(2\pi i / q)$ and $n \in \mathbb{Z}$; we have $g_m = g_n$ if and only if $m \equiv n \mod (q)$, so there are q distinct branches $g_0, g_1, \ldots, g_{q-1}$ and we therefore take q disjoint copies $E_0, E_1, \ldots, E_{q-1}$ of E lying at different levels above E, with each E_n the domain of g_n. This is illustrated in Fig. 4.35. As we continue analytically across the line $z \leqslant 0$, with $\mathrm{Im}(z)$ decreasing, we pass from g_0 to g_1, so we join the edge $\mathrm{Im}(z) > 0$ of E_0 to the edge $\mathrm{Im}(z) < 0$ of E_1 along a line L_0 from 0 to ∞. Similarly, we join E_1 to E_2 along L_1, and so on until E_{q-2} is joined to E_{q-1} along L_{q-2}. The process so far, giving us a spiral-shaped surface with q sheets, is easily visualised, but the next

Fig. 4.35

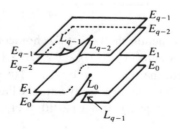

step is not: since continuation across $z \leqslant 0$ takes g_{q-1} to $g_q = g_0$, we need to join the edge $\mathrm{Im}(z) > 0$ of E_{q-1} to the edge $\mathrm{Im}(z) < 0$ of E_0 along a line L_{q-1}. We must do this without forcing the resulting surface T to pass through itself, and this is impossible in the euclidean space $\mathbb{R}^3$: however, our only reason for regarding the sheets E_n as subsets of $\mathbb{R}^3$ is to help visualisation and illustration, and there is no problem if we regard them as abstract surfaces. As in Fig. 4.36, our illustrations will show apparent self-intersections, whereas in reality the surface does not intersect itself.

Fig. 4.36

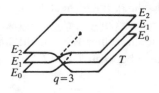

Fig. 4.37

There is an alternative method for constructing T in $\mathbb{R}^3$, which avoids self-intersections but has the disadvantage of not effectively illustrating the projections of the sheets onto E; this uses the fact that E is homeomorphic to the region E' shown in Fig. 4.37, formed by cutting the annulus $1 < |z| < 2$ along the negative real line from -1 to -2 (such a homeomorphism is given by $re^{i\theta} \mapsto (2 - e^{-r})e^{i\theta}$). We take q copies E'_n of E' $(0 \leqslant n \leqslant q - 1)$, each identified with E'_n in the above way, and join each E'_n to E'_{n+1} along a line L'_n homeomorphic to L_n, from -1 to -2, for $n = 0, 1, \ldots, q - 2$. With a little twisting and stretching, we can now join E'_{q-1} to E'_0 along a line L'_{q-1} in $\mathbb{R}^3$, and the resulting surface T' is homeomorphic to T; of course, the required twisting destroys the natural projections from the sheets. See Fig. 4.38.

Fig. 4.38

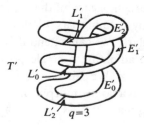

The surface

$$T = \left(\bigcup_{n=0}^{q-1} E_n \right) \cup \left(\bigcup_{n=0}^{q-1} L_n \right)$$

is an unbranched covering-surface of $\mathbb{C}\backslash\{0\}$ with q sheets; each point $s \in T$ lies over a unique point $z = \psi(s) \in \mathbb{C}\backslash\{0\}$, and each $z \in \mathbb{C}\backslash\{0\}$ lies beneath q points $s \in T$, one on each sheet. We define a function $\phi: T \to \mathbb{C}\backslash\{0\}$ by putting $\phi(s) = g_n(s)$ for all $s \in E_n$, and then extending ϕ to each line L_n by continuity; thus ϕ is locally equal to a branch of $z^{1/q}$, and we have $\phi(s)^q = \psi(s)$ for all $s \in T$ (see Fig. 4.39). Each sheet E_n of T is mapped homeomorphically by ϕ onto the region $(2n - 1)\pi/q < \arg(w) < (2n + 1)\pi/q$ of $\mathbb{C}\backslash\{0\}$, and L_n is mapped homeomorphically onto the line $\arg(w) = (2n + 1)\pi/q$, so ϕ is a homeomorphism between T and $\mathbb{C}\backslash\{0\}$, as depicted in Fig. 4.40.

Fig. 4.39

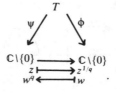

Fig. 4.40

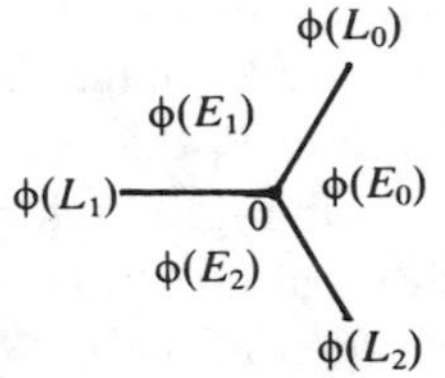

$$q = 3$$

There is no analytic (or even meromorphic) continuation of $z^{1/q}$ at $z = 0$: in any neighbourhood of 0, if we continue along a closed path winding once around 0 in the positive direction, then any branch of $z^{1/q}$ is multiplied by $\varepsilon = \exp(2\pi i/q)$, so we cannot choose a single-valued branch of $z^{1/q}$ on any region containing 0. Nevertheless, for each branch g_n of $z^{1/q}$ we have $\lim_{z \to 0} g_n(z) = 0$, so we adjoin a single point s_0 to T, lying above $0 \in \mathbb{C}$, and we regard s_0 as the origin of each of the cut planes E_n; we then extend the map $\phi: T \to \mathbb{C}\backslash\{0\}$ by defining $\phi(s_0) = 0 \in \mathbb{C}$, so that $\phi: T \cup \{s_0\} \to \mathbb{C}$ is continuous at s_0. Similarly, we can extend ϕ by continuity (though not meromorphically) at ∞, adjoining a single point s_∞ to T, lying above $\infty \in \Sigma$, and putting $\phi(s_\infty) = \infty$. We call $S = T \cup \{s_0, s_\infty\}$ the *Riemann surface* of $z^{1/q}$; by construction, ϕ is a homeomorphism from S to Σ. We can extend the unbranched covering map $\psi: T \to \mathbb{C}\backslash\{0\}$ to a branched covering map

$\psi:S\to\Sigma$ by defining $\psi(s_0)=0$ and $\psi(s_\infty)=\infty$; near s_0 and s_∞, ψ is a q-to-one mapping, so s_0 and s_∞ are branch-points of order $q-1$. (See §§1.5 and 3.15 for branch-points.)

We have seen that S is homeomorphic (under ϕ) to the sphere Σ. We may therefore regard S as a sphere, divided into regions $E_0, E_1, \ldots, E_{q-1}$ by lines $L_0, L_1, \ldots, L_{q-1}$ from s_0 to s_∞. The projection $\psi:S\to\Sigma$ wraps S q times around Σ, the region E of Σ being covered homeomorphically by each of the regions E_n. As a point $z\in\Sigma$ follows a path winding around 0 (or around ∞), any point $s\in\psi^{-1}(z)$ moves continuously above z, passing through sheets$\ldots$, $E_0, E_1, E_2, \ldots$ of S, corresponding to the way in which $\phi(s)$ passes through the branches$\ldots, g_0, g_1, g_2, \ldots$ of $z^{1/q}$. See Fig. 4.41.

Fig. 4.41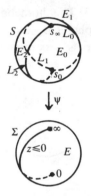

4.9 The Riemann surface of $\sqrt{p(z)}$, *p* a polynomial

Any polynomial $p(z)\in\mathbb{C}[z]$ is analytic on $\mathbb{C}$, and since we can continue the function $\sqrt{\zeta}$ analytically for $\zeta\neq0,\infty$, we can therefore continue the 2-valued function $\sqrt{p(z)}$ analytically along any path in $\mathbb{C}$ which avoids the set of zeros of p. We shall consider the Riemann surface S resulting from this continuation in various cases, depending on the number and multiplicities of the zeros of p. By absorbing the leading coefficient into z, there is no loss of generality in putting this coefficient equal to 1, so that p is monic.

(i) $p(z)=z-a$. This is the general case in which p has degree 1, and we have already considered the special case $a=0$ in §4.8 (putting $q=2$). Since the substitution $z\mapsto z-a$ is an automorphism of Σ we must expect the general case to resemble the special case.

If, as shown in Fig. 4.42, we cut Σ along a line L from a to ∞ (say, along $z-a\leqslant0$), then the resulting region $E=\Sigma\backslash L$ is simply connected and contains no zeros of $p(z) = z - a$, so by the monodromy theorem 4.5.3 we have single-valued analytic branches f_1, f_2 of $\sqrt{p}$ on E (satisfying $f_1 \equiv -f_2$); we therefore take two copies E_1, E_2 of E (regarded as lying above E), the domains of f_1, f_2 respectively. If we continue f_1 analytically across L (in either direction), then we obtain f_2, so we join the two edges of E_1 to the opposite edges of E_2 along lines L_1, L_2 lying above L. The projection of the resulting surface $T = E_1 \cup E_2 \cup L_1 \cup L_2$ onto the sphere Σ is an unbranched 2-sheeted covering map $\psi: T \to \mathbb{C} \setminus \{a\}$. By adding two branch-points s_a, s_∞ to T, lying above $a, \infty \in \Sigma$, we have a 2-sheeted branched covering map $\psi: S = T \cup \{s_a, s_\infty\} \to \Sigma$. We can define $\phi: T \to \Sigma$ by putting $\phi(s) = f_n(s)$ when $s \in E_n$, and then extending this by continuity to the rest of S. Thus $\phi: S \to \Sigma$ is a homeomorphism, locally representing a branch of $\sqrt{p}$. See Fig. 4.43. This homeomorphism is seen most easily by opening out the two cut spheres E_1 and E_2 to form hemispheres which are then joined together across a common boundary $L_1 \cup L_2$. This is illustrated in Fig. 4.44.

Fig. 4.42

Fig. 4.43

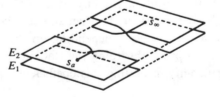

Fig. 4.44

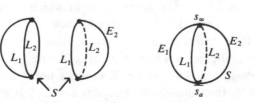

(ii) $p(z) = (z - a)(z - b), a \neq b$. If we cut Σ along a line L from a to b (say along the Euclidean line segment from a to b in $\mathbb{C}$), then the resulting region $E = \Sigma \backslash L$ is simply connected. We can continue $\sqrt{p}$ meromorphically along all paths in E: analytic continuation in $E \cap \mathbb{C}$ is straightforward, and if $\infty \in E$ then each branch of $\sqrt{p}$ is meromorphic at ∞ (with a simple pole) since each branch of $z\sqrt{(p(1/z))} = \sqrt{((1 - az)(1 - bz))}$ is analytic and non-zero at 0. As before, we may therefore use the monodromy theorem to show that there are two single-valued meromorphic branches f_1 and f_2 of $\sqrt{p}$ on E, each having a simple pole at ∞. We therefore take two copies E_1, E_2 of E, with E_n lying above E, and regard E_n as the domain of f_n. By crossing L in either direction we pass from one branch to the other, so we join E_1 and E_2 together along lines L_1, L_2 lying above L, to obtain a 2-sheeted unbranched covering surface $T = E_1 \cup E_2 \cup L_1 \cup L_2$ of $\Sigma \backslash \{a, b\}$. Finally, we include branch-points s_a, s_b lying above a, b to obtain a 2-sheeted branched covering surface $S = T \cup \{s_a, s_b\}$ of Σ. As in case (i), S is homeomorphic to a sphere. See Fig. 4.45.

Fig. 4.45

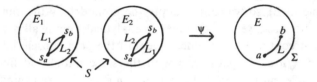

In this case, there is no branch-point on S lying above ∞, for if we continue $\sqrt{p}$ analytically along a closed path γ winding once around ∞ in Σ, as in Fig. 4.46, then provided γ is sufficiently close to ∞ it winds once around each of a and b (regarding γ as a closed path in $\mathbb{C}$), so both factors $\sqrt{(z - a)}$ and $\sqrt{(z - b)}$ of $\sqrt{p}$ are multiplied by -1 and hence $\sqrt{p}$ is invariant. Thus we do not pass from one branch to another by continuing along γ, so S is not branched above ∞.

Fig. 4.46

(iii) $p(z) = (z - a)^2$. To complete our examination of quadratic polynomials p, we consider the case where p has equal roots. Now there are two branches $f_1(z) = z - a$ and $f_2(z) = -(z - a)$ of $\sqrt{p}$, both analytic on $\mathbb{C}$ and meromorphic on Σ. We therefore take the Riemann surface S to consist of two copies E_1, E_2 of Σ, the domains of f_1, f_2 respectively. Since we cannot pass from one branch to another by meromorphic continuation of $\sqrt{p}$, we do not join E_1 and E_2, but instead we regard the surface S as a 2-sheeted unbranched covering surface of Σ consisting of two disjoint spheres. (Notice that although f_1 and f_2 take the same values at a and at ∞, they are not identically equal in neighbourhoods of these points, so there are two sheets of S lying above each of a and ∞.)

(iv) $p(z) = (z - a)(z - b)(z - c)$, a, b, c *all distinct.* As in the preceding cases, there are two branches f_1 and f_2 of $\sqrt{p}$, so the Riemann surface S of $\sqrt{p}$ is a 2-sheeted covering of Σ. By continuing $\sqrt{p}$ analytically along a small closed path around a, b or c we pass from f_1 to f_2 or vice-versa, so S has branch-points s_a, s_b, s_c lying above a, b, c. There is also a branch-point s_∞ above ∞, since continuation around ∞ multiplies each factor $\sqrt{(z-a)}$, $\sqrt{(z-b)}$ and $\sqrt{(z-c)}$ by -1, so that $\sqrt{p}$ is multiplied by $(-1)^3 = -1$.

If we cut Σ along disjoint simple paths L, M from a to b and from c to ∞, then the resulting region $E = \Sigma \backslash (L \cup M)$ is not simply connected (for example, in Fig. 4.47 a closed path winding once around L is not null-homotopic in E), so we cannot apply the monodromy theorem directly to E. Nevertheless, we can continue $\sqrt{p}$ analytically along all paths within E, and that continuation is independent of the path chosen: for if continuation of $\sqrt{p}$ along paths γ, δ with the same initial and final points produces different branches of $\sqrt{p}$, then $\sqrt{p}$ is not invariant under continuation along the closed path $\gamma^{-1}\delta$; however, the winding-numbers of $\gamma^{-1}\delta$ satisfy $n_a(\gamma^{-1}\delta) = n_b(\gamma^{-1}\delta)$ ($= n$, say) and $n_c(\gamma^{-1}\delta) = 0$, so continuation along $\gamma^{-1}\delta$ multiplies $\sqrt{p} = \sqrt{(z-a)(z-b)(z-c)}$ by $(-1)^{2n} = 1$, and $\sqrt{p}$ is invariant. Hence, by continuing along paths in E we obtain two single-valued analytic branches f_1, f_2 of $\sqrt{p}$ on E. We therefore take two copies E_1, E_2 of E, the domains of f_1 and f_2 respectively, and we join

Fig. 4.47

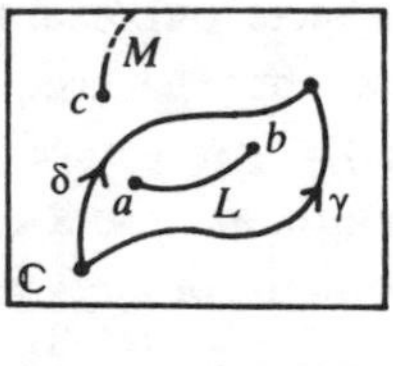

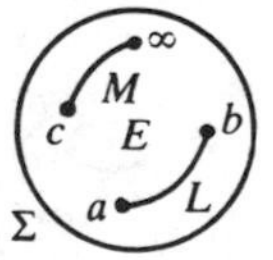

them by lines L_1, L_2 above L and M_1, M_2 above M, similar to the joins used in cases (ii) and (i).

If we open out the cuts on E_1 and E_2, then (as illustrated in Fig. 4.48) we see that E_1 and E_2 are homeomorphic to the two halves of a torus, joined across their common boundary components $L_1 \cup L_2$ and $M_1 \cup M_2$, so that S is homeomorphic to a torus.

There is a close connection here with elliptic functions. If Ω is a lattice in $\mathbb{C}$ then the Weierstrass function $\wp$ corresponding to Ω satisfies an ordinary differential equation

$$(\wp')^2 = 4\,\wp^3 - g_2\,\wp - g_3,$$

where g_2 and g_3 are constants depending on Ω (see §3.10); thus

$$\wp' = \sqrt{(p(\wp))},$$

where p is the cubic polynomial

$$p(z) = 4z^3 - g_2 z - g_3,$$

which has distinct roots by Theorem 3.10.9. Using this, we can construct a homeomorphism α from the torus $\mathbb{C}/\Omega$ to the Riemann surface S of

Fig. 4.48

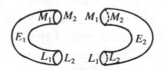

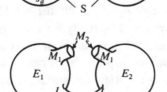

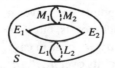

$w = \sqrt{(p(z))}$ (consult Fig. 4.49). We recall from §3.15 that $\wp$ and $\wp'$ induce branched coverings $\hat{\wp} : \mathbb{C}/\Omega \to \Sigma$ and $\hat{\wp}' : \mathbb{C}/\Omega \to \Sigma$; for each $t \in \mathbb{C}/\Omega$ we therefore define $\alpha(t)$ to be the unique point in S which lies above $z = \hat{\wp}(t) \in \Sigma$ and carries the value $w = \hat{\wp}'(t)$ of $\sqrt{p(z)}$, that is, $\psi(\alpha(t)) = \hat{\wp}(t)$ and $\phi(\alpha(t)) = \hat{\wp}'(t)$, where ψ is the covering map $S \to \Sigma$ and $\phi : S \to \Sigma$ is the single-valued function induced by analytic continuation of $\sqrt{p}$. It is straightforward to check that $\alpha : \mathbb{C}/\Omega \to S$ is a homeomorphism, so we may regard S as being parametrised by a variable $t \in \mathbb{C}/\Omega$; if we identify S with $\mathbb{C}/\Omega$ then, for example, $\psi : S \to \Sigma$ is identified with the 2-sheeted covering map $\hat{\wp} : \mathbb{C}/\Omega \to \Sigma$ which we studied in §3.15. Conversely, we will show in §6.5 that if $p(z)$ is any cubic polynomial $4z^3 - g_2 z - g_3$ with distinct roots, then there is a lattice Ω such that the associated Weierstrass function $\wp$ satisfies $(\wp')^2 = p(\wp)$, and hence the Riemann surface S of $\sqrt{p}$ may be identified with $\mathbb{C}/\Omega$ in the above way.

Fig. 4.49

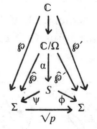

(v) $p(z) = \prod_{j=1}^{m} (z - a_j)^{e_j}$, $a_1, \ldots, a_m$ *all distinct*. Having dealt with some special cases of $\sqrt{p}$, we are now ready for the general case. If the multiplicities e_j of the roots a_j of p are all even, then as in case (iii) there are two single-valued meromorphic branches $f_1(z) = \prod (z - a_j)^{e_j/2}$ and $f_2(z) = -f_1(z)$ of $\sqrt{p}$, so the Riemann surface S of $\sqrt{p}$ is a disjoint union of two spheres, each the domain of a branch f_n. Hence we may assume that some root a_j of p has odd multiplicity e_j. Extracting repeated factors of p, we therefore have $\sqrt{p} = q\sqrt{r}$ where q and r are polynomials, and r has no repeated roots and is not constant. Since q is single-valued and meromorphic on Σ, the construction of the Riemann surface of $\sqrt{p}$ is identical to that for $\sqrt{r}$, so by replacing p by r we may assume that p has distinct roots and is not constant, that is, $m \geqslant 1$ and each $e_j = 1$.

As in previous cases, the Riemann surface S of $\sqrt{p}$ has two sheets E_1 and E_2, meeting at branch-points lying over $a_1, \ldots, a_m$. If $m = 2k$ is even, then these are the only branch-points, but if $m = 2k - 1$ is odd, then there is a branch-point lying over $a_{2k} = \infty$, as in case (iv); in either case, we have an even number of branch-points over $a_1, \ldots, a_{2k}$, so we cut Σ along simple, mutually disjoint paths from a_1 to a_2, a_3 to $a_4, \ldots, a_{2k-1}$ to a_{2k}, as shown in Fig. 4.50. The resulting region E is not simply connected (unless $k = 1$), but by the argument used in case (iv), meromorphic continuation of $\sqrt{p}$ within E is independent of the path of continuation since any closed path in E winds around the branch-points an even number of times. We can therefore form the Riemann surface S by taking two copies E_1, E_2 of E, each the domain of a meromorphic branch of $\sqrt{p}$, and by joining E_1 to E_2 across lines lying above the k cuts in Σ. This gives a 2-sheeted covering surface S of Σ, branched over $a_1, \ldots, a_{2k}$, and by opening out the cuts in E_1 and E_2, as in case (iv), we see that S is homeomorphic to a sphere with $k - 1$ handles attached. This is illustrated in Fig. 4.51. We say that such a surface S has *genus* $g = k - 1$; for example, a sphere and a torus have genus 0 and 1 respectively. We shall define and investigate the concept of genus more fully in §4.16.

Fig. 4.50

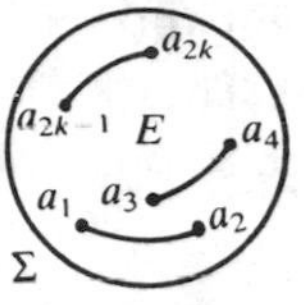

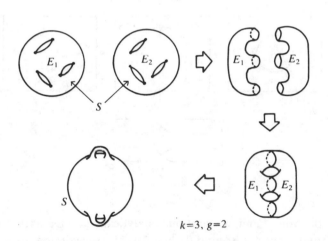

Fig. 4.51

$k=3, g=2$

4.10 Branch-points and the monodromy group

The examples considered so far may have given the impression that at each branch-point of a Riemann surface, *all* the sheets come together at a single point. However, the following example shows that this is not generally so.

The many-valued function $f(z) = \sqrt{(1 + \sqrt{z})}$ has four analytic branches on $E = \Sigma \backslash \{0, 1, \infty\}$, there being two values for $\zeta = \sqrt{z}$ and each such value determining two values for $f(z) = \sqrt{(1 + \zeta)}$. If we take α to be the positive value $+\sqrt{\frac{1}{2}}$ of $\sqrt{\frac{1}{2}}$, then we can label the branches $f_1, \ldots, f_4$ of f near $z = \frac{1}{2}$ so that they take the distinct values $+\sqrt{(1 + \alpha)}$, $+\sqrt{(1 - \alpha)}$, $-\sqrt{(1 + \alpha)}$, $-\sqrt{(1 - \alpha)}$ respectively at $z = \frac{1}{2}$ (notice that $1 \pm \alpha > 0$, so that these values are all real).

If we continue f_1 analytically along a closed path γ from $\frac{1}{2}$ to $\frac{1}{2}$ with $n_\gamma(0) = 1$ and $n_\gamma(1) = 0$ (say $\gamma(s) = \frac{1}{2}e^{2\pi i s}$), then $1 + \zeta$ is transformed from $1 + \alpha$ to $1 - \alpha$, and $f(z) = \sqrt{(1 + \zeta)}$ is transformed from $+\sqrt{(1 + \alpha)} = f_1(\frac{1}{2})$ to $+\sqrt{(1 - \alpha)} = f_2(\frac{1}{2})$, as in Fig. 4.52. Thus analytic continuation around 0 transforms f_1 to f_2, and similarly f_2 to f_1, f_3 to f_4, and f_4 to f_3, so these four branches are permuted according to the permutation $\pi(\gamma) = (12)(34)$ of their indices. We therefore join the corresponding sheets E_1 to E_2, and E_3 to E_4 at two branch-points above 0, so that as $z \in \Sigma$ rotates once around 0 each point $s \in \psi^{-1}(z)$ above z moves from E_1 to E_2, or E_2 to E_1, etc (see Fig. 4.53). Thus $\psi^{-1}(0)$ consists of two points, each a branch-point of order 1 since it joins two sheets.

Fig. 4.52

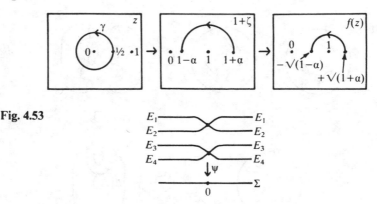

Fig. 4.53

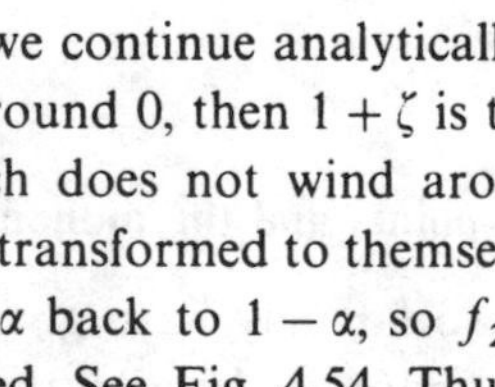

If, on the other hand, we continue analytically along a path δ winding once around 1 but not around 0, then $1 + \zeta$ is transformed from $1 + \alpha$ to $1 + \alpha$ along a path which does not wind around 0, so f_1 and f_3 (the branches of $\sqrt{(1 + \zeta)}$) are transformed to themselves; however, $1 - \zeta$ winds once around 0 from $1 - \alpha$ back to $1 - \alpha$, so f_2 and f_4 (the branches of $\sqrt{(1 - \zeta)}$) are interchanged. See Fig. 4.54. Thus the permutation of the branches induced by continuation around 1 is $\pi(\delta) = (1)(24)(3)$, so sheets E_2 and E_4 are joined at a branch-point of order 1 over $z = 1$, while sheets

E_1 and E_3, which carry analytic branches at $z = 1$, are unbranched. In this case, $\psi^{-1}(1)$ consists of three points: one branch-point where E_2 and E_4 meet, and two points on E_1 and E_3 where S is not branched. This is illustrated in Fig. 4.55.

Finally, suppose that we continue analytically along a closed path ε winding once around ∞ in Σ; this means that, as a path in $\mathbb{C}$, ε winds once around both 0 and 1 (see Fig. 4.56). Both γ and δ follow the positive

Fig. 4.54

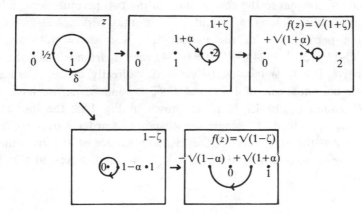

Fig. 4.55

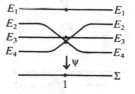

Fig. 4.56

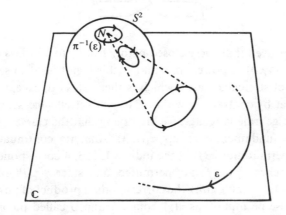

(anti-clockwise) orientation of $\mathbb{C}$, and this induces (by stereographic projection) orientations of S^2 and hence of Σ. For ε to be consistent with this orientation of Σ at ∞, we need its projection $\pi^{-1}(\varepsilon)$ to be consistent with the orientation of S^2 at N, so ε must follow the *negative* (clockwise) orientation of $\mathbb{C}$; for example, we could take

$$\varepsilon = (\gamma\delta)^{-1} = \delta^{-1}\gamma^{-1},$$

as in Fig. 4.57, or any other path in $E = \Sigma \setminus \{0, 1, \infty\}$ homotopic to $\delta^{-1}\gamma^{-1}$. It follows that the permutation of the branches, induced by continuation around ∞, is equal to the composition of the two permutations induced by continuation along δ^{-1} and γ^{-1}, in that order, so the corresponding permutation of the indices $1, \ldots, 4$ is $\pi(\varepsilon) = \pi(\delta)^{-1}\pi(\gamma)^{-1} = (24)^{-1}(34)^{-1}(12)^{-1} = (24)(34)(12) = (1234)$, reading from left to right. (It is instructive, if a little tedious, to verify this directly by considering the analytic continuations of $1 + \zeta$ and $1 - \zeta$ around ∞, as we have done above around 0 and 1.) Thus, as shown in Fig. 4.58, the four sheets $E_1, \ldots, E_4$ are joined at a single branch-point of order 3 over ∞, in the same way as the four sheets of the Riemann surface of $z^{1/4}$ are joined at ∞: we must expect this since $f(z) = \sqrt{(1 + \sqrt{z})}$ behaves like $z^{1/4}$ for large $|z|$.

Fig. 4.57

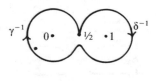

Fig. 4.58

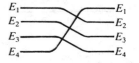

It can be proved that *every* closed path χ from $\frac{1}{2}$ to $\frac{1}{2}$ in E is homotopic to a product $P(\gamma, \delta)$ of powers of γ and δ (this is not hard to see: imagine χ as a piece of stretched elastic which is then released, except at its endpoints, so that by contracting within E it winds itself successively around 0 and 1); in algebraic language, we are saying that the classes $[\gamma]$ and $[\delta]$ generate the fundamental group $\pi_1(E, \frac{1}{2})$. Analytic continuation along χ induces a permutation $\pi(\chi)$ of the indices 1, 2, 3, 4 corresponding to the way the branches $f_1, \ldots, f_4$ are permuted, and since $\chi \simeq P(\gamma, \delta)$ we have $\pi(\chi) = \pi(P(\gamma, \delta)) = P(\pi(\gamma), \pi(\delta))$, the corresponding product of powers of $\pi(\gamma)$ and $\pi(\delta)$. These permutations $\pi(\chi)$ from a group G called the *monodromy*

group of the Riemann surface S; we have shown that G is generated by $\pi(\gamma) = (12)(34)$ and $\pi(\delta) = (24)$, and that there is a homomorphism from $\pi_1(E, \frac{1}{2})$ onto G given by $[\chi] \mapsto \pi(\chi)$. Using the relations $\pi(\gamma)^2 = \pi(\delta)^2 = 1$ and $(\pi(\gamma)\pi(\delta))^4 = \pi(\varepsilon)^{-4} = 1$ we see that G is a dihedral group of order 8 contained in S_4 (see §2.13 for dihedral groups).

In the same way, we can define the monodromy group G of the Riemann surface S of *any* many-valued function to be the group of permutations of the sheets induced by meromorphic continuation along closed paths in Σ. There are many interesting connections between algebraic properties of G and topological properties of S; for example, it is not hard to see that S is connected if and only if G permutes the sheets transitively.

4.11 Abstract Riemann surfaces

In this chapter we have seen how certain many-valued functions f may be represented by single-valued functions $\phi : S \to \Sigma$ which can be regarded as being meromorphic on a domain S, the Riemann surface of f. In earlier chapters we considered the surfaces $S = \Sigma$ and $\mathbb{C}/\Omega$ (Ω a lattice), together with the single-valued meromorphic functions defined on them. Our aim in the rest of this chapter is to present a unified theory of Riemann surfaces and their meromorphic functions, including the above examples as special cases; this will enable us to define the most general domain on which a meromorphic function can be defined.

This theory was developed during the second half of the nineteenth century. Riemann surfaces were introduced by Riemann in his doctoral dissertation *Foundations for a General Theory of Functions of a Complex Variable* in 1851 as essentially topological aids to our understanding of many-valued functions. He stated many powerful results, but his proofs, though illuminating, were not always completely rigorous since the necessary analytic techniques had not yet been fully developed. This omission was eventually repaired, mainly by the Weierstrass, and by the early twentieth century Riemann's theory had been placed on a sound basis. During this process it became clear that Riemann surfaces could be regarded as mathematical objects worthy of study in their own right; this point of view was expressed in F. Klein's book *On Riemann's Theory of Algebraic Functions and their Integrals*, and it eventually lead to the definition of an abstract Riemann surface in H. Weyl's book *The Concept of a Riemann Surface* in 1913. In the preface Weyl states

I shared his [Klein's] conviction that Riemann surfaces are not merely a device for visualizing the many-valuedness of analytic functions, but rather an indispensable essential component of the theory; not a supplement, more or less artificially distilled from the functions, but their native land, the only soil in which the functions grow and thrive.

We begin with the concept of a surface. A *surface* S is a Hausdorff topological space such that every point $s \in S$ has an open neighbourhood U homeomorphic to an open subset of $\mathbb{C}$ (or equivalently $\mathbb{R}^2$); thus S has the same local topological properties as the plane. (Some authors also require S to be connected, but we will not impose this restriction.) More generally, an *n-manifold* is a Hausdorff space in which every point has a neighbourhood homeomorphic to an open subset of $\mathbb{R}^n$; thus a surface is a 2-manifold.

Any surface S is covered by a family of open sets U_i, such that for each U_i there is a homeomorphism $\Phi_i : U_i \to W_i$, where W_i is an open subset of $\mathbb{C}$. We call such a set of pairs $\mathscr{A} = \{(U_i, \Phi_i)\}$ an *atlas* for S; if $s \in U_i$ we call (U_i, Φ_i) a *chart* at s and $z_i = \Phi_i(s)$ a *local coordinate* for s. (Geographic charts and atlases of the Earth's surface are obvious analogues, and are helpful for visualising more abstract cases.)

If (U_i, Φ_i) and (U_j, Φ_j) are charts at $s \in S$ giving local coordinates z_i and z_j for s, then $z_i = (\Phi_i \circ \Phi_j^{-1})(z_j)$ expresses the change in local coordinates for s corresponding to the two different charts. This is illustrated in Fig. 4.59. The functions

$$\Phi_i \circ \Phi_j^{-1} : \Phi_j(U_i \cap U_j) \to \Phi_i(U_i \cap U_j),$$

called the *coordinate transition functions*, are defined whenever $U_i \cap U_j \neq \varnothing$; an atlas $\mathscr{A}$ on S is called *analytic* if all its coordinate transition functions are analytic.

Fig. 4.59

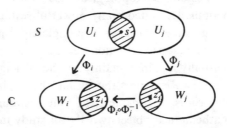

In order to define what is meant by an analytic or meromorphic function on S, without having to specify a particular atlas, we define analytic atlases $\mathscr{A} = \{(U_i, \Phi_i)\}$ and $\mathscr{B} = \{(V_j, \Psi_j)\}$ to be *compatible* if, whenever $(U_i, \Phi_i) \in \mathscr{A}$ and $(V_j, \Psi_j) \in \mathscr{B}$ satisfy $U_i \cap V_j \neq \varnothing$, then

$$\Phi_i \circ \Psi_j^{-1} : \Psi_j(U_i \cap V_j) \to \Phi_i(U_i \cap V_j)$$

is analytic; equivalently, the atlas $\mathscr{A} \cup \mathscr{B}$ is analytic. Compatibility of atlases is an equivalence relation (see Exercise 4F), and an equivalence class of atlases is called a *complex structure* on S. Finally, a surface with a complex structure is called a *Riemann surface*, or sometimes an *abstract Riemann surface* to distinguish it from the Riemann surfaces of specific functions, as constructed in §§4.7–4.10; before showing (in §4.13) that these surfaces are also abstract Riemann surfaces, we shall first give some rather simpler examples. In each case it is sufficient to specify one atlas on S, since this can be taken as a representative for its equivalence class, thus defining a complex structure on S.

(1) Let $S = \mathbb{C}$ and let $\mathscr{A} = \{(\mathbb{C}, \mathrm{id} : \mathbb{C} \to \mathbb{C})\}$, an atlas for $\mathbb{C}$ consisting of a single chart; clearly $\mathscr{A}$ is analytic. There are many compatible atlases, for example $\mathscr{B} = \{(U_i, \mathrm{id} : U_i \to U_i)\}$, where U_i ranges over all open discs in $\mathbb{C}$ of radius 1; every coordinate transition function is the identity and therefore analytic, so $\mathbb{C}$ is a Riemann surface.

(2) Any open subset T of a Riemann surface is itself a Riemann surface. For if $\{(U_i, \Phi_i)\}$ is an analytic atlas on S, and if Ψ_i is the restriction of Φ_i to $U_i \cap T$ (whenever this intersection is non-empty), then $\{(U_i \cap T, \Psi_i)\}$ is an analytic atlas on T. It is easily seen that compatible atlases on S induce compatible atlases on T, so each complex structure on S induces a unique complex structure on T.

(3) Let $S = \Sigma = \mathbb{C} \cup \{\infty\}$, with the topology defined in §1.2, so that S is homeomorphic to the 2-sphere S^2. There is an atlas $\mathscr{A}$ on S consisting of two charts $(U_i, \Phi_i)(i = 1, 2)$: we take $U_1 = \mathbb{C}$ with $\Phi_1 = \mathrm{id} : \mathbb{C} \to \mathbb{C}$, and $U_2 = \Sigma \backslash \{0\}$ with $\Phi_2 = J : \Sigma \backslash \{0\} \to \mathbb{C}$ given by $J(z) = 1/z$ for $z \in \mathbb{C}$ and $J(\infty) = 0$. Clearly $\Sigma = U_1 \cup U_2$, and Φ_1 and Φ_2 are homeomorphisms (Φ_2 is induced by a rotation of S^2). We have $(\Phi_2 \circ \Phi_1^{-1})(z) = 1/z$, which is analytic on $\Phi_1(U_1 \cap U_2) = \mathbb{C} \backslash \{0\}$, and similarly $(\Phi_1 \circ \Phi_2^{-1})(z) = 1/z$ is analytic on $\Phi_2(U_1 \cap U_2) = \mathbb{C} \backslash \{0\}$, so $\mathscr{A}$ is an analytic atlas, giving a complex structure on Σ. The resulting Riemann surface is called the *Riemann sphere*; we shall see that our definitions of concepts for Riemann surfaces are consistent with the way in which they were used in Chapters 1 and 2 for Σ.

(4) In view of its importance, we embody the next example in a theorem:

Theorem 4.11.1. *If Ω is a lattice in $\mathbb{C}$ then $\mathbb{C}/\Omega$ is a Riemann surface.*

(As shown in §3.5, $\mathbb{C}/\Omega$ is homeomorphic to a torus; as part of the following proof, we shall show that $\mathbb{C}/\Omega$ is a surface as defined above.)

Proof. First we show that $\mathbb{C}/\Omega$ is Hausdorff (this was obvious in the previous examples). Recall that $p:\mathbb{C} \to \mathbb{C}/\Omega$ is the projection map $z \mapsto [z] = z + \Omega$, and $U \subseteq \mathbb{C}/\Omega$ is defined to be open if and only if $p^{-1}(U)$ is open in $\mathbb{C}$; thus p is open and continuous.

Let $s_1 = [z_1]$ and $s_2 = [z_2]$ be distinct points in $\mathbb{C}/\Omega$. Since Ω is discrete, there exists

$$\eta = \inf_{\omega\in\Omega} |z_2 - (z_1 + \omega)| > 0,$$

so let V_1 and V_2 be open discs of radius $\eta/2$, centred at z_1 and z_2 respectively. We now show that $(V_1 + \omega) \cap V_2 = \varnothing$ for all $\omega\in\Omega$. For if not then there exist $z\in V_1$ and $\omega\in\Omega$ satisfying $z + \omega\in V_2$, and hence the triangle inequality gives

$$\begin{aligned}
|z_2 - (z_1 + \omega)| &\leqslant |z_2 - (z + \omega)| + |(z + \omega) - (z_1 + \omega)| \\
&= |z_2 - (z + \omega)| + |z - z_1| \\
&< \frac{\eta}{2} + \frac{\eta}{2} \\
&= \eta,
\end{aligned}$$

contradicting the definition of η. Since p is open, $p(V_1)$ and $p(V_2)$ are disjoint open neighbourhoods of s_1 and s_2 in $\mathbb{C}/\Omega$, so $\mathbb{C}/\Omega$ is a Hausdorff space.

We now construct an atlas on $\mathbb{C}/\Omega$. Since Ω is discrete, there exists

$$\delta = \inf_{\omega\in\Omega\setminus\{0\}} |\omega| > 0.$$

If $\mathscr{V}$ is the set of all open discs in $\mathbb{C}$ of diameter at most $\delta/2$, then

(i) $V \cap (V + \omega) = \varnothing$ for all $V\in\mathscr{V}$, $\omega\in\Omega\setminus\{0\}$;

(ii) if $V, V'\in\mathscr{V}$ then V has non-empty intersection with at most one of the translates $V' + \omega$ of $V'(\omega\in\Omega)$.

The triangle inequality immediately gives (i), while for (ii), suppose that $z_1\in V\cap(V' + \omega_1)$ and $z_2\in V\cap(V' + \omega_2)$ with $\omega_1, \omega_2\in\Omega$, say $z_1 = z'_1 + \omega_1$ and $z_2 = z'_2 + \omega_2$ with $z'_1, z'_2\in V'$; then the element $\omega_1 - \omega_2$ of Ω satisfies

$$\begin{aligned}
|\omega_1 - \omega_2| &= |(z_1 - z'_1) - (z_2 - z'_2)| \\
&= |(z_1 - z_2) - (z'_1 - z'_2)| \\
&\leqslant |z_1 - z_2| + |z'_1 - z'_2| \\
&< \frac{\delta}{2} + \frac{\delta}{2} \\
&= \delta,
\end{aligned}$$

so $\omega_1 = \omega_2$ by definition of δ.

By (i), the restriction p_V of p to V is one-to-one; since p is continuous, so is p_V. Moreover, p_V is open: if a subset A of V is relatively open in V then (since V is open) A is open in $\mathbb{C}$, so $p_V(A) = p(A)$ is open in $\mathbb{C}/\Omega$ (since p is open) and hence relatively open in $p_V(V) = p(V)$. Thus p_V is a homeomorphism of V onto its image $p(V)$ ($= U_V$, say), so there is a homeomorphism $\Phi_V = p_V^{-1} : U_V \to V$. The set of all charts (U_V, Φ_V), with $V \in \mathscr{V}$, is therefore an atlas for $\mathbb{C}/\Omega$, so $\mathbb{C}/\Omega$ is a surface.

To show that this atlas is analytic, suppose that (U_V, Φ_V) and $(U_{V'}, \Phi_{V'})$ are charts with $U_V \cap U_{V'} \neq \varnothing$; we must show that the coordinate transition function

$$\Phi_{V'} \circ \Phi_V^{-1} : \Phi_V(U_V \cap U_{V'}) \to \Phi_{V'}(U_V \cap U_{V'})$$

is analytic (see Fig. 4.60). Let $z \in \Phi_V(U_V \cap U_{V'})$ and let $z' = (\Phi_{V'} \circ \Phi_V^{-1})(z)$; then $p_{V'}(z') = \Phi_V^{-1}(z) = p_V(z)$, so $p(z') = p(z)$ and hence $z = z' + \omega$ for some $\omega \in \Omega$ (possibly depending on z). Since $z \in V$ and $z' \in V'$ we have $V \cap (V' + \omega) \neq \varnothing$, and hence ω is independent of z by (ii). Thus $z' = z - \omega$ for a constant $\omega \in \Omega$, so z' is an analytic function of z, as required. $\square$

Fig. 4.60

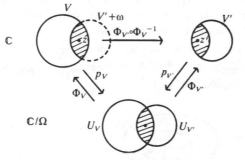

Although different lattices $\Omega \subset \mathbb{C}$ give homeomorphic surfaces $\mathbb{C}/\Omega$, we shall see in §6.1 that they give rise to infinitely many essentially distinct complex structures on the torus.

In future, when we refer to $\mathbb{C}$, Σ or $\mathbb{C}/\Omega$ (or to any of their subsets) as Riemann surfaces, then it should be understood that we are using the complex structures described in the above examples.

The following simple result will be found useful:

Theorem 4.11.2. *If $\mathscr{A} = \{(U_i, \Phi_i)\}$ is an analytic atlas, and if for each i, $\{V_{ij}\}$ is a family of open sets covering U_i, and Ψ_{ij} is the restriction of Φ_i to V_{ij}, then $\mathscr{B} = \{(V_{ij}, \Psi_{ij})\}$ is a compatible analytic atlas.*

Proof. The underlying surface is covered by the sets U_i, and hence by the sets V_{ij}. The coordinate transition functions for the atlas $\mathscr{A} \cup \mathscr{B}$ are either the identity or the restrictions of the coordinate transition functions of the analytic atlas $\mathscr{A}$, so $\mathscr{B}$ is analytic and compatible with $\mathscr{A}$. ☐

Using this result, we can always represent a given complex structure by an atlas in which the open sets are, in some sense, 'arbitrarily small'.

4.12 Analytic, meromorphic and holomorphic functions on Riemann surfaces

If S is a Riemann surface, then a function $f:S \to \mathbb{C}$ is defined to be *analytic* if, for every chart (U, Φ) on S, the function $f \circ \Phi^{-1}:\Phi(U) \to \mathbb{C}$ is analytic on $\Phi(U)$ (in the usual sense of being a differentiable function of a complex variable). See Fig. 4.61. Since Φ is a homeomorphism, it follows that f is continuous on S. If (V, Ψ) is a chart in a compatible atlas on S, with $U \cap V \neq \varnothing$, then $\Phi \circ \Psi^{-1}$ is analytic (again, in the usual sense), and so

$$f \circ \Psi^{-1} = (f \circ \Phi^{-1}) \circ (\Phi \circ \Psi^{-1})$$

is a composition of analytic functions and hence analytic. Thus the above definition of an analytic function $f:S \to \mathbb{C}$ depends only on the complex structure on S, and not on the particular atlas of charts giving rise to that structure. We say that f is analytic at some point $s \in S$ if it is analytic on some open neighbourhood of S, this neighbourhood being a Riemann surface by Example (2) of §4.11.

Fig. 4.61

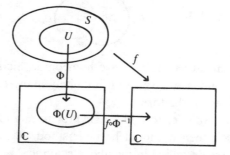

Examples. (1) Any open subset $U \subseteq \mathbb{C}$ is a Riemann surface with a single chart $(U, \mathrm{id}:U \to U)$, and the functions which are analytic on U in the sense defined above are precisely those which are analytic on U in the traditional sense; thus there is no ambiguity in referring to analytic functions on Riemann surfaces contained in $\mathbb{C}$.

(2) The Riemann sphere Σ has two charts $(U_i, \Phi_i)(i = 1, 2)$, as defined

in Example (3) of §4.11. Given $f:\Sigma \to \mathbb{C}$, the functions $f \circ \Phi_i^{-1}:\Phi_i(U_i) \to \mathbb{C}$ have the form

$$f \circ \Phi_1^{-1}:\mathbb{C} \to \mathbb{C}, \quad z \mapsto f(z),$$

and

$$f \circ \Phi_2^{-1}:\mathbb{C} \to \mathbb{C}, \quad z \mapsto f(1/z),$$

so f is analytic on Σ if and only if both $f(z)$ and $f(1/z)$ are analytic on $\mathbb{C}$. For $z \in \mathbb{C} \setminus \{0\}$, $f(1/z)$ is analytic if and only if $f(z)$ is analytic, so f is analytic on Σ if and only if $f(z)$ is analytic on $\mathbb{C}$ and $f(1/z)$ is analytic at $z = 0$. This is precisely the condition given in §1.3 for f to be analytic on Σ, so our two definitions agree for $S = \Sigma$.

Let S_1 and S_2 be Riemann surfaces. Then a continuous function $f:S_1 \to S_2$ is defined to be *holomorphic* if, whenever (U_1, Φ_1) and (U_2, Φ_2) are charts on S_1 and S_2 with $U_1 \cap f^{-1}(U_2) \neq \varnothing$, then the function

$$\Phi_2 \circ f \circ \Phi_1^{-1}:\Phi_1(U_1 \cap f^{-1}(U_2)) \to \mathbb{C}$$

is analytic. See Fig. 4.62. (Notice that since f is continuous, $\Phi_1(U_1 \cap f^{-1}(U_2))$ is an *open* subset of $\mathbb{C}$.) As in the case of analytic functions on Riemann surfaces, one can show quite easily that this definition is independent of the choices of atlases of charts for the complex structures on S_1 and S_2. Moreover, if $S_2 \subseteq \mathbb{C}$ then $f:S_1 \to S_2$ is holomorphic if and only if it is analytic. Now a word of warning: because of the preceding fact, that holomorphic functions are a generalisation of analytic functions, some authors use the word 'analytic' where we have used 'holomorphic', to describe functions between arbitrary Riemann surfaces rather than merely those with co-domain $S_2 \subseteq \mathbb{C}$; this can lead to confusion when the co-domain S_2 is Σ, since a function can then be 'analytic' while having ∞ as a value! For example, Weierstrass' elliptic function $\wp$ induces a holomorphic function $\hat{\wp}:\mathbb{C}/\Omega \to \Sigma$, with $\hat{\wp}([0]) = \infty$; for this reason we have restricted the term 'analytic' to functions $f:S_1 \to S_2 \subseteq \mathbb{C}$.

Fig. 4.62

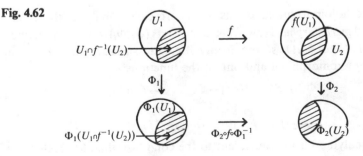

Theorem 4.12.1. *If $f:S_1 \to S_2$ and $g:S_2 \to S_3$ are holomorphic functions between Riemann surfaces, then $g \circ f:S_1 \to S_3$ is holomorphic.*

Proof. By Theorem 4.11.2 we can choose the atlases $\mathcal{A}_i$ on $S_i (i = 1, 2, 3)$ so that for each $(U_1, \Phi_1) \in \mathcal{A}_1$ there exists $(U_2, \Phi_2) \in \mathcal{A}_2$ with $f(U_1) \subseteq U_2$, and for each $(U_2, \Phi_2) \in \mathcal{A}_2$ there exists $(U_3, \Phi_3) \in \mathcal{A}_3$ with $g(U_2) \subseteq U_3$. Since $\Phi_2 \circ f \circ \Phi_1^{-1}$ and $\Phi_3 \circ g \circ \Phi_2^{-1}$ are analytic, so is

$$\Phi_3 \circ g \circ f \circ \Phi_1^{-1} = (\Phi_3 \circ g \circ \Phi_2^{-1}) \circ (\Phi_2 \circ f \circ \Phi_1^{-1}).$$

Now if (V, Ψ) is any chart on S_3 with $(g \circ f)(U_1) \cap V \neq \varnothing$, then

$$\Psi \circ (g \circ f) \circ \Phi_1^{-1} = (\Psi \circ \Phi_3^{-1}) \circ (\Phi_3 \circ g \circ f \circ \Phi_1^{-1})$$

is analytic since both the coordinate transition function $\Psi \circ \Phi_3^{-1}$ and $\Phi_3 \circ g \circ f \circ \Phi_1^{-1}$ are analytic. Clearly $g \circ f$ is continuous, so it is holomorphic. This proof is illustrated in Fig. 4.63. $\square$

Fig. 4.63

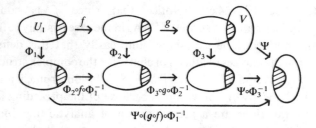

Theorem 4.12.2. *If $\mathbb{C}/\Omega$ is the Riemann surface defined in Theorem 4.11.1, then the projection $p:\mathbb{C} \to \mathbb{C}/\Omega$, $z \mapsto [z] = z + \Omega$, is holomorphic.*

Proof. $\mathbb{C}$ has a single chart $(\mathbb{C}, \Phi)$, with $\Phi = \mathrm{id}:\mathbb{C} \to \mathbb{C}$, while the charts on $\mathbb{C}/\Omega$ have the form (U_V, Φ_V), where $\Phi_V \circ p$ restricts to the identity function $\Phi_V \circ p_V$ on each $V \in \mathcal{V}$. Thus $\Phi_V \circ p \circ \Phi^{-1}:V \to V$ is the identity function, obviously analytic. Since p is continuous, it is holomorphic. $\square$

Just as the holomorphic functions $S \to \mathbb{C}$ are called analytic, those into Σ are called *meromorphic*. Now Σ has two charts (U_i, Φ_i) $(i = 1, 2)$ as defined in Example (3) of §4.11, so if S is any Riemann surface then a function $f:S \to \Sigma$ is meromorphic if and only if the functions

$$\Phi_1 \circ f:S \backslash f^{-1}(\infty) \to \mathbb{C}, \qquad s \mapsto f(s),$$

and

$$\Phi_2 \circ f:S \backslash f^{-1}(0) \to \mathbb{C}, \quad s \mapsto 1/f(s),$$

are both analytic. This is equivalent to the condition that, for each chart

(U, Φ) on S, the function

$$f \circ \Phi^{-1} : \Phi(U) \to \Sigma, \quad z \mapsto f(\Phi^{-1}(z)),$$

is meromorphic (in the usual sense). It is clear that when $S \subseteq \Sigma$, this use of the word 'meromorphic' coincides with that in Chapter 1, so we have:

Example.(1)   The meromorphic functions $f : \Sigma \to \Sigma$ are the rational functions (see Theorem 1.4.1).

(2) Since $\mathbb{C} \subseteq \Sigma$, every analytic function on a Riemann surface is meromorphic.

(3) If Ω is a lattice in $\mathbb{C}$ then the meromorphic functions $f : \mathbb{C}/\Omega \to \Sigma$ can be identified, in the obvious way, with the elliptic functions with respect to Ω. For if $g : \mathbb{C} \to \Sigma$ is such an elliptic function, then there is a well-defined function $f = \hat{g} : \mathbb{C}/\Omega \to \Sigma$, $[z] \mapsto g(z)$; if (U_V, Φ_V) is a chart on $\mathbb{C}/\Omega$ (as defined in the proof of Theorem 4.11.1), then $f \circ \Phi_V^{-1} = f \circ p = g$ on V, and so f is meromorphic. Conversely, if $f : \mathbb{C}/\Omega \to \Sigma$ is meromorphic then $g = f \circ p : z \mapsto f([z])$ is a meromorphic function $\mathbb{C} \to \Sigma$, elliptic with respect to Ω. See Fig. 4.64.

Fig. 4.64

If f and g are meromorphic functions on S then for $s \in S$ we can define

$$(f \pm g)(s) = f(s) \pm g(s),$$
$$(f \cdot g)(s) = f(s) \cdot g(s),$$
$$(f/g)(s) = f(s)/g(s)$$

if g is not identically zero. These functions $f \pm g$, $f \cdot g$ and f/g are meromorphic on S, so the meromorphic functions on S form a field; as we have just seen, the elements of this field are the rational functions if $S = \Sigma$, and they can be identified with the elliptic functions if $S = \mathbb{C}/\Omega$. The analytic functions on S are closed under addition, subtraction and multiplication, so they form a ring contained within the field of meromorphic functions; if S is compact and connected, then this ring consists of the constant functions $S \to \mathbb{C}$ (for $S = \Sigma$ see the proof of Theorem 1.3.3, for $S = \mathbb{C}/\Omega$ see Theorem 3.6.1, and for the general case see Exercise 4K).

4.13 The sheaf of germs of meromorphic functions

In §§4.7–4.10 we constructed the Riemann surfaces S of certain many-valued functions. We now show how this construction may be carried out in general, so that the resulting surface S is an abstract Riemann surface as defined in §4.11, and the maps ψ and $\phi:S\to\Sigma$ are meromorphic in the sense of §4.12. Since the technique is rather abstract, we first outline briefly how the above examples of Riemann surfaces may be given complex structures.

Let T be the unbranched Riemann surface of any of the many-valued functions $w=f(z)$ considered in §§4.7–4.10, and let $\psi:T\to\Sigma$ be the projection map. Each $s\in T$ lying above $\mathbb{C}$ has an open neighbourhood U mapped by ψ homeomorphically onto an open subset of $\mathbb{C}$, so we take $(U,\psi|U)$ to be a chart at s; if s lies above ∞ then we replace ψ by $J\circ\psi$. Whenever two such neighbourhoods intersect, the corresponding co-ordinate transition function has the form $z\mapsto z\ (z\neq\infty)$ or $z\mapsto 1/z\ (z\neq 0)$, and is therefore analytic. With this atlas of charts, T is an abstract Riemann surface, and ψ and $\phi:T\to\Sigma$ are meromorphic. Now the full Riemann surface S consists of T together with any branch-points, and we cannot use ψ to give local coordinates at a branch-point since it is not locally one-to-one. Instead, near a branch-point s_c of order $q-1$ above $c\in\Sigma$ we can represent the locally q-valued function $f(z)$ by a single-valued function $F(\zeta)$ of ζ, where $\zeta=(z-c)^{1/q}$ or $z^{-1/q}$ as $c\in\mathbb{C}$ or $c=\infty$; we then use ζ as a local coordinate at s_c, and since the coordinate transition functions (between ζ and z) are analytic on overlapping neighbourhoods, it follows that S is an abstract Riemann surface.

We now turn to the general case. Rather than construct the various Riemann surfaces individually, we shall construct a *single* abstract Riemann surface $\mathscr{S}$ which is so large that it has, among its subspaces, the Riemann surface of every many-valued meromorphic function! One of the main difficulties, which we now confront, is that of describing the points in these surfaces, that is, of defining precisely the concept of 'a single-valued branch of a many-valued function $w=f(z)$ at a point $z\in\Sigma$'. It is insufficient to define this to be a pair $(z,w)\in\Sigma\times\Sigma$ satisfying $w=f(z)$ (so that

Fig. 4.65

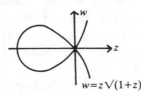

$w=z\sqrt{(1+z)}$

the Riemann surface of $w = f(z)$ is identified with the graph of f), since two branches of f may agree *at* z but not *near* z: for example, $f(z) = z\sqrt{(1 + z)}$ has two branches with the same value at $z = 0$, so the graph has one point at $z = 0$, as in Fig. 4.65, whereas the Riemann surface has two.

Given $a \in \Sigma$, let $\mathscr{F}_a$ be the set of all functions which are meromorphic in some open neighbourhood of a. If $f, g \in \mathscr{F}_a$ then we write $f \sim_a g$ if $f \equiv g$ in some neighbourhood of a; clearly this is an equivalence relation on $\mathscr{F}_a$, and the equivalence class of f, denoted by $[f]_a$, is called the *germ of* f *at* a. We call this germ *analytic* if f is analytic at a, that is, $f(a) \neq \infty$.

We can think of the germs $[f]_a (a \in \Sigma)$ as representing single-valued meromorphic functions defined near a, since by definition we have $[f]_a = [g]_b$ if and only if $a = b$ and $f \equiv g$ near a (in which case, by Lemma 4.1.3, $f \equiv g$ on any region on which both f and g are meromorphic). The set $\mathscr{M}$ of all germs $[f]_a$, for all $a \in \Sigma$, is called the *sheaf of germs of meromorphic functions*; we shall show that $\mathscr{M}$ is an abstract Riemann surface, and that the unbranched Riemann surface T of any many-valued function may be identified with an open subspace of $\mathscr{M}$, so that by Example (2) of §4.11 T is an abstract Riemann surface.

First we define a topology on $\mathscr{M}$, the germs close to $[f]_a$ being the germs $[f]_b$ at points b close to a in Σ. If $a \in \Sigma$ then a *disc centred at* a is an open disc $D = \{z \in \mathbb{C} \,|\, |z - a| < \varepsilon\}$ $(\varepsilon > 0)$ if $a \in \mathbb{C}$, or an open disc $D = \{z \in \mathbb{C} \,|\, |z| > \varepsilon\} \cup \{\infty\} (\varepsilon > 0)$ if $a = \infty$; in either case, the set $\mathscr{D}_a$ of all discs centred at a has the property that if $D, E \in \mathscr{D}_a$ then either $D \subseteq E$ or $E \subseteq D$. If $m = [f]_a$ is a germ in $\mathscr{M}$, then f is meromorphic on some $D \in \mathscr{D}_a$, and we write $m = [D, f]_a$; by Lemma 4.1.3, m determines f uniquely on D. In this case we define the *D-neighbourhood* $D(m)$ of m in $\mathscr{M}$ to be $D(m) = \{[f]_b \,|\, b \in D\}$. A subset $A \subseteq \mathscr{M}$ is defined to be *open* if for each $m \in A$ there is some D-neighbourhood $D(m) \subseteq A$. Clearly, $\mathscr{M}$ is open, as is any union of open sets, and the empty set is open since it contains no germs m which can violate the definition; to show that we have a topology on $\mathscr{M}$ it remains to show that if A and B are open then so is $A \cap B$. Any $m \in A \cap B$ has the form $m = [D, f]_a$ where $D \in \mathscr{D}_a$ and $D(m) \subseteq A$, and similarly $m = [E, g]_a$ where $E \in \mathscr{D}_a$ and $E(m) \subseteq B$; thus $[f]_b \in A$ for all $b \in D$, and $[g]_b \in B$ for all $b \in E$. Now $D \cap E$ is a disc $F \in \mathscr{D}_a$, on which both f and g are meromorphic; since $[f]_a = m = [g]_a$ we have $f \equiv g$ near a and hence $f \equiv g$ on F by Lemma 4.1.3. Thus $[f]_b = [g]_b \in A \cap B$ for all $b \in F$, so $A \cap B$ contains $F(m)$ and is therefore open.

Given $m = [f]_a \in \mathscr{M}$, we define $\psi(m) = a$. To show that $\psi : \mathscr{M} \to \Sigma$ is continuous, take any open set $U \subseteq \Sigma$ and let $m \in \psi^{-1}(U)$, so $a = \psi(m) \in U$. Since U is open, there exists $D \in \mathscr{D}_a$ such that $m = [D, f]_a$ and $D \subseteq U$.

Each element of $D(m)$ has the form $[f]_b$ for some $b \in D$, so $\psi([f]_b) = b \in U$. Thus $D(m) \subseteq \psi^{-1}(U)$, so $\psi^{-1}(U)$ is open and ψ is continuous.

Using ψ we can show that $\mathcal{M}$ is a Hausdorff space. If $m = [f]_a$ and $n = [g]_b$ are distinct elements of $\mathcal{M}$, then either $a \neq b$, or else $a = b$ but there is no $D \in \mathcal{D}_a$ on which $f \equiv g$. If $a \neq b$ we can choose disjoint open neighbourhoods U and V for a and b in Σ (which is Hausdorff), and then $\psi^{-1}(U)$ and $\psi^{-1}(V)$ are disjoint open neighbourhoods of m and n in $\mathcal{M}$, as required. If $a = b$ we can choose some $D \in \mathcal{D}_a$ on which both f and g are meromorphic, so $f \not\equiv g$ on D; then $D(m)$ and $D(n)$ are open neighbourhoods of m and n in $\mathcal{M}$. If these are not disjoint, then $[f]_c = [g]_c$ for some $c \in D$, so $f \equiv g$ in some neighbourhood U of c in D, and hence $f \equiv g$ on D by Lemma 4.1.3. This contradiction shows that $D(m) \cap D(n)$ is empty, so $\mathcal{M}$ is a Hausdorff space.

It is clear that the restriction $\psi_{D,m}$ of ψ to a D-neighbourhood $D(m)$ of m maps $D(m)$ homeomorphically onto the open set $D \subseteq \Sigma$. If $\psi(m) \neq \infty$ then $D \subseteq \mathbb{C}$ and we can use $(D(m), \psi_{D,m})$ as a chart at m. If $\psi(m) = \infty$ then $J \circ \psi_{D,m} : [f]_b \mapsto 1/b$ maps $D(m)$ homeomorphically onto an open disc in $\mathbb{C}$ (centred at 0), so we can use $(D(m), J \circ \psi_{D,m})$ as a chart. Each coordinate transition function is either the identity function on a subset of $\mathbb{C}$ or else the restriction of J to a subset of $\mathbb{C} \setminus \{0\}$; these functions are analytic, so with this atlas of charts $\mathcal{M}$ is an abstract Riemann surface, and it is easily seen that $\psi : \mathcal{M} \to \Sigma$ is meromorphic (in the sense defined in §4.12).

Given $m = [f]_a \in \mathcal{M}$, the value $f(a)$ is independent of the particular choice of f, so we have a function $\phi : \mathcal{M} \to \Sigma$ given by $\phi(m) = f(a)$. To show that ϕ is meromorphic we show that ϕ maps the various local coordinates on $\mathcal{M}$ meromorphically into Σ. First consider a chart on $\mathcal{M}$ of the form $(D(m), \psi_{D,m})$ where $m = [D, f]_a$ and $a = \psi(m) \neq \infty$, so $D \subseteq \mathbb{C}$. Since f is meromorphic on D, the transformation of coordinates $\phi \circ \psi_{D,m}^{-1} : D \to \Sigma$, $z \mapsto f(z)$ is meromorphic. See Fig. 4.66. If, on the other hand, we consider a chart $(D(m), J \circ \psi_{D,m})$ where $m = [D, f]_\infty$, then the transformation of coordinates is given by $\phi \circ (J \circ \psi_{D,m})^{-1} = \phi \circ \psi_{D,m}^{-1} \circ J : J(D) \to \Sigma$, $z \mapsto f(1/z)$, and this is meromorphic on $J(D)$ (since $0 \notin J(D)$ and f is meromorphic on D). Thus $\phi : \mathcal{M} \to \Sigma$ is meromorphic.

Using $\mathcal{M}$, we have a simple description of meromorphic continuation

Fig. 4.66

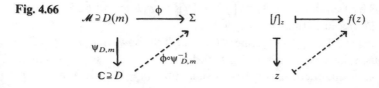

along paths. Suppose that $m \in \mathcal{M}$ and γ is a path in Σ starting at $\psi(m)$. If some meromorphic function element in m can be continued along γ, then at each point $\gamma(t)$, $t \in I$, this continuation gives a germ $\tilde{\gamma}(t)$ satisfying $\psi(\tilde{\gamma}(t)) = \gamma(t)$, as in Fig. 4.67; by Theorem 4.4.1, $\tilde{\gamma}(t)$ depends only on m, γ and t, and not on the method of continuation along γ. The function $\tilde{\gamma} : I \to \mathcal{M}$, called the *meromorphic continuation of m along γ*, is continuous, so it is a path in $\mathcal{M}$ starting at m: if A is an open subset of $\mathcal{M}$ and $t_0 \in \tilde{\gamma}^{-1}(A)$, say $\tilde{\gamma}(t_0) = n = [f]_a$ where $a = \gamma(t_0)$, then since A is open we have $D(n) \subseteq A$ for some $D \in \mathcal{D}_a$; for all t sufficiently close to t_0 we have $\gamma(t) \in D$ (as in Fig. 4.68) and $\tilde{\gamma}(t) = [f]_{\gamma(t)} \in D(n) \subseteq A$, so $t \in \tilde{\gamma}^{-1}(A)$ and hence $\tilde{\gamma}$ is continuous. Conversely, if $\tilde{\gamma}$ is any path in $\mathcal{M}$ starting at m, then there is a path $\gamma = \psi \circ \tilde{\gamma}$ in Σ starting at $\psi(m)$, and the germs $\tilde{\gamma}(t)$, $t \in I$, give a meromorphic continuation of $m = \tilde{\gamma}(0)$ along γ.

Fig. 4.67

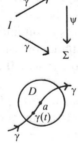

Fig. 4.68

This gives us an equivalence relation on $\mathcal{M}$, two germs m and n being related if n can be obtained by meromorphic continuation of m along some path γ in Σ, or, equivalently, if there is a path $\tilde{\gamma}$ in $\mathcal{M}$ from m to n. The equivalence classes partition $\mathcal{M}$ into disjoint sets which, being path-connected, are connected; it is easily seen that these sets are open, and hence also closed (their complements, being unions of open sets, are open), so they must be the connected components of $\mathcal{M}$, each one an abstract Riemann surface by Example (2) of §4.11. The component $\mathcal{M}(m)$ containing m is called the *unbranched Riemann surface of m*; for example, if m represents a branch of $z^{1/q}$ at some point in $\mathbb{C} \setminus \{0\}$, then $\mathcal{M}(m)$ can be identified with the unbranched Riemann surface T of $z^{1/q}$ (see §4.8), since the elements of T correspond to the different branches of $z^{1/q}$ at points in $\mathbb{C} \setminus \{0\}$, that is, the germs n which can be obtained by continuation of m.

If $A(z,w)$ is a single-valued function of two variables z and w, then the *unbranched Riemann surface $\mathcal{M}_A$* of the equation $A(z,w) = 0$ is defined to be the largest open subset of $\mathcal{M}$ on which $A(\psi(m), \phi(m)) = 0$. Thus $\mathcal{M}_A$ consists of the germs $m = [D, f]_a$ such that z and $w = f(z)$ satisfy $A(z, w) = 0$ for all $z \in D$. (Notice that the condition that $\mathcal{M}_A$ is *open* requires $A(z, w)$ to vanish not just *at* a, but also in some D-neighbourhood of a.) In §§4.7–4.10 we took $A(z, w) = e^w - z$, $w^q - z$, $w^2 - p(z)$ or $w^4 - 2w^2 + 1 - z$; then $\mathcal{M}_A$ can be identified with the unbranched Riemann surface T of $\log z$, $z^{1/q}$, $\sqrt{p(z)}$ or $\sqrt{(1 + \sqrt{z})}$ respectively, as constructed in those sections. With one exception, in each of those examples $\mathcal{M}_A$ was a single component of $\mathcal{M}$ since it could be obtained by meromorphic continuation of a single germ; the exceptional case was $\sqrt{p(z)}$ where $p(z) = (z - a)^2$, there being two components, each conformally equivalent under ψ to Σ, corresponding to the two branches $\pm(z - a)$ of $\sqrt{p(z)}$.

In general, suppose that $A(z, w)$ is such that, whenever a function $f(z)$ is meromorphic at a point $a \in \Sigma$, then the function $g(z) = A(z, f(z))$ is meromorphic at a (this is always satisfied if A is a polynomial in z and w). If $m = [f]_a$ is in $\mathcal{M}_A$ then $g \equiv 0$ near a. Now continuation of m along a path γ in Σ starting at a induces a continuation of g along γ; since $g \equiv 0$ near a, the zero function is a continuation of g along γ, and this continuation is unique by Theorem 4.4.1, so $A(z, w) \equiv 0$ along γ; this shows that $\mathcal{M}_A$ contains the entire component $\mathcal{M}(m)$, so $\mathcal{M}_A$ is a union of components of $\mathcal{M}$. This argument, known as the *principle of permanence of identical relations*, does not apply directly to the function $A(z, w) = e^w - z$, since g will not be meromorphic at poles of f; however, as we saw in §4.1, any germ of $\log z$ may be continued analytically throughout $\mathbb{C} \setminus \{0\}$, so the problem does not arise.

For later applications we need a condition on a set $\mathcal{G}$ of germs which ensures that any germ in $\mathcal{G}$ can be continued meromorphically along all paths in some region E, using only germs in $\mathcal{G}$. For any region $E \subseteq \Sigma$, we define a subset $\mathcal{G} \subseteq \mathcal{M}$ to be *sufficient for continuation within E* if, for each $a \in E$,

  (i) the set $\mathcal{G}_a = \mathcal{G} \cap \psi^{-1}(a)$ of germs in $\mathcal{G}$ at a is non-empty;

  (ii) the germs in $\mathcal{G}_a$ all have the form $[D_a, f]_a$ for some *common* $D_a \in \mathcal{D}_a, D_a \subseteq E$;

  (iii) $\bigcup_{b \in D_a} \mathcal{G}_b = \bigcup_{m \in \mathcal{G}_a} D_a(m)$, that is, the germs in $\mathcal{G}_b$ (for $b \in D_a$) are precisely the germs $[f]_b$ induced at b by the germs $[D_a, f]_a$ in $\mathcal{G}_a$.

(For examples, see the proofs of Theorems 4.14.3, 14.18.1 and 6.6.4.)

Lemma 4.13.1. *If $\mathcal{G}$ is sufficient for continuation within E, then each germ in $\mathcal{G}$ can be continued along any path γ in E, and the resulting germ $\bar{\gamma}(t)$ at each point $\gamma(t)$ is in $\mathcal{G}$.*

Proof. Let $m \in \mathcal{G}_{\gamma(0)}$, and let t_1 be the supremum of all $t \in I$ such that m can be continued along γ as far as $\gamma(t)$, using germs in $\mathcal{G}$. Clearly $t_1 > 0$ since some initial segment of γ must lie in the domain $D_{\gamma(0)}$ of m. If we put $a = \gamma(t_1)$, then by condition (ii) the germs in $\mathcal{G}_a$ have a common domain $D_a \in \mathcal{D}_a$, $D_a \subseteq E$ (consult Fig. 4.69). By continuity of γ we can choose $t_2 < t_1$ sufficiently close to t_1 that $\gamma([t_2, t_1]) \subseteq D_a$. Since $t_2 < t_1$ there is a continuation of m along γ, using germs in $\mathcal{G}$, as far as $b = \gamma(t_2)$. In particular, the resulting germ $\tilde{\gamma}(t_2) = [D_b, g]_b$ is in $\mathcal{G}$, so by condition (iii) it has the form $[f]_b$ for some germ $[D_a, f]_a \in \mathcal{G}_a$. Then $f \equiv g$ near b, so $f \equiv g$ on the region $D_a \cap D_b$ and hence we can use the function element (D_a, f) to extend the continuation of m further along γ. If $t_1 < 1$ then this continues m to points $\gamma(t) \in D_a$ with $t > t_1$, contradicting the definition of t_1. Hence $t_1 = 1$ and we have continued m as far as $\gamma(1)$. $\square$

Fig. 4.69

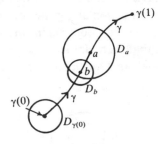

Corollary 4.13.2. *If, in addition, E is simply connected, then each $m \in \mathcal{G}$ extends to a single-valued meromorphic function f on E, with $[f]_a \in \mathcal{G}$ for all $a \in E$.*

Proof. This follows immediately from Lemma 4.13.1 and the monodromy theorem 4.5.3. $\square$

It remains for us to describe how to adjoin branch-points to $\mathcal{M}$. It is helpful to recall how we did this in §4.8 to the Riemann surface of $z^{1/q}$ at $z = 0$; we shall translate this process (typical of all branch-points) into the language of abstract Riemann surfaces. Once one has grasped the basic idea (the choice of local coordinates) there are no great difficulties; however, it is a lengthy process to verify rigorously all the details (such as the fact that the extended space is Hausdorff) so we will merely give a brief outline, in the hope that the reader can fill in the gaps. (For a more complete account, see Springer [1957].)

For convenience, we first consider branch-points at 0. Let D be a disc

$|z| < \varepsilon$, and let $a \in E = D \setminus \{0\}$. Suppose that a germ $m_0 = [f_0]_a$ can be continued analytically along all paths in E. If γ is a closed path from a to a in E, with winding number $n_0(\gamma) = 1$ about 0 (see §4.6), then by continuing m_0 analytically n times around γ ($n \in \mathbb{Z}$) we obtain a sequence of germs $m_n = [f_n]_a$ at a (see Fig. 4.70). Suppose that $m_n = m_0$ for some $n > 0$ (this happens if m_0 represents a branch of $z^{1/q}$, but not in the case of $\log(z)$); if q is the least positive integer n satisfying $m_n = m_0$, then it is easily seen that the sequence of germs has the form $\ldots, m_0, m_1, \ldots, m_{q-1}, m_0, m_1, \ldots$, repeating itself with period q, with the germs $m_0, \ldots, m_{q-1}$ all distinct. Now E is homeomorphic to $\mathbb{C} \setminus \{0\}$, and it follows easily from Theorem 4.6.1 that the fundamental group $\pi_1(E)$ is generated by the homotopy class $[\gamma]$; thus each closed path δ from a to itself in E is homotopic to a power of γ, so $m_0, \ldots, m_{q-1}$ are the only germs at a which can be obtained by continuation within E. Continuation along a path from a to some point $z \in E$ must transform distinct germs at a into distinct germs at z (otherwise continuation along the reverse path is not unique), so we obtain at least q germs at z from m_0 in this way. Interchanging the roles of a and z we see that there are *exactly* q germs $[g_0]_z, \ldots, [g_{q-1}]_z$ at z, so we have a q-valued analytic function $f(z)$ on E, the branches at z being $g_0, \ldots, g_{q-1}$.

Fig. 4.70

We now represent $f(z)$ as a *single*-valued analytic function $F(\zeta)$ of $\zeta = z^{1/q}$ near $\zeta = 0$. Let $\tilde{E} = \{\zeta \in \mathbb{C} \mid 0 < |\zeta^q| < \varepsilon\}$, define $\theta : \tilde{E} \to E$ by $\theta(\zeta) = \zeta^q$, and choose any $\tilde{a} \in \theta^{-1}(a)$. The function $F_0(\zeta) = (f_0 \circ \theta)(\zeta) = f_0(\zeta^q)$ is analytic near $\tilde{a}$, giving an analytic germ $\tilde{m}_0 = [F_0]_{\tilde{a}}$ at $\tilde{a}$. We can continue $\tilde{m}_0$ analytically along all paths in $\tilde{E}$, giving an analytic function $F(\zeta) = (f \circ \theta)(\zeta) = f(\zeta^q)$ on $\tilde{E}$ (see Fig. 4.71). To show that F is single-valued it is sufficient to consider a closed path $\tilde{\gamma}$ from $\tilde{a}$ to itself in $\tilde{E}$, with winding number $n_0(\tilde{\gamma}) = 1$. As ζ follows $\tilde{\gamma}$ once around 0 in $\tilde{E}$, $z = \theta(\zeta) = \zeta^q$ travels q times around 0 in E,

Fig. 4.71

so $f(z)$ passes successively through the germs $m_0, m_1, \ldots$ at a, starting with m_0 and finishing with $m_q = m_0$. Thus $\tilde{m}_0$ is unchanged by continuation around $\tilde{\gamma}$, and since $[\tilde{\gamma}]$ generates $\pi_1(\tilde{E})$, F is single-valued on $\tilde{E}$.

Being single-valued and analytic on the punctured disc $\tilde{E}$, F has a Laurent expansion $F(\zeta) = \sum_{r=-\infty}^{\infty} c_r \zeta^r$ on $\tilde{E}$, so putting $\zeta = z^{1/q}$ we have a *Puiseux series* (V. Puiseux, 1820–83)

$$f(z) = \sum_{r=-\infty}^{\infty} c_r z^{r/q}$$

on E, different values of $z^{1/q}$ giving the different branches of $f(z)$. Suppose that there exists N with $c_N \neq 0$ and $c_r = 0$ for all $r < N$; then F is meromorphic at 0, with $F(0) = c_0$ or ∞ as $N \geqslant 0$ or $N < 0$. If $q = 1$ then $f = F$ is single-valued and meromorphic at 0, so $\mathcal{M}$ contains a germ $[f]_0$ and there is no branch-point to be adjoined; we therefore assume that $q > 1$, in which case we adjoin a point $[f]_0$ to $\mathcal{M}$, called a *branch-point of order* $q - 1$ at 0. Similarly we can adjoin branch-points to $\mathcal{M}$ at other points $c \in \Sigma$: we put $\zeta^q = z - c$ if $c \in \mathbb{C}$, and $\zeta^q = 1/z$ if $c = \infty$; in each case, ζ is called a *local uniformising parameter* (for example in §4.10 we put $\zeta^2 = z$ near the two branch-points of order 1 for $\sqrt{(1 + \sqrt{z})}$ at 0). We define $\mathcal{S}$ to be the union of $\mathcal{M}$ and all branch-points which can be adjoined to $\mathcal{M}$ in this way, and we extend ψ and ϕ to functions $\mathcal{S} \to \Sigma$ by defining $\psi([f]_c) = c$ for a branch-point $[f]_c$ at $c \in \Sigma$, and $\phi([f]_c) = F(0)$ where $F(\zeta)$ is as defined above. If $D \in \mathcal{D}_c$ we define the D-neighbourhood $D([f]_c)$ to consist of $[f]_c$ together with the q germs $m_0, \ldots, m_{q-1}$ representing branches of f at each $a \in D \setminus \{0\}$ (this is consistent with our definition of D-neighbourhoods on $\mathcal{M}$, putting $q = 1$), and we define a subset $A \subseteq \mathcal{S}$ to be *open* if it contains a D-neighbourhood of each of its points. It is straightforward, if a little tedious, to show that $\mathcal{S}$ is a Hausdorff space, $\mathcal{M}$ is imbedded as an open subspace, and the set $\mathcal{S} \setminus \mathcal{M}$ of branch-points is discrete.

We make $\mathcal{S}$ into an abstract Riemann surface by using ζ for local coordinates near branch-points $[f]_c$. If $c = 0$ we assign the coordinate $\zeta \in \tilde{E}$ to the unique germ $[g_j]_z$ at $z = \zeta^q$ satisfying $g_j \circ \theta = F$ near ζ, while $[f]_c$ itself is given the coordinate $\zeta = 0$. Thus the q roots ζ of $\zeta^q = z$ correspond to the q germs $[g_j]_z$ at z, and it is not hard to show that the map $[g_j]_z \mapsto \zeta$ gives a homeomorphism between $D([f]_c)$ and the open disc $\tilde{D} = \tilde{E} \cup \{0\}$. The only charts overlapped by this chart are those on $\mathcal{M}$, so the coordinate transition functions have the form $\zeta \mapsto z = \zeta^q$ and $z \mapsto \zeta = z^{1/q}$, analytic since $z \neq 0$. Similarly we can use $\zeta = (z - c)^{1/q}$ or $z^{-1/q}$ for local coordinates at $c \neq 0$, so $\mathcal{S}$ is an abstract Riemann surface, and one easily shows that ψ and $\phi : \mathcal{S} \to \Sigma$ are meromorphic.

Each branch-point $[f]_c$ is attached to a *unique* connected component $\mathscr{M}(m)$ of $\mathscr{M}$, where $m \in D([f]_c)$; thus the connected components of $\mathscr{S}$ each consist of a single component of $\mathscr{M}$, together with any adjoined branch-points, so we have a bijection between components $\mathscr{S}(m)$ of $\mathscr{S}$ and components $\mathscr{M}(m)$ of $\mathscr{M}$, with $\mathscr{M}(m) \subseteq \mathscr{S}(m)$. We call $\mathscr{S}(m)$ the *complete global continuation*, or *branched Riemann surface* of m. Similarly, the (branched) *Riemann surface* $\mathscr{S}_A$ of $A(z,w) = 0$ is the largest open subset of $\mathscr{S}$ on which $A(\psi, \phi) = 0$; we have $\mathscr{M}_A = \mathscr{M} \cap \mathscr{S}_A$, so $\mathscr{S}_A$ consists of the unbranched Riemann surface $\mathscr{M}_A$ together with any attached branch-points.

4.14 The Riemann surface of an algebraic function

We say that w is an *algebraic function* of z (possibly many-valued) if the relationship between w and z has the form $A(z,w) = 0$ for some polynomial $A(z,w)$; we have already met examples of algebraic functions in §§4.8–4.10, for instance $A(z,w) = w^4 - 2w^2 + 1 - z$ in §4.10 corresponds to the many-valued function $w = \sqrt{(1 + \sqrt{z})}$. We shall prove that the Riemann surface $\mathscr{S}_A$ of an algebraic function is always compact.

We can factorise A uniquely as a product of powers of finitely many distinct irreducible polynomials $A_i(z,w)$ $(1 \leqslant i \leqslant r)$. Since $A_i = 0$ implies $A = 0$, the unbranched Riemann surface $\mathscr{M}_A$ contains each $\mathscr{M}_{A_i}$. Conversely, if some germ $m = [D, f]_a \in \mathscr{M}$ lies in $\mathscr{M}_A$, then $A(z, f(z)) \equiv 0$ on D, so each $z \in D$ satisfies $A_i(z, f(z)) = 0$ for some i depending on z; if we choose a sequence of points in D with limit a, then there is a subsequence $z_n \to a$ with $A_i(z_n, f(z_n)) = 0$ for all n and for some *fixed* i, so it follows from Theorem 1.3.1 (suitably adapted for meromorphic functions) that $A_i(z, f(z)) \equiv 0$ on D and hence $m \in \mathscr{M}_{A_i}$. Thus $\mathscr{M}_A = \bigcup_{i=1}^r \mathscr{M}_{A_i}$, and we can show that this union is disjoint: if $i \neq j$ then since A_i and A_j are co-prime, they have a common zero (z,w) for only finitely many $z \in \mathbb{C}$ (see Theorem A.14), and it follows that no germ $m = [D, f]_a$ can lie in $\mathscr{M}_{A_i} \cap \mathscr{M}_{A_j}$ since we cannot have $A_i(z, f(z)) \equiv 0 \equiv A_j(z, f(z))$ on D. Each $\mathscr{M}_{A_i}$ is a union of connected components of $\mathscr{M}$, and adding any branch-points we see that $\mathscr{S}_A$ is the disjoint union of the surfaces $\mathscr{S}_{A_i}$ $(1 \leqslant i \leqslant r)$; in fact we shall see later that each $\mathscr{S}_{A_i}$ (resp. $\mathscr{M}_{A_i}$) is connected, and is therefore a connected component of $\mathscr{S}$ (resp. $\mathscr{M}$). Now $\mathscr{S}_A$ will be compact provided each $\mathscr{S}_{A_i}$ is compact; *we therefore assume from now on that A is irreducible.*

Collecting powers of w, we can write

$$A(z, w) = a_0(z)w^n + a_1(z)w^{n-1} + \ldots + a_n(z), \qquad (4.14.1)$$

where each $a_i(z)$ is a polynomial in z, and $a_0(z) \not\equiv 0$; we call n the *degree* of A (in w). In general, for fixed $z \in \Sigma$ the equation $A(z, w) = 0$ is a polynomial equation with n distinct roots w; we call such values of z *regular points* for A. The exceptional values of z form the set C_A of *critical points*, satisfying one or more of the conditions:

(i) $z = \infty$;
(ii) $a_0(z) = 0$;
(iii) $A(z, w) = 0$ has a repeated root w.

Clearly, there are only finitely many $z \in \Sigma$ satisfying (i) or (ii). Since A is irreducible and $\partial A / \partial w$ has degree $n - 1$, A and $\partial A / \partial w$ are co-prime; it follows from Theorem A.14 that they have a common root w for only finitely many $z \in \mathbb{C}$, so C_A is finite. We shall show that the branch-points of $\mathscr{S}_A$ all lie above critical points, so they are finite in number.

At a regular point $a \in \Sigma \setminus C_A$, the equation $A(a, w) = 0$ has distinct simple roots $w = w_1, \ldots, w_n$; the next result shows that for z near a, the roots of $A(z, w) = 0$ remain simple and distinct, and vary analytically with respect to z.

Lemma 4.14.2.   *If $a \in \Sigma \setminus C_A$ then there exist $D \in \mathscr{D}_a$ and analytic function elements (D, f_i) $(1 \leqslant i \leqslant n)$ satisfying:*

(i) $f_i(a) = w_i$ for $i = 1, \ldots, n$;
(ii) *for each $z \in D$ the solutions of $A(z, w) = 0$ are $w = f_i(z)$ $(i = 1, \ldots, n)$, all simple and distinct.*

Proof.   We can choose small circular paths γ_i in $\mathbb{C}$ $(i = 1, \ldots, n)$, each γ_i winding once around w_i but not passing through or winding around any other root $w_j (j \neq i)$. For each i, the integral

$$I_i(z) = \frac{1}{2\pi i} \int_{\gamma_i} \frac{\dfrac{\partial A}{\partial w}(z, w)}{A(z, w)} \, dw$$

exists at $z = a$ (since $A(a, w)$ is analytic and non-zero on γ_i), and since we can differentiate under the integral sign $I_i(z)$ is an analytic and hence continuous function of z in some neighbourhood N_i of a. By the calculus of residues (as used in the proof of Theorem 3.6.4 for example), $I_i(z)$ is the number of zeros of $A(z, w)$ enclosed by γ_i minus the number of poles (counting multiplicities); since $A(z, w)$ is a polynomial, there are no poles, so $I_i(z)$ is the number of zeros. Thus $I_i(z)$ is integer-valued, and therefore, being continuous, it must be constant on N_i. By construction, γ_i encloses a single, simple root $w = w_i$ of $A(a, w) = 0$, so $I_i(a) = 1$ and hence $I_i(z) = 1$

for all $z \in N_i$; thus within γ_i there is a unique simple root of $A(z, w) = 0$, which we shall denote by $w = f_i(z)$. By uniqueness of the solution of $A(a, w) = 0$ enclosed by γ_i, we have $f_i(a) = w_i$. If we choose $D \in \mathcal{D}_a$ such that $D \subseteq \bigcap_i N_i$, then for any $z \in D$ the polynomial equation $A(z, w) = 0$ has at most n roots w; we have produced roots $w = f_i(z)(1 \leqslant i \leqslant n)$, mutually distinct since they lie within the disjoint interiors of the circles γ_i, so (ii) is proved. Finally, the calculus of residues (as used in the proof of Theorem 3.6.7, for instance) gives

$$f_i(z) = \frac{1}{2\pi i} \int_{\gamma_i} \frac{w \dfrac{\partial A}{\partial w}(z, w)}{A(z, w)} \, dw,$$

showing that $f_i(z)$ is analytic on N_i and hence on D.   □

Thus, at each $a \in \Sigma \setminus C_A$ we have n distinct analytic germs representing the solutions of $A(z, w) = 0$ for z near a; these germs are the elements of $\mathcal{M}_A$ projected onto a by ψ. The set $\mathcal{G}$ of all such germs (for $a \in \Sigma \setminus C_A$) is easily seen to be sufficient for continuation within $\Sigma \setminus C_A$ (in the sense of §4.13), so Corollary 4.13.2 immediately gives:

Theorem 4.14.3.   *If E is any simply connected region in $\Sigma \setminus C_A$ then there are single-valued analytic functions $f_1(z), \ldots, f_n(z)$ on E such that for any $z \in E$ the solutions of $A(z, w) = 0$ are $w = f_i(z)(1 \leqslant i \leqslant n)$, all simple and distinct.*   □

This justifies our earlier use of cuts in Σ to produce a simply connected region on which a many-valued function has single-valued branches.

We now show that above critical points $c \in C_A$, the worst that can happen is that there are branch-points or poles. In the following discussion it may be helpful to bear in mind the example $A(z, w) = w^4 - 2w^2 + (1 - z)$ corresponding to $w = \sqrt{(1 + \sqrt{z})}$ in §4.10; by considering the common roots w of A and of $\partial A/\partial w = 4w(w - 1)(w + 1)$, it is easily seen that $C_A = \{\infty, 0, 1\}$ in this case.

First we assume that $c \neq \infty$, so either $a_0(c) = 0$ or else $A(c, w) = 0$ has a repeated root. Since C_A is finite, we can choose $\varepsilon > 0$ so that the disc $D = \{z \in \mathbb{C} \mid |z - c| < \varepsilon\}$ contains no other critical points. If $a \in E = D \setminus \{c\}$ then we have germs $m_i = [f_i]_a$ $(1 \leqslant i \leqslant n)$ at a, representing the branches of $A(z, w) = 0$ near a, and these may be continued analytically, within $\mathcal{M}_A$, along any path γ in E. If γ is a closed path from a to itself, winding once around c, then continuation of any m_i along γ must produce a germ

$m_j \in \mathcal{M}_A$ at $a = \gamma(1)$, distinct germs m_i giving distinct germs m_j (by unique-ness of continuation along γ^{-1}). Since there are only finitely many germs $m_1, \ldots, m_n$ in $\mathcal{M}_A$ at a, the function $\pi(\gamma): m_i \mapsto m_j$ is a permutation of $\{m_1, \ldots, m_n\}$ (see §4.11 for examples), each germ m_i being contained in a cycle of length $q \leqslant n$. We can arrange the labelling so that m_1 is in a cycle $(m_1, m_2, \ldots, m_q)$, and then it follows from §4.13 that the corresponding branches $f_1, \ldots, f_q$ of the algebraic function $w = f(z)$ are represented by a Puiseux series

$$f(z) = F(\zeta) = \sum_{r=-\infty}^{\infty} c_r \zeta^r$$

on E, different branches corresponding to different choices of $\zeta = (z - c)^{1/q}$. If $q < n$ then there are similar series for $f_{q+1}, \ldots, f_n$, one series for each remaining cycle of $\pi(\gamma)$.

If $a_0(c) \neq 0$ then the solutions w of $A(z, w) = 0$ are bounded as $z \to c$ in E: the rational functions $a_i(z)/a_0(z)$ $(1 \leqslant i \leqslant n)$ are bounded for z near c, and if $|w| \geqslant 1$ then (4.14.1) gives

$$\left.
\begin{aligned}
|w| &= \left| \frac{a_1}{a_0} + \frac{a_2}{a_0 w} + \ldots + \frac{a_n}{a_0 w^{n-1}} \right| \\
&\leqslant \left| \frac{a_1}{a_0} \right| + \left| \frac{a_2}{a_0} \right| + \ldots + \left| \frac{a_n}{a_0} \right|,
\end{aligned}
\right\} \tag{4.14.4}$$

so $|w| \leqslant \max(1, \sum_{i=1}^{n} |a_i/a_0|)$ near c. Putting $w = f(z)$ we see that $F(\zeta)$ is bounded as $\zeta \to 0$, so $c_r = 0$ for all $r < 0$. Thus F is meromorphic (in fact, analytic) at $\zeta = 0$, so if $q > 1$ then $\mathcal{S}_A$ has a branch-point of order $q - 1$ at c (if $q = 1$ then $[f_1]_c = [f]_c \in \mathcal{M}_A$ and there is no need to add a branch-point). Similarly, any remaining cycles of $\pi(\gamma)$ give branch-points or elements of $\mathcal{M}_A$ at c.

If, on the other hand, c is a root of multiplicity $k \geqslant 1$ of $a_0(z)$, then $\lim_{z \to c} a_0(z)(z - c)^{-k} \neq 0$, so the rational functions $(z - c)^k a_i(z)/a_0(z)$ $(1 \leqslant i \leqslant n)$ are bounded near c. Multiplying (4.14.4) through by $|(z - c)^k|$, we see by a similar argument that for any solution w of $A(z, w) = 0$, $|(z - c)^k w|$ is bounded as $z \to c$. Thus, putting $w = f(z) = F(\zeta)$ we see that $\zeta^{kq} F(\zeta)$ is bounded as $\zeta \to 0$, so F is meromorphic at $\zeta = 0$, and hence $\mathcal{S}_A$ has a branch-point of order $q - 1$ at c (if $q > 1$) or else we have $[f]_c \in \mathcal{M}_A$ (if $q = 1$). Again, any remaining cycles of $\pi(\gamma)$ behave similarly.

Finally we consider $c = \infty \in C_A$. Arguments similar to those for $c \neq \infty$ give a Puiseux series

$$f(z) = F(\zeta) = \sum_{r=-\infty}^{\infty} c_r \zeta^r$$

for branches $f_1, \ldots, f_q$ of f near ∞, where $\zeta = z^{-1/q}$. For sufficiently large $k \in \mathbb{N}$ the rational functions $a_i(z)/z^k a_0(z)$ are all bounded for z near ∞, so by dividing (4.14.4) through by $|z^k|$ we see that for any solution w of $A(z, w) = 0$, $|z^{-k}w|$ is bounded as $z \to \infty$. Thus $\zeta^{kq} F(\zeta)$ is bounded as $\zeta \to 0$, so F is meromorphic at $\zeta = 0$ and as before there is a branch-point or an element of $\mathcal{M}_A$ corresponding to each cycle of $\pi(\gamma)$.

We are now able to prove our main results on algebraic functions.

Theorem 4.14.5. *If $A(z, w)$ is an irreducible polynomial then $\mathscr{S}_A$ is connected.*

Proof. By the remarks at the end of §4.13, it is sufficient to prove that the unbranched Riemann surface $\mathcal{M}_A$ is connected. To do this, it is sufficient to prove that for some $a \in \Sigma \backslash C_A$, the n germs $m_i = [f_i]_a \in \mathcal{M}_A$ at a are all in the same component of $\mathcal{M}$: for given any germ $m = [g]_b \in \mathcal{M}_A$ there is a path γ in Σ from b to a, avoiding critical points, and continuation of m along γ must give some m_i at a ($1 \leqslant i \leqslant n$), so m is in the component $\mathcal{M}(m_i)$ of $\mathcal{M}$ containing m_i.

Given $a \in \Sigma \backslash C_A$, we can label $m_1, \ldots, m_n$ so that $\mathcal{M}(m_1)$ contains $m_1, m_2, \ldots, m_k$ but not $m_{k+1}, \ldots, m_n$, where we assume for a contradiction that $k < n$. Now the symmetric functions of $f_1, \ldots, f_k$ are the coefficients $\sigma_1 = \sum_{i=1}^{k} f_i, \ldots, \sigma_k = \prod_{i=1}^{k} f_i$ in

$$\prod_{i=1}^{k} (w - f_i(z)) = w^k - \sigma_1(z) w^{k-1} + \ldots + (-1)^k \sigma_k(z); \qquad (4.14.6)$$

being polynomials in $f_1, \ldots, f_k$, they can be continued analytically along all paths in $\Sigma \backslash C_A$. By hypothesis, $f_1, \ldots, f_k$ are permuted amongst themselves by continuation along closed paths, so $\sigma_1, \ldots, \sigma_k$ are single-valued and analytic on $\Sigma \backslash C_A$. We have seen that as z approaches a critical point, each $|f_i(z)|$ grows no faster than a power of the local uniformising parameter ζ. The functions $\sigma_i(z)$ therefore have the same property, so they are meromorphic at the critical points and hence meromorphic on Σ; by Theorem 1.4.1 they are rational functions of z, so if $b(z)$ is the least common multiple of their denominators then (4.14.6) implies that the function

$$B(z, w) = b(z) \prod_{i=1}^{k} (w - f_i(z))$$

is a polynomial in z and w, of degree k in w. Now A is irreducible and of degree n, while B has degree $k < n$, so A and B are co-prime and hence there are only finitely many $z \in \mathbb{C}$ for which A and B have a common root w (see Theorem A.14). However, for all z sufficiently near a, the values z

and $w = f_1(z)$ are common zeros of A and B, so this contradiction gives $k = n$ as required. $\square$

Combining this with the argument at the beginning of this section, we immediately have:

Corollary 4.14.7. *If $A(z, w)$ is a polynomial then $\mathscr{S}_A$ consists of finitely many connected components $\mathscr{S}_{A_i}$, one for each distinct irreducible factor A_i of A.* $\square$

The following result is even more important:

Theorem 4.14.8. *If $A(z, w)$ is a polynomial then $\mathscr{S}_A$ is compact.*

Proof. We have seen that for each $a \in \Sigma$, $\mathscr{S}_A \cap \psi^{-1}(a)$ consists of between one and n points $[f]_a$ (possibly including branch-points if $a \in C_A$), and that for some $D \in \mathscr{D}_a$ the D-neighbourhoods of these points are the connected components of $\mathscr{S}_A \cap \psi^{-1}(D)$, each homeomorphic (under the appropriate chart) to an open disc in $\mathbb{C}$. (Fig. 4.72 shows a cross-section of Σ and $\mathscr{S}_A$, where $\psi^{-1}(a)$ consists of a branch-point of order 1 and a regular point.) If D_0 is a strictly smaller disc in $\mathscr{D}_a$, then the closure $\bar{D}_0$ of D_0 in Σ is contained in D, and so $\mathscr{S}_A \cap \psi^{-1}(\bar{D}_0)$ has finitely many components, each homeomorphic to a closed disc in $\mathbb{C}$; by the Heine–Borel theorem (see §1.2), these components are compact. Now Σ, being compact, is covered by just finitely many such discs D_0 (for various $a \in \Sigma$), so $\mathscr{S}_A$, being a union of finitely many compact subsets (the components of the corresponding sets $\mathscr{S}_A \cap \psi^{-1}(\bar{D}_0)$), is compact. $\square$

Fig. 4.72

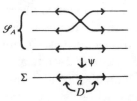

Conversely, we have the following major result:

Theorem 4.14.9. *Any compact abstract Riemann surface can be identified with the Riemann surface $\mathscr{S}_A$ of some algebraic function $A(z, w) = 0$.* $\square$

(The proof is quite difficult, requiring potential theory and harmonic functions; see Springer [1957] or Cohn [1967]. By 'can be identified with' we mean 'is conformally equivalent to', a concept we shall define in §4.17.)

These equivalences between topological and algebraic concepts (compactness corresponding to polynomials, connected components to irreducible factors) are typical of Riemann surface theory, and help to explain its central role in the development of mathematics. For example, there is a connection with Galois theory: continuation around each critical point of an algebraic function $A(z, w) = 0$ induces a permutation of the single-valued branches $w = f_i(z)$; these permutations generate the monodromy group of A (see §4.10) which we can imbed in the Galois group of A by regarding A as a polynomial in w with roots $f_i(z)$. Then the orbits of this group of permutations correspond to the components of $\mathscr{S}_A$ and to the irreducible factors of A.

Finally we mention without proof a result which can be regarded as a generalisation of Theorems 1.4.1, 3.11.1 and 3.11.2 (which refer to the Riemann surfaces Σ and $\mathbb{C}/\Omega$). By Theorem 4.14.9 and Corollary 4.14.7 we can identify any compact connected Riemann surface S with $\mathscr{S}_A$ for some irreducible polynomial $A(z, w)$; then the meromorphic functions on S form a field F containing ψ and $\phi : S \to \Sigma$, and a more precise description of F is given by:

Theorem 4.14.10. (i) *The meromorphic functions on S are the rational functions of ψ and ϕ, that is, $F = \mathbb{C}(\psi, \phi)$; (ii) if f and g are meromorphic on S then there is a non-zero irreducible polynomial $\Phi(x, y)$, with complex coefficients, such that $\Phi(f, g)$ is identically zero on S.* $\square$

4.15 Orientable and non-orientable surfaces

Surfaces can be divided into two classes, orientable and non-orientable. Roughly speaking, a surface is orientable if it is possible to choose a consistent sense of orientation (clockwise or anti-clockwise) at every point P on the surface. This means that if we take any closed path γ based at P, and cover γ by a finite number of discs (as in analytic continuation), then each disc induces an orientation on the next disc, and by following

Fig. 4.73

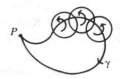

the orientation around γ we return to the original orientation at P (see Fig. 4.73). If, on the other hand, by following the orientation around some path γ we return to a different orientation at P, then the surface is non-orientable. (We will give more precise definitions shortly.)

The simplest example of a non-orientable surface is the Möbius band, constructed by taking a long strip of paper, twisting one end through an angle π, and then gluing the two ends together, as shown in Fig. 4.74. We need to delete the boundary to obtain a surface (if we do not, then we have a 'surface with boundary', a concept we have not discussed). This surface is non-compact; the simplest example of a compact non-orientable surface is the projective plane, defined in §3.5.

Fig. 4.74

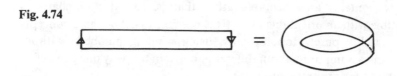

Our aim in this section is to show that all Riemann surfaces are orientable. There are many rigorous definitions of orientability of a surface, and we choose one which, although not applicable to *all* surfaces, is very convenient for Riemann surfaces. We now define the class of surfaces to which this definition applies, namely the smooth surfaces.

By identifying each $z = x + iy \in \mathbb{C}$ with $(x, y) \in \mathbb{R}^2$, we can regard the local coordinates of any surface as lying in $\mathbb{R}^2$; we say that an atlas of charts is *smooth* (or C^∞) if all its coordinate transition functions f are smooth (in the sense that the partial derivatives $\partial^n f / \partial x^k \partial y^{(n-k)}$ all exist). As with analytic atlases, two smooth atlases $\mathscr{A}$ and $\mathscr{B}$ are called *compatible* if the atlas $\mathscr{A} \cup \mathscr{B}$ is smooth. Compatibility between smooth atlases is an equivalence relation, and an equivalence class of smooth atlases is called a *smooth structure*. Finally a *smooth surface* is a surface with a smooth structure, that is, a surface on which the atlas is smooth.

Since every analytic function is smooth, it is clear that every Riemann surface is a smooth surface.

If U and V are open subsets of $\mathbb{R}^2$ and $f:(x, y) \mapsto (u, v)$ is a smooth function $U \to V$, then the *Jacobian* of f is

$$J_f = \frac{\partial u}{\partial x} \frac{\partial v}{\partial y} - \frac{\partial u}{\partial y} \frac{\partial v}{\partial x}. \tag{4.15.1}$$

If f has a smooth inverse function $g = f^{-1}: V \to U$ (as is always the case for the coordinate transition functions of a smooth surface), then the chain rule

$$J_g J_f = J_{(g \circ f)} = J_{(\mathrm{id})} = 1$$

implies that $J_f \neq 0$ at all points of U (Apostol [1963], §7.2); if $J_f > 0$ at all points of U then f is said to be *orientation-preserving*. For example, $f(x, y) = (x, -y)$ (or $f(z) = \bar{z}$ in complex coordinates) does *not* preserve orientation, since $J_f \equiv -1$ on $\mathbb{R}^2$; in fact, being a reflection, f maps any positively oriented circle in $\mathbb{R}^2$ to a circle with negative orientation. Rotations and translations of $\mathbb{R}^2$, on the other hand, have $J_f \equiv 1$, so they preserve orientation.

A smooth atlas is said to be *orientable* if all its coordinate transition functions preserve orientation; a smooth surface is *orientable* if its smooth structure contains an orientable atlas, that is, its atlas of charts is compatible with an orientable atlas. (It is possible to define these concepts of orientability purely topologically, using winding numbers, without making any assumptions about differentiability; in this way one can define orientability for arbitrary surfaces.)

Theorem 4.15.2.   *Every analytic atlas is orientable.*

Proof.   It is sufficient to show that every analytic function f, with an analytic inverse, is orientation-preserving. If

$$f(x, y) = (u, v)$$

in real coordinates, then being analytic, f satisfies the Cauchy–Riemann equations

$$\frac{\partial u}{\partial x} = \frac{\partial v}{\partial y}, \quad \frac{\partial u}{\partial y} = -\frac{\partial v}{\partial x},$$

and so (4.15.1) gives

$$J_f = \left(\frac{\partial u}{\partial x}\right)^2 + \left(\frac{\partial v}{\partial y}\right)^2 \geq 0.$$

Since f^{-1} is analytic, $J_f \neq 0$ as remarked above, so $J_f > 0$ as required.   □

Corollary 4.15.3.   *Every Riemann surface is orientable.*   □

As with smooth and analytic atlases, we define two orientable atlases $\mathscr{A}$ and $\mathscr{B}$ to be *compatible* if the atlas $\mathscr{A} \cup \mathscr{B}$ is orientable. Compatibility of orientable atlases is an equivalence relation, and an equivalence class of orientable atlases is called an *orientation*. It can be shown that every orientable surface S has just two orientations, described as follows. If $\mathscr{A} = \{U_i, \Phi_i\}$ is an orientable atlas for S, we define $\bar{\mathscr{A}} = \{(U_i, \bar{\Phi}_i)\}$ where $\bar{\Phi}_i(s) = (x, -y)$ whenever $\Phi_i(s) = (x, y) \in \mathbb{R}^2$; in complex coordinates, this is just complex conjugation. Then $\bar{\mathscr{A}}$ is easily seen to be an orientable atlas for S which is not compatible with $\mathscr{A}$. The equivalence classes containing $\mathscr{A}$ and $\bar{\mathscr{A}}$ give the two orientations for S.

4.16 The genus of a compact Riemann surface

If $A(z, w) = 0$ is an irreducible algebraic equation, then by Theorems 4.14.5 and 4.14.8 and Corollary 4.15.3 its Riemann surface $S = \mathscr{S}_A$ is a compact, connected, orientable surface. Such surfaces are classified topologically by the following result (see Massey [1967]):

Theorem 4.16.1. *Each compact, connected, orientable surface is homeomorphic to a surface S_g formed by attaching g handles to a sphere, for some unique integer $g \geqslant 0$.* $\square$

We call g the *genus* of the surface; in this section we give a method for calculating g, using polygonal subdivisions. A *polygonal subdivision M* of a surface S consists of a finite set of points of S, called *vertices*, and a finite set of simple (that is, non-self-intersecting) paths on S, called *edges*, such that

  (i) every edge has two end-points, these points being vertices;
 (ii) edges can intersect only at their end-points;
(iii) the union of the edges (which we also denote by M) is connected;
(iv) the components of the complement $S \backslash M$ are homeomorphic to open discs.

In (iv), the components are called *faces*; since each edge is incident with at most two faces, there are only finitely many faces.

Examples

(1) If we project the vertices and edges of a tetrahedron onto a sphere S enclosing the tetrahedron, then we have a polygonal subdivision of S with four vertices, six edges, and four faces, each face being three-sided.
(2) By identifying opposite sides of a square, as illustrated in Fig. 4.75, we

obtain a polygonal subdivision of a torus with five vertices (the four corners forming a single vertex), ten edges, and five faces A, B, C, D, E, each face being four-sided.

Fig. 4.75

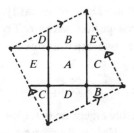

It can be shown that every compact, connected surface has a polygonal subdivision. This was first proved (for Riemann surfaces) by T. Radó in 1925; we omit the proof (which can be found in Springer [1957]) since we will calculate the genus of a Riemann surface by explicitly constructing a subdivision.

We define the *Euler characteristic* of a compact, connected surface S to be

$$\chi(S) = \chi(M) = V - E + F,$$

where M is a polygonal subdivision of S with V vertices, E edges, and F faces. It is clear that homeomorphic surfaces will have the same Euler characteristic, provided we can show that $\chi(S)$ is well defined, that is, independent of the choice of M. This, like Theorem 4.16.1, is a standard topological result, to be found in Massey [1967] or Springer [1957], for example, so we will merely sketch a proof.

We can construct new polygonal subdivisions M' from M by any of the following steps, illustrated in Fig. 4.76:

(a) place a new vertex on an existing edge, dividing that edge into two new edges;

Fig. 4.76

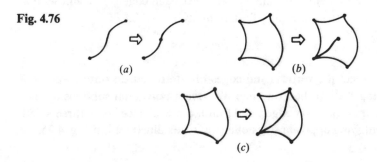

(b) place a new vertex in an existing face, and join it by a new edge to an existing vertex incident with that face;

(c) join two existing vertices, both incident with the same face, by a new edge across that face, dividing the face into two new faces.

In each case, E and either V or F are increased by 1 while the remaining parameter is unchanged, so that $\chi(M') = \chi(M)$.

Given any two polygonal subdivisions M_1 and M_2 of S, suppose that there exists a subdivision M_3 which can be obtained both from M_1 and from M_2 by finite sequences of steps (a), (b) and (c); since these steps leave χ invariant, we have $\chi(M_1) = \chi(M_3) = \chi(M_2)$, so $\chi(M_1) = \chi(M_2)$ as required. In order to construct M_3, we first take the union of M_1 and M_2, with the vertices defined to be those of M_1 and those of M_2, together with all points where M_1 meets M_2 (there is a slight technical difficulty in that an edge of M_1 may meet an edge of M_2 infinitely many times, rather as the graph of $x \sin x^{-1}$ meets the x-axis, but this can be resolved by replacing M_1 and M_2 by subdivisions with slightly 'straightened' edges, and with the same Euler characteristics, as shown in Fig. 4.77). If necessary we now add finitely many edges to $M_1 \cup M_2$ in order that the resulting subdivision M_3 should satisfy conditions (i) to (iv) (for example, $M_1 \cup M_2$ need not be connected). It is now easy to see that M_3 can be obtained from each of M_1 and M_2 by finitely many steps (a), (b) and (c), so that $\chi(S)$ is independent of the subdivision M of S.

Fig. 4.77

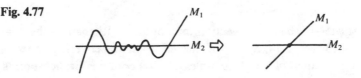

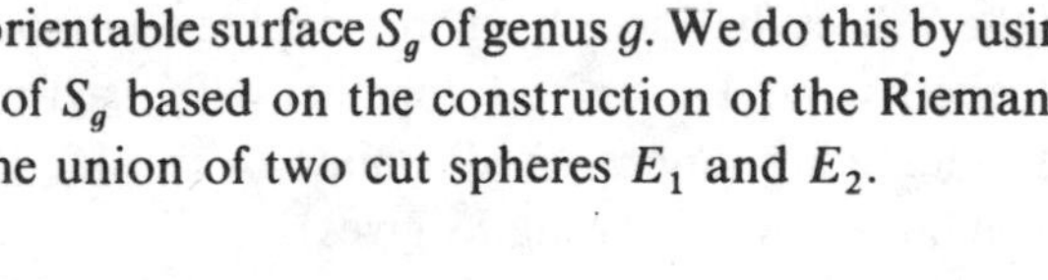

We can now calculate the Euler characteristic $\chi(S_g)$ of a compact, connected, orientable surface S_g of genus g. We do this by using a polygonal subdivision of S_g based on the construction of the Riemann surface S in §4.9 (v) as the union of two cut spheres E_1 and E_2.

Fig. 4.78

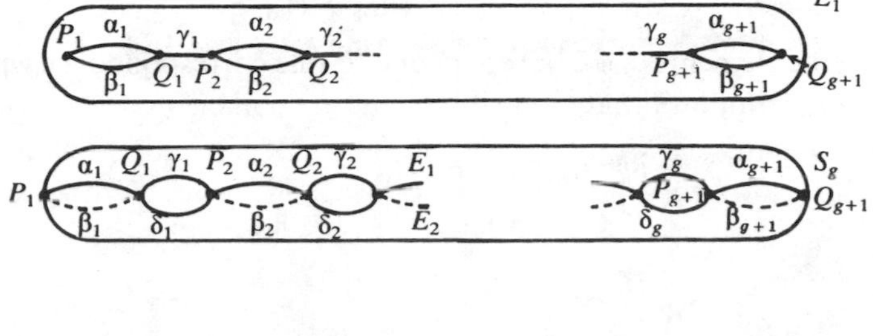

We take $g + 1$ cuts $P_1 Q_1, \ldots, P_{g+1} Q_{g+1}$ on each of the two spheres, with edges labelled $\alpha_1, \beta_1, \ldots, \alpha_{g+1}, \beta_{g+1}$ in each case; we connect the cuts by edges $Q_1 P_2, \ldots, Q_g P_{g+1}$ labelled $\gamma_1, \ldots, \gamma_g$ on E_1 and $\delta_1, \ldots, \delta_g$ on E_2, and then join E_1 and E_2 along the edges α_i and β_i to give a surface S_g of genus g. This is illustrated in Fig. 4.78. The vertices $P_1, Q_1, \ldots, P_{g+1}, Q_{g+1}$ and the edges $\alpha_1, \ldots, \alpha_{g+1}, \beta_1, \ldots, \beta_{g+1}, \gamma_1, \ldots, \gamma_g, \delta_1, \ldots, \delta_g$ form a polygonal subdivision of S_g with two faces (E_1 and E_2 with their edges removed), so $V = 2g + 2$, $E = 4g + 2$, and $F = 2$, giving $\chi(S_g) = V - E + F = 2 - 2g$. Thus we have proved:

Theorem 4.16.2. *The Euler characteristic of a compact, connected, orientable surface S_g of genus g is given by $\chi(S_g) = 2 - 2g$.* $\square$

The simplest case $g = 0$ is Euler's theorem that $V - E + F = 2$ for a sphere; a torus has $\chi = 0$ and $g = 1$ (see Example (2)).

We are now ready to prove the following result, known as the *Riemann–Hurwitz formula*, or simply the *Hurwitz formula*.

Theorem 4.16.3. *If S is the Riemann surface $\mathcal{S}_A$ of an irreducible algebraic equation $A(z, w) = 0$ of degree n in w, and if the branch-points have orders $n_1, \ldots, n_r$, then the genus g of S is given by*

$$g = 1 - n + \frac{1}{2} \sum_{i=1}^{r} n_i.$$

Proof. Let $\psi : S \to \Sigma$ be the projection map onto the Riemann sphere, as defined in §4.13. By §4.14, S is an n-sheeted branched covering space of Σ, that is, if $z \in \Sigma$ then $|\psi^{-1}(z)| = n$ unless $\psi^{-1}(z)$ contains branch-points. Since $n_i + 1$ sheets come together at a branch-point of order n_i, it follows that if $\psi^{-1}(z)$ contains branch-points of orders $n_1, \ldots, n_k$ then the permutation π_z of the n sheets (induced by meromorphic continuation around z) has cycles of lengths $n_1 + 1, \ldots, n_k + 1$, together with fixed-points corresponding to the remaining (unbranched) points in $\psi^{-1}(z)$. If there are t fixed-points, then summing cycle-lengths gives

$$n = (n_1 + 1) + \ldots + (n_k + 1) + t,$$

and since the elements of $\psi^{-1}(z)$ are in one-to-one correspondence with the cycles of π_z, it follows that

$$|\psi^{-1}(z)| = k + t$$
$$= n - (n_1 + \ldots + n_k).$$

(We can think of n_i sheets as being 'missing' at each branch-point of order n_i; in Fig. 4.72 we see an example where $n = 3$, $k = 1$, $t = 1$ and $n_1 = 1$, so $|\psi^{-1}(z)| = 2$.)

Now let $P_1, \ldots, P_r$ be the branch-points on S (there are only finitely many by compactness of S, or alternatively by the argument in §4.14), and let each P_i have order n_i. It is easily seen by induction on r that there is a polygonal subdivision M of Σ with the points $Q_i = \psi(P_i)$ $(i = 1, \ldots, r)$ included among the vertices, and it now follows that $\tilde{M} = \psi^{-1}(M)$ is a polygonal subdivision of S, the vertices and edges of $\tilde{M}$ being the points and paths of S lying above the vertices and edges of M. (We leave the detailed verification of conditions (i)–(iv) as an exercise for the reader; the important points are that ψ is open and continuous, and every branch-point is a vertex, from which it follows that the edges and faces of $\tilde{M}$ are mapped homeomorphically onto edges and faces of M.) If M has V vertices, E edges and F faces, then $V - E + F = 2$ since Σ has genus 0. Now each edge or face of M lifts to n edges or faces of $\tilde{M}$ (one on each sheet), so $\tilde{M}$ has nE edges and nF faces. By the arguments above, $\tilde{M}$ has $nV - (n_1 + \ldots + n_r)$ vertices (n_i being 'lost' at each P_i), so that

$$2 - 2g = \chi(S)$$

$$= \left(nV - \sum_{i=1}^{r} n_i \right) - nE + nF$$

$$= n(V - E + F) - \sum_{i=1}^{r} n_i$$

$$= 2n - \sum_{i=1}^{r} n_i,$$

giving

$$g = 1 - n + \frac{1}{2} \sum_{i=1}^{r} n_i. \quad \square$$

Corollary 4.16.4. $\sum_{i=1}^{r} n_i$ *is an even integer, greater than or equal to* $n - 1$. $\square$

This corollary is often useful in checking that the orders of the branch-points have been calculated correctly; $\sum_{i=1}^{r} n_i$ is called the *total order of branching*.

Examples. We now verify that the Riemann–Hurwitz formula gives the correct value for the genus when applied to some of the surfaces considered earlier.

(1) $A(z, w) = w^q - z$, as in §4.8. Here $n = q$ and there are two branch-points of order $q - 1$ at $0, \infty$; thus the total order of branching is $2q - 2$ so

$$g = 1 - q + \tfrac{1}{2}(2q - 2) = 0,$$

confirming our earlier claim that S is homeomorphic to a sphere.

(2) $A(z, w) = w^2 - (z - a_1)\ldots(z - a_m)$ with $a_1, \ldots, a_m$ distinct, as in §4.9(v). Here $n = 2$ and there are branch-points of order 1 at $a_1, \ldots, a_m$; if m is even then these are the only branch-points, so the total order of branching is m, while if m is odd then there is an additional branch-point of order 1 at ∞, so the total order of branching is now $m + 1$. We thus have

$$g = \begin{cases} 1 - 2 + \dfrac{m}{2} = \dfrac{m - 2}{2} & \text{if } m \text{ is even,} \\[2ex] 1 - 2 + \dfrac{m + 1}{2} = \dfrac{m - 1}{2} & \text{if } m \text{ is odd.} \end{cases}$$

In particular, if $p(z)$ is the polynomial $4z^3 - g_2 z - g_3$, with distinct roots, arising in the theory of elliptic functions (§3.10) then $m = 3$ gives $g = 1$, so that S is homeomorphic to a torus; we will return to this important example later.

(3) $A(z, w) = w^4 - 2w^2 + 1 - z$, or equivalently $w = \sqrt{(1 + \sqrt{z})}$, as in §4.10. Here $n = 4$, and as shown in §4.10 there are two branch-points of order 1 at $z = 0$, one branch-point of order 1 at $z = 1$, and one branch-point of order 3 at $z = \infty$. Thus the total order of branching is $2 + 1 + 3 = 6$, so $g = 1 - 4 + \tfrac{6}{2} = 0$ and S is homeomorphic to a sphere (a fact not at all apparent visually!).

4.17 Conformal equivalence and automorphisms of Riemann surfaces

If U and V are open subsets of $\mathbb{C}$, and if a homeomorphism $g: U \to V$ is analytic on U, then since g is one-to-one on U it is locally one-to-one and therefore satisfies $g'(z) \neq 0$ for all $z \in U$, by Theorem A.10, Lemma 1; it follows that the homeomorphism $g^{-1}: V \to U$ is analytic on V, with $(g^{-1})' = 1/g'$, and that both g and g^{-1} are conformal. Now if $f: S_1 \to S_2$ is a holomorphic homeomorphism between Riemann surfaces, then by applying the above argument to the induced maps between local coordinates, we see that $f^{-1}: S_2 \to S_1$ is also a holomorphic homeomorphism, and that local coordinates are transformed conformally by f and f^{-1}. We say that f is a *conformal equivalence* (or *conformal homeomorphism*), and that S_1 and S_2 are *conformally equivalent*, written $S_1 \cong S_2$. Clearly conformal equivalence is an equivalence relation: it is, in fact, the 'isomorphism' of Riemann surface theory in the sense that two conformally equivalent surfaces share the same analytic properties and are therefore indistinguishable in terms of their complex structures. (Some authors actually refer to conformal equivalence as 'isomorphism', but this can lead to confusion when the surfaces also have algebraic structures – for instance, any two tori $\mathbb{C}/\Omega_1$ and $\mathbb{C}/\Omega_2$ are isomorphic as groups, but not necessarily as Riemann surfaces.)

Examples. (1) We shall show that the disc $\mathcal{D} = \{z \in \mathbb{C} \mid |z| < 1\}$ and the upper half-plane $\mathcal{U} = \{z \in \mathbb{C} \mid \mathrm{Im}(z) > 0\}$ are conformally equivalent. We define $T : \mathcal{U} \to \mathbb{C}$ by

$$T(z) = \frac{z - i}{z + i};$$

since $T \in PGL(2, \mathbb{C})$ and $-i \notin \mathcal{U}$, T defines a holomorphic homeomorphism of $\mathcal{U}$ onto $T(\mathcal{U})$. For all $z \in \mathcal{U}$ we have

$$\begin{aligned}
|T(z)|^2 &= \frac{(z - i)(\bar{z} + i)}{(z + i)(\bar{z} - i)} \\
&= \frac{|z|^2 - 2\,\mathrm{Im}(z) + 1}{|z|^2 + 2\,\mathrm{Im}(z) + 1} \\
&< 1,
\end{aligned}$$

so T maps $\mathcal{U}$ into $\mathcal{D}$; similarly, putting $\mathrm{Im}(z) = 0$ we see that T maps the boundary $\mathbb{R} \cup \{\infty\}$ of $\mathcal{U}$ to the unit circle, which bounds $\mathcal{D}$, so $T(\mathcal{U})$, which is a disc by Theorem 2.4.1, must coincide with $\mathcal{D}$, as required. (Alternatively, we could take T to be the rotation of Σ by an angle $\pi/2$ about the axis through ± 1, mapping $\mathcal{U}$ onto $\mathcal{D}$.)

(2) Although they are homeomorphic, $\mathbb{C}$ and $\mathcal{D}$ are *not* conformally equivalent: if $f : \mathbb{C} \to \mathcal{D}$ is holomorphic (that is, analytic), then being bounded f is constant by Liouville's theorem, so f cannot be a homeomorphism.

In this context, we have the following important result, the *Riemann mapping theorem*, which classifies all simply connected open subsets of $\mathbb{C}$ up to conformal equivalence. The proof, which we shall omit, can be found in many textbooks on complex analysis, e.g. Ahlfors [1966] and Rudin [1974].

Theorem 4.17.1. *If S is a simply connected open subset of $\mathbb{C}$, then either $S = \mathbb{C}$ or else $S \cong \mathcal{D}$.*

The following related result is sometimes called the *Uniformisation*

Theorem (due to F. Klein, H. Poincaré and P. Koebe), since it allows us to uniformise (that is, parametrise) compact Riemann surfaces by means of single-valued functions. For a proof, see Beardon [1984], and for further comments on uniformisation, see §5.12.

Theorem 4.17.2.   *Every simply connected Riemann surface is conformally equivalent to just one of*:

(i) *the Riemann sphere* Σ;
(ii) *the complex plane* $\mathbb{C}$;
(iii) *the open unit disc* $\mathscr{D}$ (*or equivalently* $\mathscr{U}$, *by Example* (1)). $\square$

As shown in (2) above, $\mathbb{C} \not\cong \mathscr{D}$; moreover Σ, being compact, cannot be conformally equivalent (or even homeomorphic) to $\mathbb{C}$ or $\mathscr{D}$. Thus Theorem 4.17.2 implies that a Riemann surface homeomorphic to Σ must be conformally equivalent to Σ; in other words, each topological sphere has just one complex structure (when we pass from genus 0 to genus 1, the corresponding situation is much more complicated, as we shall see in §4.18).

A conformal homeomorphism $f : S \to S$ is called an *automorphism* of S. (There is no conflict with the definition of an automorphism of Σ given in §2.1, since the meromorphic bijections $f : \Sigma \to \Sigma$ are precisely the conformal homeomorphisms with respect to the analytic atlas introduced in §4.11.) It follows from Theorem 4.12.1 and the remarks at the beginning of this section that the set Aut S of automorphisms of S is a group under composition. As we shall see in §4.19, it is important to determine the automorphism groups of the three simply connected Riemann surfaces:

Theorem 4.17.3.

(i) Aut $\Sigma = PSL(2, \mathbb{C})$;
(ii) Aut $\mathbb{C} = \{z \mapsto az + b \mid a, b \in \mathbb{C}, a \neq 0\}$;
(iii) Aut $\mathscr{U} = PSL(2, \mathbb{R})$.

Proof.   (i) This was proved in Theorem 2.1.3.
(ii) If $f : \mathbb{C} \to \mathbb{C}$ is an automorphism then f is analytic on $\mathbb{C}$, so

$$f(z) = a_0 + a_1 z + a_2 z^2 + \dots$$

for all $z \in \mathbb{C}$, this power series having infinite radius of convergence by Theorem 4.3.3 since f has no singular points in $\mathbb{C}$. Hence the function $g = f \circ J$ has an expansion

$$\begin{aligned} g(z) &= f(1/z) \\ &= a_0 + a_1 z^{-1} + a_2 z^{-2} + \dots \end{aligned}$$

for all $z \neq 0$. There are now two possibilities: either f is a polynomial (that is, $a_n = 0$ for all sufficiently large n), or else g has an essential singularity at $z = 0$. We shall show that the latter is impossible. If g has an essential singularity at 0, then by Weierstrass' theorem (Theorem A.9) in any neighbourhood of 0 the values of $g(z)$ come arbitrarily close to any given complex number. Thus if $\mathcal{D}$ is the open unit disc then $g(\mathcal{D} \backslash \{0\})$ is dense in $\mathbb{C}$. However, if V is any non-empty open subset of $\mathbb{C} \backslash \mathcal{D}$, then since g is a non-constant analytic function, $g(V)$ is open and hence has non-empty intersection with the dense set $g(\mathcal{D} \backslash \{0\})$; this contradicts the fact that g is one-to-one (since f and J are), so f must be a polynomial. Since f is one-to-one on $\mathbb{C}$, the fundamental theorem of algebra implies that f has degree 1, so that $f(z) = az + b$ with $a \neq 0$. Conversely, it is obvious that any transformation of this form is an analytic homeomorphism $\mathbb{C} \to \mathbb{C}$, and is therefore an automorphism.

(iii) We shall first consider the automorphisms of $\mathcal{D}$, and then use Example (1) to pass from $\mathcal{D}$ to $\mathcal{U}$.

First, we show that every automorphism f of $\mathcal{D}$ which fixes 0 has the form $z \mapsto e^{i\theta} z (\theta \in \mathbb{R})$, and is therefore a Möbius transformation (this result, of wider importance for complex function theory, is often referred to as Schwarz's Lemma, after H. A. Schwarz, 1843–1921). Since f is analytic on $\mathcal{D}$ and $f(0) = 0$, we have

$$f(z) = a_1 z + a_2 z^2 + \dots$$

and so $f(z)/z$ is analytic on $\mathcal{D}$. If C is any circle, centred at 0, of radius $\rho < 1$, then since $C \subseteq \mathcal{D}$ and $|f(z)| < 1$ on $\mathcal{D}$ we have $|f(z)/z| < 1/\rho$ on C, and hence $|f(z)/z| < 1/\rho$ inside C by the maximum-modulus principle (Theorem A.11). As ρ can be made arbitrarily close to 1 it follows that $|f(z)| \leqslant |z|$ for all $z \in \mathcal{D}$. Now f^{-1} is also an automorphism of $\mathcal{D}$ fixing 0, so the same argument gives $|f^{-1}(z)| \leqslant |z|$ for all $z \in \mathcal{D}$. Replacing z by $f(z)$ we then have $|f^{-1}(f(z))| \leqslant |f(z)|$ and so $|f(z)| \geqslant |z|$ for all $z \in \mathcal{D}$. Thus $|f(z)/z| = 1$ and so $f(z) = e^{i\theta} z$ for some real θ, possibly depending on z. Now $f(z)/z$, being analytic on the region $\mathcal{D}$, is either an open mapping or else constant; since the image of $\mathcal{D}$ is not open (being a subset of the unit circle), $f(z)/z$ is constant and hence θ can be chosen to be constant. Thus f is a Möbius transformation $R_\theta : z \mapsto e^{i\theta} z$ (in the notation of §2.3).

By Example (1) the Möbius transformation

$$T(z) = \frac{z - i}{z + i}$$

is a conformal homeomorphism from $\mathcal{U}$ to $\mathcal{D}$; since $T(i) = 0$, it follows that every automorphism of $\mathcal{U}$ fixing i has the form $g = T^{-1} f T = T^{-1} R_\theta T$, and is therefore a Möbius transformation. Now if $h \in \operatorname{Aut} \mathcal{U}$ maps i to $a + bi$ $(a, b \in \mathbb{R})$ then $k : z \mapsto a + bz$ is both a Möbius transformation and an automorphism of $\mathcal{U}$ mapping i to $a + bi$, so $g = k^{-1}h$ is an automorphism of $\mathcal{U}$ fixing i; as we have seen, g must be a Möbius transformation and hence so is $h = kg$. Thus $\operatorname{Aut} \mathcal{U} \subseteq PSL(2, \mathbb{C})$ and now Theorem 2.8.1(ii) gives $\operatorname{Aut} \mathcal{U} = PSL(2, \mathbb{R})$. $\square$

Notice that in each of the three cases in Theorem 4.17.3, the automorphism group consists entirely of Möbius transformations, so we can make use of the information established in Chapter 2.

4.18 Conformal equivalence of tori

In this section we consider the following problem: if Ω and Ω' are lattices in $\mathbb{C}$, then when are the tori $\mathbb{C}/\Omega$ and $\mathbb{C}/\Omega'$ conformally equivalent? In §5.7 we shall see that every compact Riemann surface of genus 1 is conformally equivalent to $\mathbb{C}/\Omega$ for some lattice Ω, so that by answering this question we shall have a classification of such surfaces.

If Ω is a lattice and $a \in \mathbb{C}$, then we define $a\Omega$ to be $\{a\omega \,|\, \omega \in \Omega\}$; this is a lattice if and only if $a \neq 0$. Lattices Ω and Ω' are *similar* (in the sense of Euclidean geometry) if $\Omega' = a\Omega$ for some $a \neq 0$; this is an equivalence relation. If Ω and Ω' are lattices, then we will denote the elements of $\mathbb{C}/\Omega$ and $\mathbb{C}/\Omega'$ by $[z]$ and $[z]'$ respectively.

Theorem 4.18.1.   *The holomorphic functions $f : \mathbb{C}/\Omega \to \mathbb{C}/\Omega'$ are the transformations $f_{a,b} : [z] \mapsto [az + b]'$, where $a, b \in \mathbb{C}$ and $a\Omega \subseteq \Omega'$; $f_{a,b}$ is a conformal homeomorphism if and only if $a\Omega = \Omega'$, so $\mathbb{C}/\Omega$ and $\mathbb{C}/\Omega'$ are conformally equivalent if and only if Ω and Ω' are similar.*

(In §6.1 we will give necessary and sufficient conditions, in terms of their bases, for Ω and Ω' to be similar.)

Proof.   The main step in the proof is to show that if $f : \mathbb{C}/\Omega \to \mathbb{C}/\Omega'$ is a holomorphic function, then there is an automorphism $\tilde{f}$ of $\mathbb{C}$ such that

Fig. 4.79

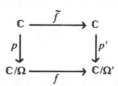

$f \circ p = p' \circ \tilde{f}$, where p and p' are the natural projections $\mathbb{C} \rightarrow \mathbb{C}/\Omega$ and $\mathbb{C} \rightarrow \mathbb{C}/\Omega'$ (see Fig. 4.79).

If $f([z]) = [z']'$ then as in the proof of Theorem 4.11.1 there exist open discs V and V' in $\mathbb{C}$, centred at z and z', mapped homeomorphically by p and p' onto neighbourhoods U and U' of $[z]$ and $[z']'$. The inverse homeomorphisms $q:U \rightarrow V$ and $q':U' \rightarrow V'$ are the charts described in §4.11, and since f is holomorphic the map

$$F = q' \circ f \circ p : q(U \cap f^{-1}(U')) \rightarrow q'(U' \cap f(U))$$

(representing the change of local coordinates induced by f) is analytic (consult Fig. 4.80). Notice that F is *not* uniquely determined by z: we could replace V' by any of its translates $V' + \omega' (\omega' \in \Omega')$, and this has the effect of replacing F by $F + \omega'$. Thus at each $z \in \mathbb{C}$ we have a set of analytic germs $[F + \omega']_z$, and it is easily seen that as z ranges over $\mathbb{C}$ the set $\mathscr{G}$ of all such germs is sufficient for continuation in $\mathbb{C}$; then Corollary 4.13.2 implies that, since $\mathbb{C}$ is simply connected, any one of these germs extends to a single-valued analytic function $\tilde{f}:\mathbb{C} \rightarrow \mathbb{C}$ satisfying $f \circ p = p' \circ \tilde{f}$ (since each local branch F of $\tilde{f}$ satisfies $f \circ p = p' \circ F$ on its domain). We call $\tilde{f}$ a *lift* of f (its existence follows in more general situations from covering space theory: see §4.19 and Massey [1967], Theorem 5.1).

Fig. 4.80

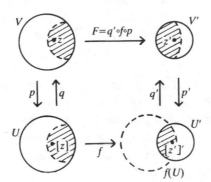

Since $f \circ p = p' \circ \tilde{f}$ we have $f([z]) = [\tilde{f}(z)]'$ for all $z \in \mathbb{C}$, and so for any fixed $\omega \in \Omega$ we have $\tilde{f}(z + \omega) = \tilde{f}(z) + \omega'_z$ where ω'_z is an element of Ω' possibly depending on z. Now the function

$$z \mapsto \omega'_z = \tilde{f}(z + \omega) - \tilde{f}(z)$$

is continuous (since $\tilde{f}$ is analytic), and it maps a connected space $\mathbb{C}$ into a discrete space Ω', so it must be constant. Thus ω'_z depends only on ω, and not on z, so dropping the suffix z we have

$$\tilde{f}(z + \omega) = \tilde{f}(z) + \omega'$$

for all $z \in \mathbb{C}$, where $\omega' \in \Omega'$ depends on $\omega \in \Omega$. Differentiating with respect to z, we see that $d\tilde{f}/dz$ is elliptic with respect to Ω; it is also analytic, since $\tilde{f}$ is, so it must be constant by Theorem 3.6.1. Thus

$$\tilde{f}(z) = az + b$$

for constants $a, b \in \mathbb{C}$, so

$$f([z]) = [az + b]',$$

and $f = f_{a,b}$. For each $\omega \in \Omega$, $f([z + \omega]) = f([z])$ and hence $[a(z + \omega) + b]' = [az + b]'$; thus $a\omega \in \Omega'$ and so $a\Omega \subseteq \Omega'$. Conversely, it is easily seen that if $a\Omega \subseteq \Omega'$ then this transformation $f_{a,b} : \mathbb{C}/\Omega \to \mathbb{C}/\Omega'$ is holomorphic.

If $f = f_{a,b}$ is a conformal homeomorphism, then its holomorphic inverse must have the form $[z]' \mapsto [(z - b)/a]$ where $a^{-1}\Omega' \subseteq \Omega$, so that $\Omega' \subseteq a\Omega$ and hence $\Omega' = a\Omega$; conversely, if $\Omega' = a\Omega$ then this defines a holomorphic inverse, so f is a conformal homeomorphism. $\square$

It is easily seen that there are infinitely many similarity classes of lattices in $\mathbb{C}$, and hence infinitely many conformal equivalence classes of tori; in other words, a compact orientable surface of genus 1 admits infinitely many distinct complex structures. We will return to this topic in §6.1.

The above proof of Theorem 4.18.1 depends heavily on the compactness of $\mathbb{C}/\Omega$ (needed to prove Theorem 3.6.1, which shows that $d\tilde{f}/dz$ is constant). We now give an alternative argument, independent of compactness, which determines the conformal equivalences between tori and which we will generalise in Theorem 5.9.3 to apply to other Riemann surfaces, not necessarily compact.

As in the above proof, we can lift any conformal equivalence $f : \mathbb{C}/\Omega \to \mathbb{C}/\Omega'$ to an analytic map $\tilde{f} : \mathbb{C} \to \mathbb{C}$, and similarly we can lift $g = f^{-1} : \mathbb{C}/\Omega' \to \mathbb{C}/\Omega$ to $\tilde{g} : \mathbb{C} \to \mathbb{C}$, again analytic (see Fig. 4.81). Then $p \circ \tilde{g} \circ \tilde{f} = g \circ f \circ p = p$, so that $(\tilde{g} \circ \tilde{f})(z) = z + \omega_z (\omega_z \in \Omega)$ for all $z \in \mathbb{C}$. As before, the function $\mathbb{C} \to \Omega$, $z \mapsto \omega_z = (\tilde{g} \circ \tilde{f})(z) - z$ is continuous and hence constant, so $(\tilde{g} \circ \tilde{f})(z) = z + \omega$ for all $z \in \mathbb{C}$ and for some fixed $\omega \in \Omega$. This immediately implies that $\tilde{f}$ is one-to-one, and a similar argument involving $\tilde{f} \circ \tilde{g}$ shows that $\tilde{f}$ is onto; thus $\tilde{f}$ is an automorphism of $\mathbb{C}$, so it has the form $z \mapsto az + b$ by Theorem 4.17.3(ii). Finally, we argue as before to show that $a\Omega = \Omega'$ and that every such map $\tilde{f}$ induces a conformal equivalence.

Fig. 4.81

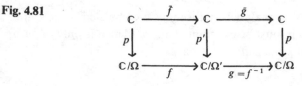

Putting $\Omega' = \Omega$ in Theorem 4.18.1 we have

Theorem 4.18.2. *The automorphisms of the Riemann surface* $\mathbb{C}/\Omega$ *are the transformations* $f_{a,b}:[z] \mapsto [az + b]$ *such that* $a, b \in \mathbb{C}$ *and* $a\Omega = \Omega$. $\square$

It is straightforward to verify that the group $G = \mathrm{Aut}(\mathbb{C}/\Omega)$ has a normal subgroup N, isomorphic (as a group!) to $\mathbb{C}/\Omega$, consisting of the transformations $f_{1,b}$, and that G/N is a cyclic group of order 2, 4 or 6 (see Exercise 4Q). Notice that G acts transitively on $\mathbb{C}/\Omega$, so all points of $\mathbb{C}/\Omega$ have the same analytic properties (as is the case with Σ).

Finally, we show that for any lattice Ω, $\mathbb{C}/\Omega$ is conformally equivalent to the Riemann surface S of $\sqrt{p(z)}$, where p is the cubic polynomial appearing in the differential equation $\wp' = \sqrt{p(\wp)}$ satisfied by the Weierstrass function $\wp$ associated with Ω (see §3.10); for a form of converse, see Theorem 6.5.11.

Theorem 4.18.3. *Let* $p(z) = 4z^3 - g_2 z - g_3$, *where* $g_2 = 60\Sigma'\omega^{-4}$ *and* $g_3 = 140\Sigma'\omega^{-6}$ *(summations over all non-zero* $\omega \in \Omega$*), and let* S *be the Riemann surface of* $\sqrt{p(z)}$*; then* $S \cong \mathbb{C}/\Omega$.

Proof. As shown in §4.9(iv) there is a homeomorphism $\alpha:\mathbb{C}/\Omega \to S$, mapping each $t \in \mathbb{C}/\Omega$ to the germ $[f]_c \in S$ where $c = \wp(t)$ and f is the local branch of $\sqrt{p}$ near c satisfying $f(c) = \wp'(t)$; putting $t = [a] \in \mathbb{C}/\Omega (a \in \mathbb{C})$ we have $c = \wp(a)$ and $f(c) = \wp'(a)$.

It is sufficient to show that α induces a conformal transformation of local coordinates from $\mathbb{C}/\Omega$ to S. The charts on $\mathbb{C}/\Omega$ at t have the form (U_V, Φ_V) described in the proof of Theorem 4.11.1, where V is a small disc in $\mathbb{C}$ containing a, and Φ_V is the inverse of the projection p_V of V onto its image $U_V \subset \mathbb{C}/\Omega$; on S, provided c is not one of the four branch-points, we use the projection $\psi:S \to \Sigma$, $[f]_z \mapsto z$, to give local coordinates near c, as in §4.13. It follows that away from the branch-points, α induces the transformation $\psi \circ \alpha \circ \Phi_V^{-1} = \psi \circ \alpha \circ p_V = \wp \circ p_V$ of local coordinates, and this is just the restriction of $\wp$ to V. Now the branch-points on S, lying above ∞ and the roots e_1, e_2 and e_3 of p, correspond under α to the points $[0]$, $[\frac{1}{2}\omega_1]$, $[\frac{1}{2}\omega_2]$ and $[\frac{1}{2}\omega_3]$ of $\mathbb{C}/\Omega$; since $\wp$ has non-zero derivative on $\mathbb{C}\backslash\frac{1}{2}\Omega$ (by Theorem 3.10.7), α transforms the local coordinates conformally away from the branch-points. Each branch-point has order 1, so (as in §4.13) we use the local coordinates $\zeta = z^{-1/2}$ (near ∞) or $\zeta = (z - e_i)^{1/2}$ (near e_i); by Corollary 3.10.8 $\wp$ takes the values ∞ and e_i with multiplicity 2, so the functions $\wp^{-1/2}$ and $(\wp - e_i)^{1/2}$, representing the transformations of local coordinates at the branch-points, are conformal. $\square$

4.19 Covering surfaces of Riemann surfaces

Now that we have precise definitions of surfaces and of Riemann surfaces, we can return a little more rigorously to the concept of a covering surface, which we introduced rather briefly in §1.5. A *covering surface* of a surface S is a pair $(\tilde{S}, p)$ where $\tilde{S}$ is a surface and p is a continuous function from $\tilde{S}$ onto S with the following property: each $s \in S$ has an open neighbourhood U, homeomorphic to the open disc $\mathscr{D}$, such that each connected component V of $p^{-1}(U)$ is mapped homeomorphically by p onto U. We will refer to U and V as *elementary neighbourhoods* (it is standard to apply this to U, but not V; however, it is convenient and should cause no confusion, to apply it also to V). When p (called a *covering map*) is understood, we will simply refer to $\tilde{S}$ as a covering surface of S; the phrase 'a covering of S by $\tilde{S}$' is also used. We call $p^{-1}(s)(s \in S)$ the *fibre* above s; since $|p^{-1}(s) \cap V| = 1$ for all V, $|p^{-1}(s)|$ is the number (possibly infinite) of connected components of $p^{-1}(U)$, so $|p^{-1}(s)| = |p^{-1}(s')|$ for all $s' \in U$. Thus $|p^{-1}(s)|$ is locally constant on S, so if S is connected then $|p^{-1}(s)|$ is a constant, called the *number of sheets* of the covering.

We will pay particular attention to coverings of Riemann surfaces, but it should be noted that this is part of a more general topological theory of covering spaces, and that many of our results can be generalised to manifolds of arbitrary dimension; see, for example Massey [1967].

Examples. (1) If Ω is a lattice in $\mathbb{C}$ then the projection $p:\mathbb{C} \to \mathbb{C}/\Omega$ is a covering map with infinitely many sheets. To see this, we cover $\mathbb{C}$ by open discs V satisfying $V \cap (V + \omega) = \varnothing$ for all $\omega \in \Omega \setminus \{0\}$, as in the proof of Theorem 4.11.1; then these, and their images $U_V = p(V)$, form the required elementary neighbourhoods in $\mathbb{C}$ and $\mathbb{C}/\Omega$, since the components $V + \omega$ of $p^{-1}(U_V)$ are mapped homeomorphically by p onto U_V. Each fibre is a coset of Ω, and hence infinite.

(2) We saw in §1.5 that if $f: \Sigma \to \Sigma$ is a non-constant rational function and B its set of branch-points, then f restricts to a covering map $\Sigma \setminus f^{-1}(f(B)) \to \Sigma \setminus f(B)$, the number of sheets being the degree of f. (To obtain a covering map from f, it is *not* sufficient simply to remove all branch-points and their images; as shown in Example (2) of §1.5, whenever $f^{-1}(s)$ contains a branch-point then *all* of $f^{-1}(s)$ must be removed, together with s. Similar considerations apply to the next two examples.)

(3) By analogy with (2), it follows from §3.15 that non-constant elliptic functions induce coverings of a punctured sphere by a punctured torus, the number of sheets being the order of the function; for instance, the

Weierstrass function $\wp:\mathbb{C}\to\Sigma$ induces a 2-sheeted covering map

$$\hat{\wp}:\mathbb{C}/\Omega\setminus\{[0],[\tfrac{1}{2}\omega_1],[\tfrac{1}{2}\omega_2],[\tfrac{1}{2}\omega_3]\}\to\Sigma\setminus\{\infty,e_1,e_2,e_3\}.$$

(4) It follows from §4.14 (and in particular Lemma 4.14.2) that the unbranched Riemann surface $\mathscr{M}_A$ of an algebraic function A is a covering surface of a punctured sphere, the punctures being the finitely many images in Σ of the branch-points; the number of sheets is equal to the degree of A. For specific examples, see §§4.8–4.10.

(5) Similarly, §4.7 shows that the Riemann surface of $\log(z)$ is a covering surface of $\Sigma\setminus\{0,\infty\}$, with infinitely many sheets. It should, however, be pointed out that transcendental functions do not always give rise to covering surfaces in this way: for instance, a point $s\in\Sigma$ might be on the natural boundary for one branch of the function, but not for a second branch, so that s would not have a suitable elementary neighbourhood.

Although we will not pursue this idea very far, we briefly mention here how to extend our definition of a covering surface to include branch-points. The basic idea is that a typical branch-point of order $n-1$ is given by the behaviour, at the origin, of the function $\pi_n:\mathscr{D}\to\mathscr{D}$, $z\mapsto z^n$, so all we need do is transfer this model to a more general setting. We say that a continuous surjection $p:\tilde{S}\to S$ is a *branched* (or *ramified*) covering map if each $s\in S$ has an open neighbourhood U and a homeomorphism $\Phi:U\to\mathscr{D}$ such that for each connected component V of $p^{-1}(U)$ there is a homeomorphism $\Psi:V\to\mathscr{D}$ satisfying $\Phi\circ p=\pi_n\circ\Psi$ for some integer $n\geqslant 1$ (see Fig. 4.82). We have $n=1$ if and only if $p(=\Phi^{-1}\circ\pi_n\circ\Psi)$ is a homeomorphism $V\to U$, as happens for (unbranched) covering maps; if $n>1$ for some V then we say that the unique element $\tilde{s}$ of $V\cap p^{-1}(s)$ is a *branch-point* of *order* $n-1$ since p is (like π_n) locally n-to-one near $\tilde{s}$. For instance, if we include the branch-points in Examples (2), (3) and (4) above then we have a branched covering in each case.

Fig. 4.82

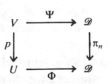

For the remainder of this section we will assume that $(\tilde{S},p)$ is a covering surface of a surface S (with no branch-points).

A covering transformation of $(\tilde{S},p)$ is a homeomorphism $g:\tilde{S}\to\tilde{S}$ such that $p\circ g=p$, see Fig. 4.83 (equivalently, g maps each fibre $p^{-1}(s)$ to itself); these form a group under composition. For instance in Example (1) above, the covering transformations are the translations of $\mathbb{C}$ by elements of Ω, forming a group isomorphic to Ω. In Example (4), if we take $\tilde{S}$ to be the unbranched Riemann surface of $z^{1/q}$ (see §4.8), and $S = \Sigma \backslash \{0, \infty\}$, so that $p = \psi : \tilde{S} \to S$ maps each germ $[f]_a \in \tilde{S}$ to $a \in S$, then the covering transformations are $[f]_a \mapsto [\varepsilon^r f]_a$ for $r = 0, 1, \dots, q - 1$, where $\varepsilon = \exp(2\pi i/q)$; these form a cyclic group of order q.

Fig. 4.83

Lemma 4.19.1. *The set F_g of fixed-points of a covering transformation g is both open and closed.*

Proof. Let $\tilde{s} \in F_g$, so that $g(\tilde{s}) = \tilde{s}$, and let V be an elementary neighbourhood of $\tilde{s}$ in $\tilde{S}$. Since g is a homeomorphism, $g^{-1}(V)$ is a neighbourhood of $\tilde{s}$, and hence so is $V' = V \cap g^{-1}(V)$. If $v \in V'$ then since $p = p \circ g$ we have $p(v) = p(g(v))$; but v and $g(v)$ both lie in V, on which p is one-to-one, so $v = g(v)$. Thus $V' \subseteq F_g$, so F_g is open.

Now let $\tilde{s} \in \tilde{S} \backslash F_g$, so that $g(\tilde{s}) \neq \tilde{s}$. Since $\tilde{S}$ is a Hausdorff space, there exist disjoint neighbourhoods A and B of $\tilde{s}$ and $g(\tilde{s})$ respectively. Let V be an elementary neighbourhood of $\tilde{s}$, and let $V' = V \cap A \cap g^{-1}(B)$, a neighbourhood of $\tilde{s}$. Then $V' \subseteq A$ while $g(V') \subseteq B$, so $V' \cap g(V') = \varnothing$ and hence $V' \subseteq \tilde{S} \backslash F_g$, showing that F_g is closed. $\quad\square$

Thus F_g is a union of connected components of $\tilde{S}$. For example, let S be connected, and $\tilde{S}$ a disjoint union of any number of copies of S, each mapped homeomorphically onto S by p (see §4.9(iii)). Then the covering transformations are the permutations of the copies of S (that is, the components of $\tilde{S}$), and F_g is just the union of those copies mapped to themselves by g.

In fact, the proof of Lemma 4.19.1 gives a little more than we have stated. If G is a group of homeomorphisms of a topological space X onto itself, then G acts *discontinuously* on X if every $x \in X$ has a neighbourhood V such that $V \cap g(V) = \varnothing$ for all non-identity $g \in G$. For instance, a lattice Ω acts discontinuously on $\mathbb{C}$ as a group of translations, by Example (1) above. The next result generalises this:

Theorem 4.19.2. *If $\tilde{S}$ is connected, then the group G of covering transform-*

ations of $p:\tilde{S}\to S$ acts discontinuously on $\tilde{S}$; in particular, a non-identity element of G can have no fixed-points on $\tilde{S}$.

Proof. Let V be an elementary neighbourhood of some $\tilde{s}\in\tilde{S}$, and suppose that $V\cap g(V)\neq\varnothing$ for some $g\in G$, so that there exists $v\in V$ with $g(v)\in V$; then since $p = p\circ g$ and p is one-to-one on V, we have $v = g(v)$, so F_g is non-empty. Since $\tilde{S}$ is connected, we must have $F_g = \tilde{S}$, so that g is the identity. $\quad\square$

We say that $(\tilde{S}, p)$ is a *regular covering space* of S if, for each $s\in S$, the group G of covering transformations acts transitively on the fibre $p^{-1}(s)$; thus any two elements of $p^{-1}(s)$ 'look alike'. Examples include $p:\mathbb{C}\to\mathbb{C}/\Omega$ (with $G = \Omega$) and the covering of $S = \Sigma\backslash\{0, \infty\}$ by the unbranched Riemann surface of $z^{1/q}$, as shown above. For an example of a non-regular covering, take the function $w = \sqrt{(1 + \sqrt{z})}$ considered in §4.10: this has branch-points at $z = 0, 1$ and ∞, so we take $S = \Sigma\backslash\{0, 1, \infty\}$ and $\tilde{S} = \psi^{-1}(S)$, a 4-sheeted unbranched covering of S. We now divide $\tilde{S}$ into four sheets $E_1,\ldots, E_4$ as in §4.10, take $s\in S$ close to 1, and let s_i be the unique element of $\psi^{-1}(s)$ in E_i. At $z = 1, E_1$ and E_3 are unbranched, while E_2 and E_4 meet at a branch-point of order 1; it follows that if δ is a closed path from s to itself, winding once around 1 (and not around 0 or ∞) then $\psi^{-1}(\delta)$ consists of two closed paths (from s_1 and s_3 to themselves, on E_1 and E_3 respectively) and two paths which are not closed (from s_2 to s_4 and vice versa). Now the covering transformations must permute these four paths amongst themselves, mapping closed paths to closed paths, so they cannot, for example, map s_1 to s_2.

The following result illustrates the importance of regular coverings.

Theorem 4.19.3. *If $(\tilde{S}, p)$ is a regular covering surface of S, and if G is the group of covering transformations, then there is a homeomorphism $q:S\to\tilde{S}/G$ given by $q(s) = [\tilde{s}]_G$ where $s\in S$ and $\tilde{s}\in p^{-1}(s)$.*

Proof. Let $\pi:\tilde{S}\to\tilde{S}/G$ be the natural projection, sending each $\tilde{s}\in\tilde{S}$ to its orbit $[\tilde{s}]_G$, so that q is defined by $q(s) = \pi(\tilde{s})$, where $p(\tilde{s}) = s$, as shown in Fig. 4.84. To show that this depends only on s, and not on $\tilde{s}$, let $\tilde{s}_1, \tilde{s}_2\in p^{-1}(s)$;

Fig. 4.84

since $\tilde{S}$ is a regular covering surface, there exists $g \in G$ with $g(\tilde{s}_1) = \tilde{s}_2$, and hence $\pi(\tilde{s}_1) = \pi(g(\tilde{s}_1)) = \pi(\tilde{s}_2)$, as required. Thus we have a well-defined function $q : S \to \tilde{S}/G$, which is clearly onto. If $s_1, s_2 \in S$ satisfy $q(s_1) = q(s_2)$, then any $\tilde{s}_1 \in p^{-1}(s_1)$ and $\tilde{s}_2 \in p^{-1}(s_2)$ satisfy $\pi(\tilde{s}_1) = \pi(\tilde{s}_2)$, so $\tilde{s}_2 = g(\tilde{s}_1)$ for some $g \in G$; then $s_1 = p(\tilde{s}_1) = (p \circ g)(\tilde{s}_1) = p(\tilde{s}_2) = s_2$, so q is one-to-one and hence a bijection.

It follows easily from the definitions of a covering map and of the quotient topology on $\tilde{S}/G$ that p and π are both open and continuous; then q also has both these properties and is therefore a homeomorphism. $\square$

The above result gives us a natural way of identifying S with $\tilde{S}/G$, and p with π. We now quote one of the most important results on covering surfaces; a proof can be found in most topology textbooks, for example in Massey [1967].

Theorem 4.19.4. *Every connected surface S has a covering surface $(\hat{S}, p)$ such that $\hat{S}$ is simply connected.* $\square$

For instance, if $S = \mathbb{C}/\Omega$ then we can take $\hat{S} = \mathbb{C}$ as in Example (1) above; if $S = \mathbb{C} \setminus \{0\}$ we can take $\hat{S} = \mathbb{C}$ and $p = \exp : \mathbb{C} \to \mathbb{C} \setminus \{0\}$; if S is simply connected we can take $\hat{S} = S$ and $p = \mathrm{id} : \hat{S} \to S$.

We call $(\hat{S}, p)$ (or simply $\hat{S}$) a *universal covering surface* for S; it can be shown to be unique in the sense that if $(\tilde{S}, q)$ is any other simply connected covering surface of S then there is a homeomorphism $f : \hat{S} \to \tilde{S}$ with $p = q \circ f$ (see Fig. 4.85). More generally, if $(\tilde{S}, q)$ is *any* connected covering surface of S then there is a covering map $f : \hat{S} \to \tilde{S}$ with $p = q \circ f$, so that $\hat{S}$ covers all connected covering surfaces of S.

Fig. 4.85

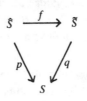

Our aim is to show that $\hat{S}$ is always a regular covering surface of S; while this can be proved for arbitrary surfaces, we shall find it most convenient to restrict our attention to Riemann surfaces, so that we can apply the principle of meromorphic continuation. First we show that if S has a complex structure, then this induces a complex structure on each covering surface $\tilde{S}$ of S.

Theorem 4.19.5. *If S is a Riemann surface with covering surface $\tilde{S}$ then there is a unique complex structure on $\tilde{S}$ such that $p:\tilde{S}\to S$ is holomorphic.*

Proof. Using Theorem 4.11.2 to give a suitable atlas on S, we can choose the elementary neighbourhoods V on $\tilde{S}$ sufficiently small that each is mapped by p homeomorphically onto some $U \subseteq S$, where (U, Φ) is a chart on S (see Fig. 4.86). We then define an atlas on $\tilde{S}$ to consist of all charts of the form $(V, \Phi \circ p)$; this atlas is analytic, since if $(V_i, \Phi_i \circ p)$ and $(V_j, \Phi_j \circ p)$ are charts with $V_i \cap V_j \neq \varnothing$, then the coordinate transition function is

$$(\Phi_j \circ p) \circ (\Phi_i \circ p)^{-1} = \Phi_j \circ \Phi_i^{-1},$$

which is a coordinate transition function on S and is therefore analytic. A similar argument shows that compatible atlases on S induce compatible atlases on $\tilde{S}$, so the complex structure on $\tilde{S}$ is well defined.

Fig. 4.86

The transformation of local coordinates induced by p has the form $\Phi \circ p \circ (\Phi \circ p)^{-1}$, which is the identity and hence analytic, so p is holomorphic. It is straightforward to check that any other analytic atlas on $\tilde{S}$, with respect to which p is holomorphic, must be compatible with that defined above, so the complex structure on $\tilde{S}$ is unique. $\square$

Corollary 4.19.6. *If S and $\tilde{S}$ are as in Theorem 4.19.5, then each covering transformation of $(\tilde{S}, p)$ is an automorphism of $\tilde{S}$.*

Proof. If g is a covering transformation, and $(V_i, \Phi_i \circ p)$ and $(V_j, \Phi_j \circ p)$ are charts at $\tilde{s}$ and at $g(\tilde{s})$, where $\tilde{s} \in \tilde{S}$, then g induces the transformation of local coordinates

$$(\Phi_j \circ p) \circ g \circ (\Phi_i \circ p)^{-1} = \Phi_j \circ (p \circ g) \circ p^{-1} \circ \Phi_i^{-1}$$
$$= \Phi_j \circ p \circ p^{-1} \circ \Phi_i^{-1}$$
$$= \Phi_j \circ \Phi_i^{-1},$$

which is analytic. Thus g is holomorphic, and the same applies to g^{-1}, so g is an automorphism of $\tilde{S}$. $\square$

By Theorem 4.19.5, if S is any connected Riemann surface, then its universal covering surface $\hat{S}$ has a unique complex structure with respect to which the covering map $p:\hat{S}\to S$ is holomorphic. Thus $\hat{S}$ is a simply connected Riemann surface, so by Theorem 4.17.2 it is conformally equivalent to Σ, $\mathbb{C}$ or $\mathscr{U}$; without loss of generality we can take $\hat{S}$ to be one of these three surfaces, so that in all cases we have $\hat{S}\subseteq\Sigma$. By Corollary 4.19.6 the covering transformations are automorphisms of $\hat{S}$, and these are completely described in Theorem 4.17.3. It only remains to show that $\hat{S}$ is a regular covering surface of S, so that we can apply Theorem 4.19.3 to identify S with $\hat{S}/G$ for some $G\leqslant\operatorname{Aut}\hat{S}$.

Theorem 4.19.7. *If S is a connected Riemann surface then the universal covering surface $(\hat{S}, p)$ is a regular covering surface of S.*

Proof. We must show that whenever $\hat{s}_1$ and $\hat{s}_2$ lie in the same fibre $p^{-1}(s)$ ($s\in S$) then there is a covering transformation $f:\hat{S}\to\hat{S}$ with $f(\hat{s}_1)=\hat{s}_2$. Our method (similar to that used in Theorem 4.18.1) is first to construct f locally near $\hat{s}_1$, and then use meromorphic continuation to extend to a global transformation.

We can choose elementary neighbourhoods V_i ($i=1,2$) of $\hat{s}_i$ such that the restriction p_i of p to V_i maps V_i homeomorphically onto an elementary neighbourhood U of s. Then the function $F=p_2^{-1}\circ p_1:V_1\to V_2$ is holomorphic (since each p_i is), satisfies $p\circ F=p$ on its domain V_1, and maps $\hat{s}_1$ to $\hat{s}_2$. If (as above) we identify $\hat{S}$ with a subset of Σ, then as s ranges over S and $\hat{s}_1,\hat{s}_2$ over $p^{-1}(s)$, the corresponding function elements (V_1,F) give a set $\mathscr{G}$ of meromorphic germs. It is clear that $\mathscr{G}$ is sufficient for meromorphic continuation within $\hat{S}$ (in the sense of §4.13), so because $\hat{S}$ is simply connected, Corollary 4.13.2 implies that any one of these germs extends to a single-valued meromorphic function f on $\hat{S}$; moreover, f maps $\hat{S}$ into itself and satisfies $p\circ f=p$, since each function element (V_1,F) has this property locally. If we choose F to map a particular $\hat{s}_1\in p^{-1}(s)$ to $\hat{s}_2$, then f will also have this property since it is a meromorphic continuation of the germ $[F]_{\hat{s}_1}$.

Applying a similar argument to $F^{-1}=p_1^{-1}\circ p_2:V_2\to V_1$ we have a meromorphic function $g:\hat{S}\to\hat{S}$, equal to F^{-1} near $\hat{s}_2$, satisfying $p\circ g=p$. Then $g\circ f:\hat{S}\to\hat{S}$ is meromorphic, and near $\hat{s}_1$ it coincides with $F^{-1}\circ F=\operatorname{id}:V_1\to V_1$, so $g\circ f$ is the identity on $\hat{S}$ by the uniqueness of meromorphic continuation; similarly $f\circ g$ is the identity, so $f:\hat{S}\to\hat{S}$ is a homeomorphism and hence a covering transformation of $\hat{S}$. $\square$

The above proof appears to need the uniformisation theorem (4.17.2) which, by identifying $\hat{S}$ with a subset of Σ, enables us to apply meromorphic continuation to F. In fact, we can avoid this deep result: it is not difficult to check that the arguments leading up to Corollary 4.13.2 can be applied to *any* simply connected Riemann surface (such as $\hat{S}$), and not just to subsets $E \subseteq \Sigma$ as stated.

Theorem 4.19.8. *If S is a connected Riemann surface not conformally equivalent to the sphere Σ, the plane $\mathbb{C}$, the punctured plane $\mathbb{C} \backslash \{0\}$, or a torus $\mathbb{C}/\Omega$, then S has universal covering space $\hat{S} = \mathcal{U}$, the upper half-plane, and S is conformally equivalent to $\mathcal{U}/G$ for some subgroup G of $PSL(2, \mathbb{R})$ acting discontinuously on $\mathcal{U}$.*

Proof. By Theorems 4.19.3 and 4.19.7 there is a homeomorphism $q : S \to \hat{S}/G$, where $\hat{S} = \Sigma$, $\mathbb{C}$ or $\mathcal{U}$ by Theorem 4.17.2, and $G \leqslant \text{Aut}\,\hat{S}$ by Corollary 4.19.6. Using q, we can carry the complex structure on S over to $\hat{S}/G$: we define the charts on $\hat{S}/G$ to have the form $(q(W), \Phi \circ q^{-1})$ where (W, Φ) ranges over the charts on S. Then $q : S \to \hat{S}/G$ is a conformal equivalence, and the projection $\pi : \hat{S} \to \hat{S}/G$ is holomorphic.

By Theorem 4.19.2 G acts discontinuously on $\hat{S}$, the non-identity elements having no fixed-points; we shall see that when $\hat{S} = \Sigma$ or $\mathbb{C}$, this restricts the possibilities for G and hence for S.

If $\hat{S} = \Sigma$ then $\text{Aut}\,\hat{S} = PSL(2, \mathbb{C})$ by Theorem 4.17.3(i); by Theorem 2.9.1 every non-identity element of $PSL(2, \mathbb{C})$ has at least one fixed-point, so G is the trivial group and $S \cong \Sigma/G = \Sigma$.

If $\hat{S} = \mathbb{C}$ then $\text{Aut}\,\hat{S}$ consists of the transformations $z \mapsto az + b$ ($a, b \in \mathbb{C}, a \neq 0$) by Theorem 4.17.3(ii); if $a \neq 1$ then $b/(1 - a)$ is a fixed-point, so G must consist of translations $z \mapsto z + b$. Now such a group G acts discontinuously on $\mathbb{C}$ if and only if the corresponding elements b form a discrete subgroup of $\mathbb{C}$, so by Theorem 3.1.3 G is the trivial group, an infinite cyclic group, or a lattice. In the first case $S \cong \mathbb{C}/G = \mathbb{C}$, and in the third case $\mathbb{C}/G$ is a torus. In the second case, if ω is a generator for G then the function $z \mapsto \exp(2\pi i z/\omega)$, $\mathbb{C} \to \mathbb{C} \backslash \{0\}$, induces a conformal equivalence $\mathbb{C}/G \to \mathbb{C} \backslash \{0\}$, so $S \cong \mathbb{C} \backslash \{0\}$ (see §3.3 for an example of this in connection with simply periodic functions).

If we exclude the above four possibilities for S, then $\hat{S} = \mathcal{U}$ and $S \cong \mathcal{U}/G$ where G is a discontinuous subgroup of $\text{Aut}\,\hat{S} = PSL(2, \mathbb{R})$, by Theorem 4.17.3(iii). $\quad\square$

This theorem shows that every connected Riemann surface S may be obtained from one of the three simply connected Riemann surfaces $\hat{S} = \Sigma, \mathbb{C}$ or $\mathcal{U}$ by factoring out a discontinuous subgroup G of $\text{Aut}\,\hat{S}$ (itself a known group); moreover, in all except four simple cases, the universal covering space $\hat{S}$ is $\mathscr{U}$, rather than Σ or $\mathbb{C}$. This leads naturally to the subject of the next chapter, the upper half-plane $\mathscr{U}$ and the action on $\mathscr{U}$ of various subgroups of its automorphism group $PSL(2, \mathbb{R})$.

In this section we have deliberately confined ourselves to a rather abbreviated account of covering surfaces, partly through lack of space and partly because the extra structure of a Riemann surface enables us to take certain short-cuts not available in the general theory. One particular topic we have ignored is the important connection between covering surfaces and fundamental groups; for example, if $\hat{S}$ is the universal covering space of a connected surface S, then the group G of covering transformations can be identified (in a natural way) with $\pi_1(S)$, and each connected covering surface of S can be obtained from $\hat{S}$ by factoring out a subgroup of $\pi_1(S)$. For details, see Massey [1967].

EXERCISES

4A. Prove that the series $\sum_{n=1}^{\infty} z^{2^n}$ has the unit circle as its natural boundary.

4B. Prove that the point $z = 1$ is a singular point for the power series $\sum_{n=1}^{\infty}(z^n/n^2)$.

4C. Construct the Riemann surface of $\sin^{-1} z$.

4D. Find the position and order of the branch-points of the following many-valued functions. In each case construct a cut plane in which it is possible to define a branch of the function:

(a) $\sqrt{((z-1)(z-2)(z-3)(z-4)(z-5))}$;
(b) $(1 - z^n)^{1/m}$ (m, n positive integers);
(c) $\log \sin z$.

4E. Compute the monodromy groups for the Riemann surfaces of the following many-valued functions:

(a) $z^{1/n}$ (n a positive integer);
(b) $\sqrt{(z + \sqrt{z})}$;
(c) $z^{1/3} + ((z-1)/(z-2))^{1/2}$.

4F. Show that compatibility of atlases is an equivalence relation.

4G. Let S be a Riemann surface, p a point of S and let $f : S \to \Sigma$ be a meromorphic function. If (U_i, Φ_i) is a chart at p then $f \circ \Phi_i^{-1} : \mathbb{C} \to \Sigma$ is a meromorphic function on $\Phi_i(U_i)$ and so can be expanded in a power series about the point $\zeta = \Phi_i(p)$,

$$f \circ \Phi_i^{-1}(z) = \sum_{n=N}^{\infty} a_n(z - \zeta)^n.$$

Prove that the integer N is independent of the chart (U_i, Φ_i) at p. (The integer N is called the *order* of f at p.) How would you define the order of a zero or a pole of a meromorphic function on a Riemann surface?

4H. Let S_1 be a connected Riemann surface and $f: S_1 \to S_2$, $g: S_1 \to S_2$ be two holomorphic functions which coincide on some sequence of points with a limit point in S_1. Prove that $f = g$. (Hint: let $X = \{q \in S_1 \mid f \text{ and } g \text{ coincide in a neighbourhood of } q\}$. X is open by definition; show that X is non-empty and closed.)

4I. Let S_1 be a connected Riemann surface and $f: S_1 \to S_2$ be a non-constant holomorphic function. Prove that f is an open mapping. (Use the corresponding result in the complex plane, Theorem A.10.)

4J. Let S be a connected Riemann surface and $f: S \to \mathbb{C}$ be a non-constant analytic function. Prove that $|f|$ has no relative maximum in S. (Use the maximum modulus principle in $\mathbb{C}$, Theorem A.11.)

4K. Show that if S is a compact connected Riemann surface the only analytic functions $f: S \to \mathbb{C}$ are the constant functions.

4L. Use the Riemann–Hurwitz formula to find the genus of the Riemann surfaces of the many-valued functions given in (b) and (c) of Exercise 4E.

4M. Find the genus of the Riemann surfaces of the following algebraic functions:

(a) $w^8 + z^8 - 1 = 0$;

(b) $w^2 - z^4(z - 1) = 0$;

(c) $w^3 - w + z = 0$.

4N. Let B be a finite subset of the sphere Σ consisting of $r \geqslant 2$ points and let $n \geqslant 1$ be an integer. Show that there exists a compact orientable surface S and a branched covering $p: S \to \Sigma$ with branch points of order $(n - 1)$ at each of the points of B if and only if either

(i) r is even; or

(ii) r and n are both odd.

(Hint: to construct such a covering consider an equation of the form $w^n = f(z)$ where f is a suitable rational function.)

4P. Prove that the projective plane contains a subset homeomorphic to the Möbius band and deduce that the projective plane is non-orientable.

4Q. Prove the statement made after Theorem 4.18.2 that if Ω is a lattice then $G = \mathrm{Aut}(\mathbb{C}/\Omega)$ has a normal subgroup N isomorphic to $\mathbb{C}/\Omega$ and that G/N is a cyclic group of order 2, 4 or 6. (Use Exercise 3H.) Describe the lattices Ω for which G/N has order 4 or 6.

4R. (a) Let T_λ denote the cyclic group of automorphism of the plane generated by $z \mapsto z + \lambda$ ($\lambda \in \mathbb{C} \setminus \{0\}$). Define an analytic atlas on $\mathbb{C}/T_\lambda$ such that the natural projection $p: \mathbb{C} \mapsto \mathbb{C}/T_\lambda$ is holomorphic.

(b) Show that $\mathbb{C}/T_\lambda$ is conformally equivalent to $\mathbb{C}/T_1$.

(c) Give an example of a non-constant analytic function

$$f : \mathbb{C}/T_\lambda \to \mathbb{C}.$$

4S. Prove that the sphere is a two-sheeted covering surface of the projective plane. Describe the covering transformations.

4T. Let S_1 be a compact surface on which there is a polygonal subdivision M. Let S_2 be another compact surface and suppose that $p : S_2 \to S_1$ is an n-sheeted covering map. Use the method of proof of Theorem 4.16.3 to show that

$$\chi(S_2) = n\chi(S_1),$$

and deduce that a finite-sheeted covering surface of a compact orientable surface of genus 1 is itself of genus 1.

4U. Prove that if G is a discontinuous group of homeomorphisms of a surface S, then S is a covering surface of S/G.

5

PSL(2, ℝ) *and its discrete subgroups*

At the end of the previous chapter we noted how important the group *PSL*(2, ℝ) and its subgroups are to Riemann surface theory. In this chapter we give an introduction to this topic. The plan of the chapter is as follows.

In §§5.1 and 5.2 we discuss basic algebraic properties of *PSL*(2, ℝ). Much of this is similar to some of the work in Chapter 2 on the group *PSL*(2, ℂ). In §§5.3 and 5.4 we discuss the hyperbolic metric on the upper half-plane 𝒰. With this metric 𝒰 becomes a model of the hyperbolic plane and *PSL*(2, ℝ) now acts as a group of isometries. In §§5.6 and 5.7 we introduce Fuchsian groups; these are discrete subgroups of *PSL*(2, ℝ) and all subgroups of *PSL*(2, ℝ) which act discontinuously on 𝒰 come into this class. Thus the groups G of Theorem 4.19.8 are all Fuchsian groups. As these groups are discontinuous groups of hyperbolic isometries they are comparable with lattices which are discontinuous groups of Euclidean isometries. As with lattices, the discontinuity implies the existence of a fundamental region and this is studied in §5.8. As before, we can use Dirichlet regions but now these are hyperbolic polygons and not Euclidean polygons as they were in §2.4. In §5.9, the quotient-spaces of the upper half-plane by Fuchsian groups are proved to be Riemann surfaces. In §5.10, the hyperbolic area of a fundamental region is shown to be an important invariant and it is used in §5.11 to give results about automorphism groups of compact Riemann surfaces of genus $g > 1$. These are necessarily finite groups, unlike the cases $g = 0, 1$ treated in earlier chapters, where infinite groups may occur.

5.1 The transformations of *PSL*(2, ℝ)

Let

$$T(z) = \frac{az + b}{cz + d}, \quad a, b, c, d \in \mathbb{R}, \quad \Delta = ad - bc > 0; \tag{5.1.1}$$

then dividing the numerator and denominator by $\sqrt{\Delta}$ we obtain

$$T(z) = \frac{(a/\sqrt{\Delta})z + (b/\sqrt{\Delta})}{(c/\sqrt{\Delta})z + (d/\sqrt{\Delta})}$$

and as $(a/\sqrt{\Delta})(d/\sqrt{\Delta}) - (b/\sqrt{\Delta})(c/\sqrt{\Delta}) = 1$, this shows that $T \in PSL(2, \mathbb{R})$ (see §2.1). In particular $PSL(2, \mathbb{R})$ contains all transformations of the form

$$z \mapsto az + b = \frac{(\sqrt{a})z + b/\sqrt{a}}{0 \cdot z + 1/\sqrt{a}} \quad (a, b \in \mathbb{R}, a > 0),$$

and hence all transformations of the form $z \mapsto az, (a > 0)$. Another important transformation of $PSL(2, \mathbb{R})$ is

$$z \to -1/z = \frac{0 \cdot z - 1}{1 \cdot z + 0}.$$

If $T(z)$ is the transformation in (5.1.1) then

$$T(z) = \frac{(az + b)(c\bar{z} + d)}{|cz + d|^2} = \frac{(acz\bar{z} + bd) + (adz + bc\bar{z})}{|cz + d|^2},$$

so that if $z = x + iy$, and $T(z) = u + iv$, then

$$v = \frac{(ad - bc)y}{|cz + d|^2} \tag{5.1.2}$$

and as $ad - bc > 0$ this confirms that T is an automorphism of $\mathcal{U}$. In Theorem 4.17.3, the stronger result that $\operatorname{Aut} \mathcal{U} = PSL(2, \mathbb{R})$ is proved.

On the other hand, if $ad - bc < 0$ then T maps $\mathcal{U}$ conformally and bijectively onto the lower half-plane.

5.2 Transitivity, conjugacy and centralisers

Theorem 5.2.1.

(i) $PSL(2, \mathbb{R})$ *is transitive on* $\mathcal{U}$.
(ii) $PSL(2, \mathbb{R})$ *is doubly transitive on* $\mathbb{R} \cup \{\infty\}$.

Proof. (i) Let $ai + b \in \mathcal{U}$, so that $a > 0$. Then if $T(z) = az + b$, $T \in PSL(2, \mathbb{R})$ and $T(i) = ai + b$. Thus the orbit of i under the action of $PSL(2, \mathbb{R})$ is $\mathcal{U}$ and so $PSL(2, \mathbb{R})$ is transitive on $\mathcal{U}$.

(ii) If $a, b \in \mathbb{R}$, $a > b$ then if $S(z) = (z - a)/(z - b)$, $S \in PSL(2, \mathbb{R})$ maps the ordered pair (a, b) to $(0, \infty)$. Also $z \mapsto -1/z$ maps $(0, \infty)$ to $(\infty, 0)$ and $z \mapsto z + b$ maps $(0, \infty)$ to (b, ∞). It follows that the orbit of $(0, \infty)$ under the action of $PSL(2, \mathbb{R})$ consists of all ordered pairs (a, b), $(a, b \in \mathbb{R} \cup \{\infty\}, a \neq b)$, so that $PSL(2, \mathbb{R})$ is doubly transitive on $\mathbb{R} \cup \{\infty\}$. $\square$

We now discuss the conjugacy classes in $PSL(2, \mathbb{R})$. In §§2.9 and 2.10 we obtained the conjugacy classes in $PSL(2, \mathbb{C})$. The treatment here is similar but a little care is needed in that it does not necessarily follow that if two transformations of $PSL(2, \mathbb{R})$ are conjugate in $PSL(2, \mathbb{C})$ then they need be conjugate in $PSL(2, \mathbb{R})$. We can still classify the elements as parabolic, hyperbolic, elliptic and loxodromic according to the trace as we did in §2.10. Also, the fixed-points of the transformations still play an important role. The fixed-points are found by solving

$$z = \frac{az + b}{cz + d}, \quad a, b, c, d \in \mathbb{R}, \quad ad - bc = 1,$$

and as a, b, c, d are real we see that this transformation has either two fixed-points in $\mathbb{R} \cup \{\infty\}$, one fixed point in $\mathbb{R} \cup \{\infty\}$, or a pair of complex conjugate fixed-points.

Parabolic elements $(|a + d| = 2)$. Let T be a parabolic element with the single fixed-point $\alpha \in \mathbb{R} \cup \{\infty\}$. By Theorem 5.2.1 (ii) there exists $S \in PSL(2, \mathbb{R})$ such that $S(\alpha) = \infty$. Hence STS^{-1} is parabolic with fixed-point ∞ and therefore

$$W = STS^{-1} : z \mapsto z + t \qquad (t \in \mathbb{R} \setminus \{0\}).$$

Let $V(z) = (1/|t|)z$. Then $VWV^{-1} : z \mapsto z \pm 1$, the sign depending on whether $t > 0$ or $t < 0$. A simple calculation shows that $z \mapsto z + 1$ is not conjugate to $z \mapsto z - 1$ in $PSL(2, \mathbb{R})$ so that there are two conjugacy classes of parabolic elements in $PSL(2, \mathbb{R})$.

Hyperbolic elements $(|a + d| > 2)$. Let T be hyperbolic with fixed-points $\alpha, \beta \in \mathbb{R} \cup \{\infty\}$. By Theorem 5.2.1 (ii), there exists $S \in PSL(2, \mathbb{R})$ such that S maps α to 0 and β to ∞. Thus

$$STS^{-1} = U_\lambda \qquad \lambda \in \mathbb{R} \setminus \{0, 1\},$$

in the notation of §2.9. If $B(z) = -1/z$ then $BU_\lambda B^{-1} = U_{\lambda^{-1}}$ so that U_λ is conjugate to $U_{\lambda^{-1}}$ in $PSL(2, \mathbb{R})$. However, by Theorem 2.9.3, U_λ is not conjugate to U_κ if $\kappa \neq \lambda$ or λ^{-1}. Thus every hyperbolic element of $PSL(2, \mathbb{R})$ is conjugate to a unique element of the form $U_\lambda, (\lambda > 1)$.

Elliptic elements $(|a + d| < 2)$. Let T be an elliptic element with fixed-point $\xi \in \mathcal{U}$. By Theorem 5.2.1 (i) there exists $S \in PSL(2, \mathbb{R})$ such that $S(\xi) = i$. Then $S(\bar{\xi}) = \bar{i} = -i$ so that $W = STS^{-1}$ is an elliptic element with fixed-points i and $-i$. Therefore

$$\frac{W(z) - i}{W(z) + i} = \lambda \left(\frac{z - i}{z + i} \right).$$

As W maps $\mathbb{R} \cup \{\infty\}$ to itself we can find $\alpha \in \mathbb{R}$ such that $W(\alpha) \in \mathbb{R}$. Hence

$$\left| \frac{W(\alpha) - i}{W(\alpha) + i} \right| = \left| \frac{\alpha - i}{\alpha + i} \right| = 1,$$

and so $\lambda = e^{i\theta}$ $(0 \leqslant \theta < 2\pi)$. Therefore T is conjugate to W, where

$$\frac{W(z) - i}{W(z) + i} = e^{i\theta} \left(\frac{z - i}{z + i} \right), \qquad (0 \leqslant \theta < 2\pi). \tag{5.2.2}$$

We cannot get a simpler form in $PSL(2, \mathbb{R})$, but it is worth noting that if

$$\frac{z - i}{z + i} = z', \qquad \frac{W(z) - i}{W(z) + i} = w'$$

then $z', w' \in \mathscr{D}$, the unit disc, and

$$w' = e^{i\theta} z'.$$

Thus, inside $PSL(2, \mathbb{C})$, T is conjugate to a rotation of the unit disc through θ. We note that, as in $PSL(2, \mathbb{C})$, all elements of finite order are elliptic.

Loxodromic elements $(a + d$ not real$)$. As all elements of $PSL(2, \mathbb{R})$ have real trace there are no loxodromic elements in $PSL(2, \mathbb{R})$.

If G is any group and $g \in G$, then the *centraliser* of g in G is defined by

$$C_G(g) = \{h \in G \mid hg = gh\}.$$

$C_G(g)$ is a subgroup of G and if $k \in G$ then

$$C_G(kgk^{-1}) = kC_G(g)k^{-1}$$

so that to calculate the structure of the centralisers of elements of $PSL(2, \mathbb{R})$ we need only look at the centralisers of the representatives of the conjugacy classes found above. The calculations are made easier by the following lemma which is true in any group of transformations.

Lemma 5.2.3.   *If $ST = TS$ then S maps the fixed-point set of T to itself.*

Proof.   Suppose that T fixes p. Then

$$S(p) = ST(p) = TS(p) = T(S(p)),$$

so that $S(p)$ is fixed by T. $\square$

Now suppose that $T(z) = z + 1$. Then if $S \in C_{PSL(2, \mathbb{R})}(T)$ then S fixes ∞. Thus $S(z) = az + b$ and $ST = TS$ gives $a = 1$. Hence

$$C_{PSL(2,\mathbb{R})}(T) = \{z \mapsto z + k \,|\, k \in \mathbb{R}\}$$

with the same result for the centraliser of $z \mapsto z - 1$.

In a similar way we show that the centraliser of $z \mapsto \lambda z$ $(\lambda > 0, \lambda \neq 1)$ in $PSL(2, \mathbb{R})$ consists of all transformations of the form $z \mapsto \mu z$ $(\mu > 0)$ and the centraliser of the elliptic element (5.2.2) consists of all transformations

$$\frac{W(z) - i}{W(z) + i} = e^{i\phi}\left(\frac{z - i}{z + i}\right) \qquad (0 \leqslant \phi < 2\pi).$$

From these calculations we deduce the following two results.

Theorem 5.2.4. *Two non-identity elements of PSL(2, $\mathbb{R}$) commute if and only if they have the same fixed-point set.* $\square$

Theorem 5.2.5. *The centraliser in PSL(2, $\mathbb{R}$) of a hyperbolic (resp. parabolic, elliptic) element of PSL(2, $\mathbb{R}$) consists of all hyperbolic (resp. parabolic, elliptic) elements with the same fixed-point set, together with the identity element.* $\square$

5.3. The hyperbolic metric

The connection between the group $PSL(2, \mathbb{R})$ and hyperbolic geometry was discovered by Henri Poincaré (1854–1912) and published in 1882. (Poincaré used this discovery to illustrate the sometimes spontaneous nature of mathematical creativity. He relates how, when boarding a bus, the idea suddenly came to him that the transformations he had used to define Fuchsian functions were identical with those of non-Euclidean geometry. For a more detailed account see the chapter on Poincaré in Bell [1953].)

The transformations used to define Fuchsian functions are real Möbius transformations and the non-Euclidean geometry referred to is the hyperbolic geometry of Bolyai and Lobatchewsky. In this geometry, Euclid's parallel postulate is replaced by the postulate that given a line L and a point P not on L, then through P there are an infinite number of lines not meeting L. In this section we shall show that with 'line' defined appropriately, the upper half-plane gives a model of hyperbolic geometry.

We start by defining the hyperbolic length of a piecewise differentiable path γ. First, recall how we define the Euclidean length of a piecewise differentiable path β in $\mathbb{R}^2$. Suppose that $\beta: I \to \mathbb{R}^2$, where $I = [0, 1]$, is given by $\beta(t) = (x(t), y(t))$ where x and y are piecewise differentiable functions. The

Euclidean length $e(\beta)$ is defined by means of the formula

$$ds^2 = dx^2 + dy^2,$$

or, more precisely,

$$e(\beta) = \int_0^1 \sqrt{\left(\left(\frac{dx}{dt}\right)^2 + \left(\frac{dy}{dt}\right)^2\right)} \, dt,$$

where, of course, the positive square root is to be taken.

(If we change the parametrisation by letting $t = \alpha(s)$, where $\alpha:[0, 1] \to [0, 1]$ is a monotonic function then $(x(t), y(t)) = (x_1(s), y_1(s))$, where $x_1 = x \circ \alpha$, $y_1 = y \circ \alpha$, and we get the same value for the length on replacing x, y by x_1, y_1.)

We define the *hyperbolic length* in $\mathcal{U}$ by means of the formula

$$ds^2 = \frac{dx^2 + dy^2}{y^2} = \frac{|dz|^2}{y^2} \qquad (z = x + iy).$$

More precisely, if $\gamma: I \to \mathcal{U}$ is a piecewise differentiable path with $\gamma(t) = x(t) + iy(t) = z(t)$ then its hyperbolic length $h(\gamma)$ is given by

$$h(\gamma) = \int_0^1 \frac{\sqrt{\left(\left(\frac{dx}{dt}\right)^2 + \left(\frac{dy}{dt}\right)^2\right)}}{y} \, dt = \int_0^1 \frac{\left|\frac{dz}{dt}\right|}{y} \, dt.$$

Theorem 5.3.1. *If $T \in PSL(2, \mathbb{R})$ then $h(T(\gamma)) = h(\gamma)$. Thus hyperbolic length is invariant under transformations of $PSL(2, \mathbb{R})$.*

Proof.   Let

$$T(z) = \frac{az + b}{cz + d} \qquad (a, b, c, d \in \mathbb{R}, ad - bc = 1).$$

Then

$$\frac{dT}{dz} = \frac{a(cz + d) - c(az + b)}{(cz + d)^2} = \frac{1}{(cz + d)^2}.$$

Also, if $z = x + iy$, $T(z) = u + iv$, then by (5.1.2)

$$v = \frac{y}{|cz + d|^2},$$

and hence

$$\left|\frac{dT}{dz}\right| = \frac{v}{y}. \tag{5.3.2}$$

Thus

$$h(T(\gamma)) = \int_0^1 \frac{\left|\frac{dT}{dt}\right| dt}{v} = \int_0^1 \frac{\left|\frac{dT}{dz}\frac{dz}{dt}\right| dt}{v} = \int_0^1 \frac{v\left|\frac{dz}{dt}\right| dt}{yv}$$

$$= \int_0^1 \frac{\left|\frac{dz}{dt}\right| dt}{y} = h(\gamma). \quad \square$$

We now wish to show that between any two points in $\mathcal{U}$ there is a unique path of shortest hyperbolic length and to determine this path. Such paths will be called *hyperbolic line segments* or *H-line segments* and from Theorem 5.3.1 it follows that elements of $PSL(2, \mathbb{R})$ map H-line segments to H-line segments. (The prefix H always denotes hyperbolic.)

We first show that the H-line segment joining two points ia, ib ($b > a$) on the imaginary axis is the segment of the imaginary axis joining them. Let $\kappa: I \to \mathcal{U}$ denote this segment so that $\kappa(t) = (0, y(t))$ where $dy/dt > 0$, $y(0) = a$, $y(1) = b$. Hence

$$h(\kappa) = \int_0^1 \frac{\sqrt{\left(\left(\frac{dy}{dt}\right)^2\right)} dt}{y(t)} = \int_0^1 \frac{\left|\frac{dy}{dt}\right| dt}{y(t)} = \int_0^1 \frac{\frac{dy}{dt} dt}{y(t)} = \int_a^b \frac{dy}{y} = \ln \frac{b}{a}.$$

If $\tilde{\kappa}: I \to \mathcal{U}$ is any other piecewise differentiable path joining ia to ib, with $\tilde{\kappa}(t) = (\tilde{x}(t), \tilde{y}(t))$, then

$$h(\tilde{\kappa}) = \int_0^1 \frac{\sqrt{\left(\left(\frac{d\tilde{x}}{dt}\right)^2 + \left(\frac{d\tilde{y}}{dt}\right)^2\right)} dt}{\tilde{y}(t)} \geqslant \int_0^1 \frac{\left|\frac{d\tilde{y}}{dt}\right| dt}{\tilde{y}(t)} \geqslant \int_0^1 \frac{\frac{d\tilde{y}}{dt} dt}{\tilde{y}(t)} = h(\kappa).$$

Equality holds if and only if $d\tilde{x}/dt = 0$ and $d\tilde{y}/dt \geqslant 0$, where they exist. As $\tilde{\kappa}$ is piecewise differentiable, it follows that $\tilde{\kappa}$ is the Euclidean line segment joining ia and ib.

Exactly the same calculation shows that if z_1 and z_2 have the same real part, then the unique H-line segment joining them is the unique Euclidean line segment joining them.

Fig. 5.1

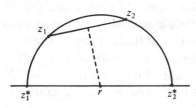

Now suppose that z_1 and z_2 do not have the same real part. Then the perpendicular bisector of the Euclidean line segment joining them cuts the real axis in a point r which is the centre of the unique Euclidean circle Q through z_1 and z_2 orthogonal to $\mathbb{R}$, as illustrated in Fig. 5.1.

Suppose that Q intersects $\mathbb{R}$ at z_1^*, z_2^*. By Theorem 5.2.1 (ii), there exists $T \in PSL(2, \mathbb{R})$ such that $T(z_1^*) = 0$, $T(z_2^*) = \infty$, and by Theorems 2.4.1, 2.11.3 it follows that $T(Q)$ is the imaginary axis so that the H-line segment joining $T(z_1)$ and $T(z_2)$ is the segment of the imaginary axis joining them. Hence, by Theorem 5.3.1, there is a unique H-line segment joining z_1 and z_2, namely the arc of Q in $\mathcal{U}$ which joins z_1 and z_2. We have now determined all the H-line segments.

Theorem 5.3.3. *The H-line segments in $\mathcal{U}$ are arcs of semi-circles with centre on the real axis or segments of Euclidean lines perpendicular to the real axis.* □

We call circles whose centres lie on the real axis or Euclidean lines perpendicular to the real axis *hyperbolic lines* or *H-lines*. We regard vertical H-lines as having one end-point at ∞, so that every H-line has two end-points in $\mathbb{R} \cup \{\infty\}$. H-lines are illustrated in Fig. 5.2.

Fig. 5.2

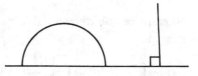

Theorem 5.3.4. *$PSL(2, \mathbb{R})$ acts transitively on the set of all H-lines.*

Proof. Let Q, Q' be two H-lines. If Q has end-points $s, t \in \mathbb{R} \cup \{\infty\}$ and Q' has end-points $s', t' \in \mathbb{R} \cup \{\infty\}$ then by Theorem 5.2.1 (ii) there exists $T \in PSL(2, \mathbb{R})$ such that $T(s) = s'$, $T(t) = t'$. Clearly, the end-points of an H-line determine it uniquely, so that $T(Q) = Q'$. □

We define the *hyperbolic distance* ρ between two points z, w of $\mathcal{U}$ to be the H-length of the H-line segment joining them. This clearly defines $\mathcal{U}$ as a metric space, the triangle inequality following from the result that an H-line segment is the unique shortest path between two points. The upper half-plane with the ρ-metric is a model of the hyperbolic plane; if we define two H-lines to be parallel if they do not intersect then given any H-line Q and a point p not on Q then there are infinitely many H-lines through

p not intersecting Q. (For a more detailed account of the axiomatic approach to hyperbolic geometry in this context the reader is referred to Magnus [1974].)

We note the following important result which follows immediately from the work in this section.

Theorem 5.3.5. $\quad \rho(T(z), T(w)) = \rho(z, w)$ *for all* $T \in PSL(2, \mathbb{R})$ *and all* $z, w \in \mathcal{U}$. $\quad \square$

5.4 Computation of $\rho(z, w)$
(The work in this section is based on Alan Beardon's treatment in Harvey [1977]. Also see Beardon [1983].)

If ia, ib $(b > a)$ are two points on the imaginary axis then in the previous section we showed that $\rho(ib, ia) = \ln(b/a)$. If z, w are any two points in $\mathcal{U}$ then it is usually more difficult to compute $\rho(z, w)$ directly so in this section we derive some simple formulae for the hyperbolic distance.

Lemma 5.4.1. *Let* $z, w \in \mathcal{U}$ $(z \neq w)$ *and let the H-line Q joining z and w have end-points z^*, w^* in $\mathbb{R} \cup \{\infty\}$, chosen in such a way that z lies between z^* and w. Then there exists a unique element $T \in PSL(2, \mathbb{R})$ such that $T(z^*) = 0$, $T(w^*) = \infty$ and $T(z) = i$. Also $T(w) = ri$ $(r > 1)$ and $\rho(z, w) = \ln r$.*

Proof. Assume that neither z^* nor w^* is ∞. We may suppose that $z^* > w^*$, otherwise we just relabel.

If we let

$$S(\zeta) = \frac{\zeta - z^*}{\zeta - w^*},$$

then $S \in PSL(2, \mathbb{R})$, $S(z^*) = 0$, $S(w^*) = \infty$, so that S maps Q to the imaginary axis. If $S(z) = ki$ then

$$T = U_{1/k} \circ S$$

is the required transformation (recall that $U_\lambda(z) = \lambda z$). It is unique by Theorem 2.5.1, and as z lies between z^* and w, $T(z) = i$ lies between $T(z^*) = 0$ and $T(w)$, so that $T(w) = ri, r > 1$. By Theorem 5.3.5, $\rho(z, w) = \rho(T(z), T(w)) = \rho(i, ri) = \ln r$. $\quad \square$

We define an *H-invariant* to be a function $g(z_1, \ldots, z_n)$ $(z_i \in \mathcal{U})$ for which $g(T(z_1), \ldots, T(z_n)) = g(z_1, \ldots, z_n)$ for all $T \in PSL(2, \mathbb{R})$. For example, $\rho(z, w)$ is an H-invariant by Theorem 5.3.5

Lemma 5.4.2.

(i) If z, w, z^*, w^* are as defined in Lemma 5.4.1 then the cross-ratio

$$\eta(z, w) = (w, z^*; z, w^*)$$

is an H-invariant.

(ii) For all $z, w \in \mathcal{U}$,

$$\tau(z, w) = \left| \frac{z - w}{z - \bar{w}} \right|$$

is an H-invariant.

Proof.

(i) As *PSL*(2, ℝ) preserves H-lines, the H-line joining $T(z)$ and $T(w)$ has end-points $T(z^*)$ and $T(w^*)$ and so the result follows from the invariance of the cross-ratio under Möbius transformations (Theorem 2.5.5).

(ii) This follows from the formula

$$|T(z) - T(w)| = |z - w| |T'(z) T'(w)|^{1/2},$$

which is easily proved by direct computation. □

Hence if z, w, r are as defined in Lemma 5.4.1, then

$$\eta(z, w) = (ri, 0; i, \infty) = r,$$

and therefore

$$\rho(z, w) = \ln \eta(z, w).$$

Also

$$\tau(z, w) = \tau(i, ri) = \frac{r - 1}{r + 1} = \frac{e^{\rho(z,w)} - 1}{e^{\rho(z,w)} + 1}, \tag{5.4.3}$$

and therefore

$$\rho(z, w) = \ln \left\{ \frac{1 + \tau(z, w)}{1 - \tau(z, w)} \right\} = \ln \left\{ \frac{|z - \bar{w}| + |z - w|}{|z - \bar{w}| - |z - w|} \right\},$$

which is an explicit formula for the hyperbolic distance.

We can obtain an attractive and useful formula from (5.4.3) by using the identities

$$\frac{e^u - 1}{e^u + 1} = \tanh \frac{u}{2},$$

and

$$\sinh^2 \frac{u}{2} = \frac{\tanh^2(u/2)}{1 - \tanh^2(u/2)}.$$

We obtain

$$\sinh^2 \tfrac{1}{2}\rho(z, w) = \frac{\tau(z, w)^2}{1 - \tau(z, w)^2} = \frac{|z - w|^2}{|z - \bar{w}|^2 - |z - w|^2}.$$

Now

$$|z - \bar{w}|^2 - |z - w|^2 = (z - \bar{w})(\bar{z} - w) - (z - w)(\bar{z} - \bar{w}) = -(z - \bar{z})(w - \bar{w})$$
$$= 4 \operatorname{Im}(z) \operatorname{Im}(w).$$

Thus

$$\sinh^2 \tfrac{1}{2}\rho(z, w) = \frac{|z - w|^2}{4 \operatorname{Im}(z) \operatorname{Im}(w)}. \tag{5.4.4}$$

For example, suppose we wish to find the hyperbolic circle with centre i and radius δ. This is

$$\{z \,|\, \sinh^2 \tfrac{1}{2}\rho(z, i) = \sinh^2 \tfrac{1}{2}\delta\}$$
$$= \{z \,|\, |z - i|^2 = 4y \sinh^2 \tfrac{1}{2}\delta\} \qquad \text{(where } z = x + iy)$$
$$= \{z \,|\, x^2 + y^2 + 1 = 2y(2 \sinh^2 \tfrac{1}{2}\delta + 1) = 2y \cosh \delta\}$$
$$= \{z \,|\, x^2 + (y - \cosh \delta)^2 = \cosh^2 \delta - 1 = \sinh^2 \delta\},$$

which is a Euclidean circle with centre $(0, \cosh \delta)$ and radius $\sinh \delta$, as pictured in Fig. 5.3.

Fig. 5.3

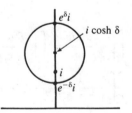

Now by Theorem 2.4.1, $PSL(2, \mathbb{R})$ maps Euclidean circles in $\mathcal{U}$ to Euclidean circles in $\mathcal{U}$, and clearly maps H-circles to H-circles. By Theorem 5.2.1(i), we see that every H-circle is a Euclidean circle. Also, every Euclidean circle is an H-circle. As the family of all open Euclidean discs coincides with the family of open hyperbolic discs we obtain the following result.

Theorem 5.4.5. *The topology induced by the hyperbolic metric is the same as the topology induced by the Euclidean metric.* $\square$

5.5 Hyperbolic area and the Gauss–Bonnet formula

If $E \subseteq \mathscr{U}$ we define $\mu(E)$, the hyperbolic area (H-area) of E by

$$\mu(E) = \iint_E \frac{dxdy}{y^2}$$

if this integral exists.

Theorem 5.5.1. $\mu(T(E)) = \mu(E)$ *for all* $T \in PSL(2, \mathbb{R})$. *Thus the H-area is invariant under all transformations of* $PSL(2, \mathbb{R})$.

Proof. Let $z = x + iy$,

$$T(z) = \frac{az+b}{cz+d}, \qquad a, b, c, d \in \mathbb{R}, \quad ad - bc = 1,$$

and $w = T(z) = u + iv$. Then using the Cauchy–Riemann equations we calculate the Jacobian

$$\frac{\partial(u, v)}{\partial(x, y)} = \frac{\partial u}{\partial x}\frac{\partial v}{\partial y} - \frac{\partial u}{\partial y}\frac{\partial v}{\partial x}$$

$$= \left(\frac{\partial u}{\partial x}\right)^2 + \left(\frac{\partial v}{\partial x}\right)^2 = \left|\frac{dT}{dx}\right|^2 = \left|\frac{dT}{dz}\right|^2$$

$$= \frac{1}{|cz+d|^4}.$$

Thus

$$\mu(T(E)) = \iint_{T(E)} \frac{dudv}{v^2} = \iint_E \frac{\partial(u, v)}{\partial(x, y)} \frac{dxdy}{v^2}$$

$$= \iint_E \frac{1}{|cz+d|^4} \frac{|cz+d|^4}{y^2} dxdy = \mu(E)$$

using (5.3.2). □

A *hyperbolic n-sided polygon* is a closed set in the closure of $\mathscr{U}$ in Σ bounded by n hyperbolic line segments. If two line segments intersect then

Fig. 5.4

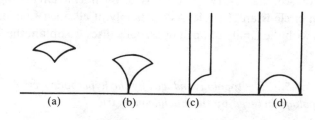

(a)   (b)   (c)   (d)

the point of intersection is called a *vertex* of the polygon. We allow vertices on $\mathbb{R} \cup \{\infty\}$ although no segment of the real axis can belong to a hyperbolic polygon. For example, in Fig. 5.4 we illustrate the four types of hyperbolic triangle, depending on whether 0, 1, 2 or 3 vertices of the triangle belong to $\mathbb{R} \cup \{\infty\}$.

The Gauss–Bonnet formula shows that the H-area of an H-triangle depends only on its angles. It bears a remarkable resemblance to the formula for the area of a spherical triangle in Theorem 2.14.1. The *angle* between two H-lines in $\mathscr{U}$ is defined to be the angle between their tangents at their point of intersection; two H-lines intersect at a point of $\mathbb{R} \cup \{\infty\}$ with angle zero.

Theorem 5.5.2. (*Gauss–Bonnet*). *Let Δ be a hyperbolic triangle with angles* α, β, γ. *Then*

$$\mu(\Delta) = \pi - \alpha - \beta - \gamma.$$

Proof.
Case 1. We first consider the case where two sides of Δ are vertical H-lines. The base of Δ is then a segment of a Euclidean semi-circle. By applying transformations of the form $z \mapsto z + \kappa$ ($\kappa \in \mathbb{R}$), $z \mapsto \lambda z$ ($\lambda > 0$) we can assume that the semi-circle has centre 0 and radius 1; these transformations will not change the H-area by Theorem 5.5.1, and as vertical sides remain vertical will preserve the zero angles, the other angles being preserved by conformality.

We can thus assume that Δ is the triangle depicted in Fig. 5.5 (where the semi-circle has radius 1).

Fig. 5.5

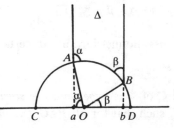

(By simple geometry the angles AOC, BOD are α, β respectively.) Assume that the vertical H-lines through A and B are $x = a$ and $x = b$ respectively. We now calculate

$$\mu(\Delta) = \iint_{\Delta} \frac{dx\,dy}{y^2} = \int_a^b dx \int_{\sqrt{(1-x^2)}}^{\infty} \frac{dy}{y^2} = \int_a^b \frac{dx}{\sqrt{(1-x^2)}}.$$

Make the substitution $x = \cos\theta$ $(0 \leqslant \theta \leqslant \pi)$; then

$$\mu(\Delta) = \int_{\pi-\alpha}^{\beta} \frac{-\sin\theta \, d\theta}{\sin\theta} = \pi - \alpha - \beta.$$

Case 2. We assume that Δ has a vertex on the real line as in (b) of Fig. 5.4. By applying a transformation of $PSL(2, \mathbb{R})$ we can map this vertex to ∞ without altering the H-area or the angles. Thus if the non-zero angles of Δ are α and β then $\mu(\Delta) = \pi - \alpha - \beta$ by Case 1.

Case 3. Δ has no vertices in $\mathbb{R} \cup \{\infty\}$. Suppose that Δ has vertices A, B, C and that (the H-line segment) AB produced cuts $\mathbb{R}$ at D. (We can apply a transformation of $PSL(2, \mathbb{R})$ to make sure that no sides of ABC are vertical H-lines.) Then we have the situation of Fig. 5.6. Here, $\Delta = \Delta_1 \setminus \Delta_2$ where Δ_1 is the H-triangle with vertices A, C, D and Δ_2 is the H-triangle with vertices B, C, D.

Fig. 5.6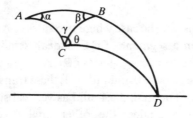

Then

$$\mu(\Delta) = \mu(\Delta_1) - \mu(\Delta_2)$$
$$= \pi - \alpha - (\gamma + \theta) - [\pi - \theta - (\pi - \beta)]$$
$$= \pi - \alpha - \beta - \gamma. \quad \square$$

We can extend the Gauss–Bonnet formula to certain types of hyperbolic polygons.

Definition. A subset C of $\mathcal{U}$ is *hyperbolically starlike* if there is a point O in the interior of C such that for all $P \in C$, the H-line segment joining O and P lies in C.

The most important examples of hyperbolically starlike sets are *hyperbolically convex sets*; C is hyperbolically convex if for any two points $P, Q \in C$, the H-line segment joining P and Q belongs to C.

Corollary 5.5.6. *Let Π be an n-sided hyperbolically starlike polygon with* *angles* $\alpha_1, \alpha_2, \ldots, \alpha_n$. *Then*

$$\mu(\Pi) = (n - 2)\pi - \alpha_1 - \alpha_2 - \cdots - \alpha_n.$$

Proof. Let the vertices of Π be $A_1, \ldots, A_n$ and let O be a point in the interior of Π such that the H-line segments $OA_1, \ldots, OA_n$ all lie in Π. The result follows by adding the H-areas of the triangles OA_1A_2, $OA_2A_3, \ldots, OA_nA_1$. $\square$

5.6 Fuchsian groups

PSL$(2, \mathbb{R})$, besides being a group, is also a topological space in that the transformation $z \mapsto (az + b)/(cz + d)$ can be identified with the point $(a, b, c, d) \in \mathbb{R}^4$. More precisely, as a topological space, $SL(2, \mathbb{R})$ can be identified with the subset of $\mathbb{R}^4$,

$$X = \{(a, b, c, d) \in \mathbb{R}^4 \mid ad - bc = 1\},$$

and if we define $\delta(a, b, c, d) = (-a, -b, -c, -d)$ then $\delta : X \to X$ is a homeomorphism and δ together with the identity forms a cyclic group of order 2 acting on X. *PSL*$(2, \mathbb{R})$ can be topologised as the quotient-space. (In the exercises it is shown that *PSL*$(2, \mathbb{R})$ is homeomorphic to $\mathbb{R}^2 \times S^1$ so that it is a 3-dimensional manifold. It is also shown that the group multiplication and taking of inverses are continuous with this topology so that *PSL*$(2, \mathbb{R})$ is a topological group.)

Definition. A *Fuchsian group* is a discrete subgroup of *PSL*$(2, \mathbb{R})$.

Fuchsian groups were first studied systematically by Poincaré in 1880, although some particular examples such as the modular group and triangle groups had been investigated before that. Poincaré was led to Fuchsian groups after reading a paper by L. Fuchs on differential equations.

In many respects, Fuchsian groups are related to the lattices of Chapter 3. Lattices are discrete groups of Euclidean isometries and their quotients are compact Riemann surfaces homeomorphic to the torus. Fuchsian groups are discrete groups of hyperbolic isometries and their quotient-spaces are also Riemann surfaces (see Theorem 5.9.1). Functions which are invariant under lattices, namely the elliptic functions, form an important family of functions; similarly there are important functions, the *automorphic functions*, which are invariant under Fuchsian groups. Some examples of automorphic functions will be studied in the next chapter. There is, however, an important difference between lattices and Fuchsian groups. Even though there are an infinite number of distinct lattices, they are all topologically similar in that their quotient-spaces are all tori. On the other hand, it follows from Theorem 4.19.8 that all orientable surfaces other than the sphere, torus, plane or punctured plane are quotients of Fuchsian groups acting on $\mathcal{U}$ without fixed-points. For this reason, the geometric and algebraic structures of Fuchsian groups can vary widely and for the rest of this chapter we can only begin to describe their theory.

Even though a wide variety of Fuchsian groups exists there are only a few examples which can be explicitly written down. First of all there are the cyclic groups. *Hyperbolic cyclic groups* are generated by a hyperbolic element, for example $z \mapsto \lambda z$ ($\lambda > 1$) clearly generates a Fuchsian group consisting of hyperbolic elements together with the identity. *Parabolic cyclic groups* are generated by a parabolic element, for example $z \mapsto z + 1$ generates such a group. *Elliptic cyclic groups*, those generated by an elliptic element, are Fuchsian if and only if they are finite (see §5.7).

An example of a Fuchsian group with a more complicated structure is the *modular group PSL*(2, ℤ). This consists of all transformations

$$z \mapsto \frac{az + b}{cz + d} \qquad (a, b, c, d \in \mathbb{Z}, ad - bc = 1).$$

This is clearly discrete and hence a Fuchsian group. It will be studied in more detail later in this and in the next chapter.

Lattices have the important property that their action on $\mathbb{C}$ is discontinuous in the sense that every point of $\mathbb{C}$ has a neighbourhood which is carried outside itself by all elements of the lattice except for the identity (see §4.19). In general, Fuchsian groups do not have such a discontinuous behaviour for if elliptic elements are present then they fix points and these fixed-points cannot have such a neighbourhood. However, Fuchsian groups turn out to be discontinuous in the following slightly weaker sense.

Definition. Let G be a group of homeomorphisms of a topological space Y. Then G acts *properly discontinuously* on Y if each point $y \in Y$ has a neighbourhood V such that if $g(V) \cap V \neq \varnothing$ for $g \in G$, then $g(y) = y$. (In some books it is further required that the stabiliser of every point is finite. This condition holds for all Fuchsian groups – see Theorem 5.7.3.)

Every discontinuous group acts properly discontinuously; the finite group of homeomorphisms of $\mathbb{C}$ generated by $z \mapsto e^{2\pi i/n} z$ ($n = 2, 3, \ldots$) acts properly discontinuously but not discontinuously. Thus, despite its name proper discontinuity is a weaker condition than discontinuity. However, the term 'proper discontinuity' is now widely used in the literature.

In order to prove that Fuchsian groups act properly discontinuously on $\mathcal{U}$ we need the following lemma which will also be useful in proving other results.

Lemma 5.6.1. *Let $w \in \mathcal{U}$ be given and let K be a compact subset of $\mathcal{U}$. Then the set*

$$E = \{T \in PSL(2, \mathbb{R}) \mid T(w) \in K\}$$

is compact.

Proof. $PSL(2, \mathbb{R})$ is topologised as a quotient-space of $SL(2, \mathbb{R})$. Thus we have a continuous map $q : SL(2, \mathbb{R}) \to PSL(2, \mathbb{R})$ defined by

$$q\begin{pmatrix} a & b \\ c & d \end{pmatrix} = T, \qquad \text{where } T(z) = \frac{az + b}{cz + d}.$$

If we show that

$$E_1 = \left\{ \begin{pmatrix} a & b \\ c & d \end{pmatrix} \in SL(2, \mathbb{R}) \,\middle|\, \frac{aw + b}{cw + d} \in K \right\}$$

is compact then it follows that $E = q(E_1)$ is compact. We prove that E_1 is compact by showing it is closed and bounded when regarded as a subset of $\mathbb{R}^4$ (identifying $\begin{pmatrix} a & b \\ c & d \end{pmatrix}$ with (a, b, c, d)). We have a continuous map $\beta : SL(2, \mathbb{R}) \to \mathcal{U}$ defined by $\beta(A) = q(A)(w)$, and as $E_1 = \beta^{-1}(K)$ it follows that E_1 is closed being the inverse image of the closed set K.

We now show that E_1 is bounded. As K is bounded there exists $M_1 \in \mathbb{R}$ such that

$$\left| \frac{aw + b}{cw + d} \right| < M_1$$

for all $\begin{pmatrix} a & b \\ c & d \end{pmatrix} \in E_1$.

Also, as K is compact in $\mathcal{U}$, there exists $M_2 > 0$ such that

$$\text{Im}\left(\frac{aw + b}{cw + d} \right) \geqslant M_2.$$

As $ad - bc = 1$, (5.1.2) implies that the left-hand side of this inequality is $\text{Im}(w)/|cw + d|^2$ so that

$$|cw + d| \leqslant \sqrt{\left(\frac{\text{Im}(w)}{M_2} \right)},$$

and thus

$$|aw + b| \leqslant M_1 \sqrt{\left(\frac{\text{Im}(w)}{M_2} \right)},$$

and we deduce that a, b, c, d are bounded. $\square$

Corollary 5.6.2.   *Let* $w \in \mathcal{U}$ *and let* K *be a compact subset of* $\mathcal{U}$. *If* Γ *is a Fuchsian group then*

$$\{ T \in \Gamma \mid T(w) \in K \}$$

is finite.

Proof.   By Lemma 5.6.1 this set is compact. However, being a subset of a Fuchsian group it is also discrete. Hence by Theorem 3.2.2 it is finite.   □

Theorem 5.6.3.

(i) *Let* Γ *be a subgroup of* $PSL(2, \mathbb{R})$. *Then* Γ *is a Fuchsian group if and only if* Γ *acts properly discontinuously on* $\mathcal{U}$.

(ii) *Let* Γ *be a Fuchsian group and let* $p \in \mathcal{U}$ *be fixed by some element of* Γ. *Then there is a neighbourhood* W *of* p *such that no other point of* W *is fixed by an element of* Γ *other than the identity.*

Proof.   We first show that a Fuchsian group acts properly discontinuously on $\mathcal{U}$. Let $z_0 \in \mathcal{U}$ and let $\overline{B_\varepsilon(z_0)}$ be a closed hyperbolic disc, centre z_0, radius $\varepsilon > 0$. As the topology induced by the hyperbolic metric coincides with the Euclidean topology by Theorem 5.4.5, $\overline{B_\varepsilon(z_0)}$ is compact. Hence the set

$$\{ T \in \Gamma \mid T(z_0) \in \overline{B_\varepsilon(z_0)} \}$$

is finite by Corollary 5.6.2. Thus there exists $0 < \delta < \varepsilon$ such that $\overline{B_\delta(z_0)}$ contains no other point in the Γ-orbit of z_0. Put $V = \overline{B_{\delta/2}(z_0)}$. Then if $V \cap S(V) \neq \varnothing$ for some $S \in \Gamma$ there exists $z \in V$ such that $S(z) \in V$. Hence $\rho(z, z_0) < \delta/2$, $\rho(S(z), z_0) < \delta/2$ and so

$$\rho(z_0, S(z_0)) \leqslant \rho(z_0, S(z)) + \rho(S(z), S(z_0))$$

$$= \rho(z_0, S(z)) + \rho(z, z_0) < \delta,$$

and by the definition of δ we must have $S(z_0) = z_0$. Therefore, Γ acts properly discontinuously on $\mathcal{U}$.

Before we consider the converse we prove part (ii). Suppose that p is fixed by $S \neq I$. Then, by what we have just proved, there is a neighbourhood W of p such that $W \cap S(W) \neq \varnothing$ implies that $S(p) = p$. If $q \in W$ is fixed by $T \neq I$ then $T(W) \cap W \neq \varnothing$ and hence $T(p) = p$; as an element of $PSL(2, \mathbb{R})$ other than the identity can fix at most one point of $\mathcal{U}$, $q = p$.

We now prove the converse of part (i), that is we show that a subgroup of $PSL(2, \mathbb{R})$ which acts properly discontinuously on $\mathcal{U}$ must be discrete. Suppose not, and choose some point $s \in \mathcal{U}$ not fixed by any non-identity element of Γ: such points do exist by part (ii). As we are supposing that Γ is not discrete, there exists a sequence (T_k) of distinct elements of Γ such that $T_k \to I$ as $k \to \infty$. Hence $T_k(s) \to s$ as $k \to \infty$ and as s is not fixed by any non-identity element of Γ, $(T_k(s))$ is a sequence of distinct points. Hence every neighbourhood of s contains other points in the Γ-orbit of s and so Γ does not act properly discontinuously. $\square$

Corollary 5.6.4. *Let Γ be a subgroup of $PSL(2, \mathbb{R})$. Then Γ is a Fuchsian group if and only if for all $z \in \mathcal{U}$, Γz, the Γ-orbit of z, is a discrete subset of $\mathcal{U}$.*

Proof. Suppose that Γz is a discrete subset of $\mathcal{U}$. Then there exists $\varepsilon > 0$ such that the open hyperbolic disc $B_\varepsilon(z)$, centre z, radius ε, contains no other point of Γz.

Hence if $V \subseteq B_{\varepsilon/2}(z)$, then by the same argument that we used to prove part (i) of the previous theorem, we can show that $V \cap S(V) \neq \emptyset$ implies that $S(z) = z$. Thus, by that theorem, Γ is a Fuchsian group. Conversely, if Γ is Fuchsian group, then it acts properly discontinuously on $\mathcal{U}$ and hence every orbit Γz $(z \in \mathcal{U})$ is discrete in $\mathcal{U}$. $\square$

Thus if $z \in \mathcal{U}$ and (T_n) is a sequence of distinct elements in Γ, then if $(T_n(z))$ has a limit point $\alpha \in \mathbb{C} \cup \{\infty\}$ then $\alpha \in \mathbb{R} \cup \{\infty\}$. The set of all possible limit points is called the *limit set* of Γ and denoted by $L(\Gamma)$. Thus, for all Fuchsian groups Γ, $L(\Gamma) \subseteq \mathbb{R} \cup \{\infty\}$.

Examples.

(i) If Γ is the modular group then $L(\Gamma) = \mathbb{R} \cup \{\infty\}$.

(ii) If Γ is the cyclic group generated by $z \mapsto 2z$, then $L(\Gamma) = \{0, \infty\}$.

In general, whether a discrete group acts discontinuously or not depends very much on the space on which the group acts. For example, the modular group does not act properly discontinuously on $\mathbb{R} \cup \{\infty\}$ as the orbit of 0 is the set $\mathbb{Q} \cup \{\infty\}$ ($\mathbb{Q}$ is the set of rationals) which is dense in $\mathbb{R} \cup \{\infty\}$. Similarly, consider the subgroup $PSL(2, \mathbb{Z}[i])$ of $PSL(2, \mathbb{C})$. This consists of transformations

$$z \mapsto \frac{az + b}{cz + d}, \quad a, b, c, d \in \mathbb{Z}[i], \quad ad - bc = 1,$$

where $\mathbb{Z}[i] = \{m + ni \mid m, n \in \mathbb{Z}\}$ is the ring of Gaussian integers. $PSL(2, \mathbb{Z}[i])$ is a discrete subgroup of $PSL(2, \mathbb{C})$ but its action on the Riemann sphere is not discontinuous for the orbit of 0 is the dense set $\{r + si \mid r, s \in \mathbb{Q}\} \cup \{\infty\}$. The group $PSL(2, \mathbb{Z}[i])$ is called the *Picard modular group* and it does have a properly discontinuous action on hyperbolic 3-space. (See Appendix 4.)

A common method of constructing Fuchsian groups is to use geometric techniques to construct groups which act properly discontinuously on the upper half-plane. This is described in §5.10; for now we indicate how this method is used to construct triangle groups. From the Gauss–Bonnet theorem we know that the sum of the angles of a hyperbolic triangle is less than π. We show that, subject to this condition, hyperbolic triangles exist with prescribed angles.

Lemma 5.6.5. *Let* α, β, γ *be non-negative real numbers such that* $\alpha + \beta + \gamma < \pi$. *Then there exists a hyperbolic triangle with angles* α, β, γ.

Proof. It is easy to construct hyperbolic triangles with some zero angles as in Fig. 5.4 so assume no angles are zero. As the sum of the angles is less than π, we can assume that $0 < \alpha < \pi/2$. We will choose one vertex of the triangle to be at i and one edge of the triangle to be a segment of the imaginary axis lying above i. Let M be that segment of the H-line lying to the right of the imaginary axis and intersecting it at an angle α. For each point $P_1 \in M$ consider the hyperbolic triangle with vertices at P_1, i and ∞. The angle at P_1 in this triangle varies continuously from $\pi - \alpha$ to 0 as P_1 moves along M from i to the real axis (see Fig. 5.7). Hence for some point Q this angle is $\beta < \pi - \alpha$. For each point P between i and Q consider the H-line intersecting M at P at an angle β. This H-line also intersects the imaginary axis, at a point above i, for otherwise we would have a hyperbolic triangle whose angle sum is not less than π. Suppose this H-line intersects the imaginary axis at $R(P)$ at an angle $\gamma(P)$. Then $\gamma(P) \to 0$ as $P \to Q$ and as $P \to i$ the H-area of the hyperbolic triangle with vertices $i, P, R(P)$ tends to 0 and so by the Gauss–Bonnet theorem the angle $\gamma(P) \to \pi - \alpha - \beta$. Hence for some point P, $\gamma(P) = \gamma$ and so we have constructed a hyperbolic triangle with angles α, β, γ. ☐

Fig. 5.7

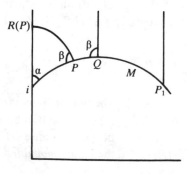

Definition. Let Q be an H-line. Then an *H-reflection* in Q is an H-isometry of $\mathscr{U}$, other than the identity, which fixes every point of Q.

If Q_0 is the imaginary axis then equation (5.4.4) shows that the map $R_0: z \to -\bar{z}$, that is Euclidean reflection in Q_0, is an H-reflection. If Q is another H-line, then by Theorem 5.3.4, there exists $T \in PSL(2, \mathbb{R})$ such that $T(Q) = Q_0$. As T is an H-isometry, $T^{-1} R_0 T$ is the H-reflection fixing Q. As R_0 has order 2, every H-reflection has order 2.

By Lemma 2.7.2 it is easy to prove the following result.

Theorem 5.6.6. *An H-reflection in Q is the restriction of a Euclidean inversion in Q to the upper half-plane.* □

Every H-reflection is an anti-conformal homeomorphism of $\mathscr{U}$; that is, it preserves angles but reverses orientation. If B is an anti-conformal homeomorphism of $\mathscr{U}$ then $R_0 B = T$ is a conformal homeomorphism of $\mathscr{U}$, and thus an element of $PSL(2, \mathbb{R})$. Hence $B = R_0 T$ and it follows that every anti-conformal homeomorphism of $\mathscr{U}$, and, in particular every H-reflection, has the form

$$z \mapsto \frac{a\bar{z} + b}{c\bar{z} + d}, \quad a, b, c, d \in \mathbb{R}, \quad ad - bc = -1.$$

(See Exercise 5C.)

Let τ be an H-triangle with vertices v_1, v_2, v_3, angles $\pi/m_1, \pi/m_2, \pi/m_3$ at these vertices and sides M_1, M_2, M_3 opposite these vertices, as illustrated in Fig. 5.8. (Here, m_1, m_2, m_3 are positive integers.)

Fig. 5.8

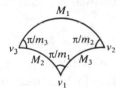

Let R_i be the H-reflection in the H-line containing M_i, $(i = 1, 2, 3)$, and let Γ^* be the group generated by the reflections R_1, R_2, R_3. As $R_i \notin PSL(2, \mathbb{R})$, Γ^* is not a Fuchsian group. However, consider $\Gamma = \Gamma^* \cap PSL(2, \mathbb{R})$; Γ^* is the union of two Γ-cosets, for example $\Gamma^* = \Gamma \cup \Gamma R_1$, for if $S \in \Gamma^* \backslash \Gamma$ then $S R_1$ is the composition of two anti-conformal homeomorphisms, so it is conformal and thus $S R_1 \in PSL(2, \mathbb{R})$. Also, $S R_1 \in \Gamma^*$ so that $S R_1 \in \Gamma$, and $S = (S R_1) R_1 \in \Gamma R_1$.

The image of τ under the H-reflection R_1 is the hyperbolic triangle $R_1(\tau)$

with sides $R_1(M_1) = M_1, R_1(M_2), R_1(M_3)$. As $R_1 R_2 R_1^{-1}$ fixes $R_1(M_2)$ pointwise, it is the H-reflection in $R_1(M_2)$. By this reflection $R_1(\tau)$ is transformed to $R_1 R_2 R_1^{-1}(R_1(\tau)) = R_1 R_2(\tau)$, as shown in Fig. 5.9.

 Fig. 5.9

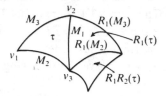

Continuing, we see that the hyperbolic triangles surrounding the vertex v_3 are $\tau, R_1(\tau), R_1 R_2(\tau), R_1 R_2 R_1(\tau), \ldots, (R_1 R_2)^{m_3 - 1} R_1(\tau)$. ($R_1 R_2$, being a product of two H-reflections fixing v_3, can be considered as a hyperbolic rotation about v_3 through an angle $2\pi/m_3$ and so $(R_1 R_2)^{m_3} = I$.)

It can be shown that $\{T(\tau) | T \in \Gamma^*\}$ forms a tessellation of $\mathcal{U}$, that is, no two Γ^*-images of τ overlap and every point of $\mathcal{U}$ belongs to some Γ^*-image of τ. For a discussion of a proof of this statement, the reader is referred to Magnus [1974], p. 81.

Now let p be any point of τ. Then the Γ^*-images of p are the corresponding points of the other triangles of the tessellation and hence they form a discrete set. Thus the Γ-orbit of each point of $\mathcal{U}$ is a discrete set, and so by Corollary 5.6.4 Γ is a Fuchsian group. A Fuchsian group constructed in this way is called a *triangle group*.

By the arguments of the previous paragraph it can be seen that every triangle of the tessellation is of the form $T(\tau)$ where T is a 'word' in R_1, R_2, R_3. (That is, T can be written as a finite product of the $R_i, i = 1, 2, 3$.) Clearly we have the relations

$$R_1^2 = R_2^2 = R_3^3 = (R_1 R_2)^{m_3} = (R_2 R_3)^{m_1} = (R_1 R_3)^{m_2} = I,$$

and it can be shown that all other relations in the group can be deduced from these. (See Magnus [1974].) It is then easy to deduce that Γ is generated by $X = R_1 R_2, Y = R_2 R_3$, that

$$X^{m_3} = Y^{m_1} = (XY)^{m_2} = I,$$

and that all relations in Γ can be deduced from these. In group theory terminology we say that

$$\Gamma = \langle X, Y | X^{m_3} = Y^{m_1} = (XY)^{m_2} = I \rangle$$

is a *presentation* of Γ (see Appendix 2).

As we are working in the hyperbolic plane, $1/m_1 + 1/m_2 + 1/m_3 < 1$. In a similar way triangular tessellations of the sphere and plane give spherical or Euclidean triangle groups for which $1/m_1 + 1/m_2 + 1/m_3 > 1$ or $= 1$ respectively. The spherical triangle groups are finite and were discussed in §2.13.

5.7 Elementary algebraic properties of Fuchsian groups

Lemma 5.7.1.

(i) *A non-trivial discrete subgroup of* $\mathbb{R}$, *the additive group of real numbers, is infinite cyclic.*
(ii) *A discrete subgroup of* S^1, *the multiplicative group of complex numbers of modulus* 1, *is finite cyclic.*

Proof. These statements follow from Theorem 3.1.3 and the proof of Lemma 2.13.3 respectively. □

Theorem 5.7.2. Let Λ be Fuchsian group, all of whose non-identity elements have the same fixed-point set. Then Λ is cyclic.

Proof. Suppose that $S \in \Lambda$ is hyperbolic. Then by choosing a conjugate group if necessary, we may assume that S fixes 0 and ∞ (see §5.2). Hence all transformations of Λ are hyperbolic and fix 0 and ∞. Thus Λ is a discrete subgroup of $H = \{z \mapsto \lambda z | \lambda > 0\}$. Now H is isomorphic as a topological group to $\mathbb{R}^*$, the multiplicative group of positive real numbers; $\mathbb{R}^*$ is isomorphic as a topological group to $\mathbb{R}$, via the isomorphism $x \mapsto \ln x$. Hence by Lemma 5.7.1(i), Λ is infinite cyclic.

Similarly, if Λ contains a parabolic element then Λ is an infinite cyclic group containing only parabolic elements. If Λ contains an elliptic element then as all transformations of the form (5.2.2) form a group isomorphic to S^1, Lemma 5.7.1 (ii) implies that Λ is finite cyclic. □

The last sentence of this proof implies

Theorem 5.7.3. An elliptic element of a Fuchsian group has finite order. □

Theorem 5.7.4. Every abelian Fuchsian group is cyclic.

Proof. This follows immediately from Theorems 5.2.5 and 5.7.2. □

Note. In particular, there is no Fuchsian group isomorphic to $\mathbb{Z} \times \mathbb{Z}$. In §4.19 we showed that every Riemann surface X is conformally equivalent to $\hat{X}/G$, where G, the group of covering transformations, is a discontinuous group of automorphisms of $\hat{X}$, the universal covering space of X. It is known (see Massey [1967], Chapter 5, Corollary 7.5) that $\pi_1(X)$ is isomorphic to G. As the fundamental group of a torus is isomorphic to $\mathbb{Z} \times \mathbb{Z}$ it follows that no Riemann surface of genus 1 has its universal covering space conformally equivalent to $\mathcal{U}$. As no subgroup of Aut Σ acts without fixed-points it follows that the universal covering space of every Riemann surface of genus 1 is conformally equivalent to $\mathbb{C}$ so that every such surface is conformally equivalent to $\mathbb{C}/\Omega$ for some lattice Ω. Another proof of this will be given in §5.10.

If G is a group and H is a subgroup of G then the *normaliser* $N_G(H)$ of H in G is

$$N_G(H) = \{g \in G \,|\, gHg^{-1} = H\};$$

$N_G(H)$ is the largest subgroup of G in which H is normal.

Theorem 5.7.5. *Let Γ be a non-cyclic Fuchsian group. Then the normaliser of Γ in $PSL(2, \mathbb{R})$ is a Fuchsian group.*

Proof. Suppose that the normaliser of Γ in $PSL(2, \mathbb{R})$ is not Fuchsian. Then it contains an infinite sequence (T_i) of distinct elements such that $T_i \to I$ as $i \to \infty$. Thus if $S \in \Gamma (S \neq I)$, then $T_i S T_i^{-1} \to S$ as $i \to \infty$. As Γ is discrete there exists an integer m such that $T_i S T_i^{-1} = S$ for all $i > m$, and so, for these values of i, Theorem 5.2.4 implies that T_i has the same fixed-point set as S. Now, as Γ is not cyclic it is not abelian by the previous theorem and so Theorem 5.2.4 implies that there exists $S' \in \Gamma$ with a different fixed-point set from that of S. However, by the same argument T_i has the same fixed-point set as S' for sufficiently large i and hence S' has the same fixed-point set as S, a contradiction. □

5.8 Fundamental regions

We define a fundamental region for a Fuchsian group Γ in the same way as we defined a fundamental region for a lattice Ω in Chapter 3. That is, F is a *fundamental region* for Γ if F is a closed set such that

(i) $\displaystyle\bigcup_{T \in \Gamma} T(F) = \mathcal{U}$,

(ii) $\mathring{F} \cap T(\mathring{F}) = \varnothing$ for all $T \in \Gamma \backslash \{I\}$,   where $\mathring{F}$ is the interior of F.

(For technical reasons we shall not require that F be connected, though this will usually be the case.)

Example. Let Γ be a triangle group obtained from a hyperbolic triangle τ as described in §5.6. Then $\tau \cup R_1(\tau)$ is easily seen to be a fundamental region for Γ.

The Dirichlet region. Let Γ be an arbitrary Fuchsian group and let $p \in \mathscr{U}$ be not fixed by any element of $\Gamma \backslash \{I\}$. Such points exist by Theorem 5.6.3(ii). Define the *Dirichlet region for* Γ *centred at* p to be the set

$$D_p(\Gamma) = \{z \in \mathscr{U} \mid \rho(z, p) \leqslant \rho(z, T(p)) \text{ for all } T \in \Gamma\}. \tag{5.8.1}$$

By the invariance of the hyperbolic metric under $PSL(2, \mathbb{R})$ this region can also be defined as

$$\{z \in \mathscr{U} \mid \rho(z, p) \leqslant \rho(T(z), p) \text{ for all } T \in \Gamma\}. \tag{5.8.2}$$

For each fixed $T_1 \in \Gamma$, $\rho(z, p) < \rho(z, T_1(p))$ describes the points z which are closer in the hyperbolic metric to p than to $T_1(p)$. Clearly, $p \in D_p(\Gamma)$ and as the Γ-orbit of p is discrete, $D_p(\Gamma)$ contains a neighbourhood of p.

The hyperbolic perpendicular bisector of the segment of the H-line joining p to $T_1(p)$ determines a hyperbolic half-plane containing p. Thus $D_p(\Gamma)$ is an intersection of hyperbolic half-planes and is thus a hyperbolically convex region. If, as is often the case, $D_p(\Gamma)$ is the intersection of finitely many hyperbolic half-planes then $D_p(\Gamma)$ is a convex hyperbolic polygon.

In §3.4 we considered Dirichlet regions for lattices. These are centred at O and the metric is Euclidean. As p lies in a neighbourhood containing no other points of the Γ-orbit of p, by Corollary 5.6.4, the proof of Theorem 3.4.5 goes through unchanged if we replace O by p and the Euclidean metric by the hyperbolic metric.

Theorem 5.8.3. *If p is not fixed by any element of $\Gamma \backslash \{I\}$, then $D_p(\Gamma)$ is a connected fundamental region for Γ.* $\square$

We now show how the Dirichlet region can be defined using the Euclidean metric. As $\sinh^2 \alpha$ is a monotonically increasing function of α, for $\alpha > 0$, equation (5.4.4) and (5.8.2) imply that

$$D_p(\Gamma) = \left\{z \in \mathscr{U} \,\middle|\, \frac{|z - p|^2}{\operatorname{Im}(z)} \leqslant \frac{|T(z) - p|^2}{\operatorname{Im} T(z)} \text{ for all } T \in \Gamma\right\}.$$

If $T(z) = (az + b)/(cz + d)$, $a, b, c, d \in \mathbb{R}$, $ad - bc = 1$, then $\operatorname{Im} T(z) =$

$\text{Im}(z)/|cz+d|^2$ so that

$$D_p(\Gamma) = \left\{ z \in \mathscr{U} \,\middle|\, \left|\frac{T(z)-p}{z-p}\right| \geq \frac{1}{|cz+d|} \text{ for all } T \in \Gamma \right\}. \quad (5.8.4)$$

Example. $\Gamma = $ the modular group. It is easily verified that ki $(k>1)$ is not fixed by any non-identity element of the modular group so choose $p = ki$, where $k > 1$. Put $T(z) = z + 1$ or $T(z) = z - 1$ in (5.8.4). Then $c = 0$, $d = 1$ so that $|z \pm 1 - ki| \geq |z - ki|$ which shows that z is closer (in the Euclidean metric) to ki than to $ki \pm 1$. Thus $D_{ki}(\Gamma)$ lies in the strip $\{z \in \mathscr{U} \mid -\frac{1}{2} \leq \text{Re}(z) \leq \frac{1}{2}\}$.

Now let $T(z) = -1/z = (0z - 1)/(1z + 0)$, so that $c = 1$, $d = 0$. Then for all $z \in D_{ki}(\Gamma)$ we have

$$\frac{\left|-\dfrac{1}{z} - ki\right|}{|z - ki|} \geq \frac{1}{|z|},$$

and so

$$|1 + kiz|^2 \geq |z - ki|^2.$$

After writing $|1 + kiz|^2 = (1 + kiz)(1 - ki\bar{z})$, etc., a simple calculation shows that $|z| \geq 1$. Thus $D_{ki}(\Gamma)$ lies outside the interior of the unit circle. We have shown that $D_{ki}(\Gamma) \subseteq F$ where

$$F = \{z \in \mathscr{U} \mid |z| \geq 1, |\text{Re}(z)| \leq \tfrac{1}{2}\}$$

(illustrated in Fig. 5.10). We wish to show that $D_{ki}(\Gamma) = F$. In order to prove this we need two lemmas.

Fig. 5.10

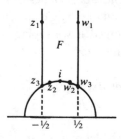

Lemma 1. $D_{ki}(\Gamma)$ *is symmetric with respect to the imaginary axis, that is, if* $A(z) = -\bar{z}$, *the reflection in the imaginary axis, then* $z \in D_{ki}(\Gamma)$ *implies that* $A(z) \in D_{ki}(\Gamma)$.

Proof. A is an H-reflection fixing all points of the imaginary axis. As A

is an H-isometry and as $A^{-1}TA \in \Gamma$ for all $T \in \Gamma$, we have

$$\rho(A(z), ki) = \rho(z, ki) \leqslant \rho(z, A^{-1}TA(ki)) \doteq \rho(A(z), TA(ki))$$
$$= \rho(A(z), T(ki)),$$

and thus $A(z) \in D_{ki}(\Gamma)$. $\quad\square$

Lemma 2. *Let* $z \in F$, $S \in \Gamma \setminus \{I\}$ *and* $w = S(z) \in F$. *Then* $z, w \in \partial F$, *and either* $z = w$ *or* z *and* w *are symmetric with respect to the imaginary axis.*

Proof. Let

$$S(z) = \frac{\alpha z + \beta}{\gamma z + \delta}, \quad \alpha, \beta, \gamma, \delta \in \mathbb{Z}, \quad \alpha\delta - \beta\gamma = 1.$$

Then

$$|\gamma z + \delta|^2 = (\gamma z + \delta)(\gamma \bar{z} + \delta) = \gamma^2 |z|^2 + 2\gamma\delta \operatorname{Re}(z) + \delta^2$$
$$\geqslant \gamma^2 - \gamma\delta + \delta^2$$
$$\geqslant 1.$$

(The last inequality is obvious if $\gamma\delta \leqslant 0$. If $\gamma\delta > 0$, write $\gamma^2 - \gamma\delta + \delta^2 = (\gamma - \delta)^2 + \gamma\delta \geqslant 1$.) Hence

$$\operatorname{Im}(w) = \frac{\operatorname{Im}(z)}{|\gamma z + \delta|^2} \leqslant \operatorname{Im}(z).$$

By interchanging the roles of z and w (using S^{-1} instead of S), we obtain $\operatorname{Im}(z) \leqslant \operatorname{Im}(w)$. Hence $\operatorname{Im}(z) = \operatorname{Im}(w)$ and $|\gamma z + \delta|^2 = 1$, and therefore each of the above inequalities is an equality. Therefore

$$(\gamma - \delta)^2 + \gamma\delta = 1 \tag{i}$$

and

$$\gamma^2(|z|^2 - 1) + \gamma\delta(2\operatorname{Re}(z) + 1) = 0. \tag{ii}$$

We have three possibilities.

(a) $\gamma = 0$, $\delta = \pm 1$ in which case $S(z) = z \pm 1$ and so z, w lie on the vertical boundary of F (like z_1, w_1 in the diagram).

(b) $\gamma = \pm 1$, $\delta = 0$ so equation (ii) implies that $|z| = 1$. In this case $S(z) = (\alpha z \mp 1)/\pm z = \pm \alpha - (1/z) = \pm \alpha - \bar{z}$. As $S(z) \in F$, $\alpha = 0$, -1 or $+1$. If $\alpha = 0$ then $S(z) = -(1/z)$ and $z, S(z)$ are like z_2, w_2 in the diagram (possibly $z_2 = w_2 = i$). If $\alpha = \pm 1$ then $z = (\pm 1 + i\sqrt{3})/2$ ($z = z_3$ or w_3 in the diagram) and $S(z) = z$.

(c) $\gamma = \delta = \pm 1$ and equation (ii) implies that $|z| = 1$, $\operatorname{Re}(z) = -\frac{1}{2}$ so that $z = z_3$. As $S(z_3) \in F$ and $S(z_3)$ has the same imaginary part as z_3, either $S(z_3) = z_3$ or $S(z_3) = w_3$.

In all cases, $S(z) = z$ or $S(z)$ and z are symmetric with respect to the imaginary axis. □

Now suppose that $z_0 \in F$. Then as $D_{ki}(\Gamma)$ is a fundamental region there exists $T \in \Gamma$ such that $T(z_0) \in D_{ki}(\Gamma) \subseteq F$. By Lemma 2, $z_0 = T(z_0)$ or z_0 and $T(z_0)$ are symmetric with respect to the imaginary axis and so by Lemma 1, $z_0 \in D_{ki}(\Gamma)$. Thus $F = D_{ki}(\Gamma)$.

We have thus proved

Theorem 5.8.4. $F = \{z \in \mathcal{U} \mid |z| \geqslant 1 \text{ and } |\mathrm{Re}(z)| \leqslant \frac{1}{2}\}$ *is a fundamental region for the modular group.* □

In the Euclidean case treated in §3.4, the Dirichlet region was a polygon with four or six sides. For Fuchsian groups, Dirichlet regions can be quite complicated. They are bounded by H-lines and possibly by sections of the real axis. If two such H-lines intersect in $\mathcal{U}$ then their point of intersection is called a *vertex* of the Dirichlet region. It can be shown that the vertices are isolated (see Exercise 5P) so that the boundary of a Dirichlet region consists of a union of (possibly infinitely many) H-lines and possibly sections of the real axis. (Later on we shall define some other vertices which are not points of intersection of distinct bounding H-lines.)

We shall be interested in the tessellation of $\mathcal{U}$ formed by a fundamental region and all its images. If this fundamental region is a Dirichlet region then this is referred to as a *Dirichlet tessellation*. (See Fig. 6.6 for a Dirichlet tessellation for the modular group.) Even though a Dirichlet region can be quite complicated, the next theorem shows that the Dirichlet tessellation has nice local properties. We first need a new concept.

Definition A fundamental region F for a Fuchsian group Γ is called *locally finite* if every point $a \in F$ has a neighbourhood $V(a)$ such that $V(a) \cap T(F) \neq \varnothing$ for only finitely many $T \in \Gamma$.

Theorem 5.8.5. *A Dirichlet region is locally finite.*

Proof. Let $F = D_p(\Gamma)$ where p is not fixed by any element of $\Gamma \backslash \{I\}$. Let $a \in F$ and let K be a compact neighbourhood of a. Suppose that $K \cap T_i(F) \neq \varnothing$ for some infinite sequence $T_1, T_2, \ldots$ of distinct elements of Γ. Let $\sigma = \sup_{z \in K} \rho(p, z)$. Then $\sigma \leqslant \rho(p, a) + \rho(a, z)$ for all $z \in K$, and as K is bounded, σ is finite. Let $w_j \in K \cap T_j(F)$. Then $w_j = T_j(z_j)$ for $z_j \in F$ and hence by the triangle inequality,

$$\rho(p, T_j(p)) \leqslant \rho(p, w_j) + \rho(w_j, T_j(p))$$
$$= \rho(p, w_j) + \rho(z_j, p)$$
$$\leqslant \rho(p, w_j) + \rho(w_j, p) \qquad \text{(as } z_j \in D_p(\Gamma))$$
$$\leqslant 2\sigma.$$

Thus the infinite set of points $T_1(p), T_2(p), \ldots$ belongs to the compact H-ball with centre p and radius 2σ. This contradicts Corollary 5.6.2. $\square$

Let F be a Dirichlet region for a Fuchsian group Γ and let u, v be vertices of F. Then u and v are called *congruent* if there exists $T \in \Gamma$ such that $T(u) = v$. Congruence is an equivalence relation on the vertices of F and the equivalence classes are called *cycles*. If u is fixed by an elliptic element S then v is fixed by the elliptic element TST^{-1}. Thus if one vertex of a cycle is fixed by an elliptic element then so are all vertices of that cycle. Such a cycle is called an *elliptic cycle* and the vertices of that cycle are called *elliptic vertices*.

As the Dirichlet region F is a fundamental region it is clear that every point w of $\mathcal{U}$ fixed by an elliptic element S' of Γ lies on the boundary of $T(F)$ for some $T \in \Gamma$. Hence $u = T^{-1}(w)$ lies on the boundary of F and is fixed by the elliptic element $S = T^{-1}S'T$. By Theorem 5.7.3, S has finite order k. Suppose first that $k \geqslant 3$; then as S is a hyperbolic isometry fixing u which maps H-lines to H-lines, u must be a vertex of F whose angle θ is at most $2\pi/k$, as illustrated in Fig. 5.11. The hyperbolically convex region F is bounded by a union of H-lines. The intersection of F with one of these H-lines is either a single point or a segment of an H-line. These segments are called *sides* of F.

Fig. 5.11

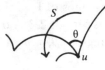

If S has order 2 then its fixed-point might lie on the interior of a side of F. In this case, S interchanges the two segments of this side separated by the fixed-point. We will include such elliptic fixed-points as vertices of F, the angle at such vertices being π. Thus a *vertex* of F is a point of intersection in $\mathcal{U}$ of two bounding H-lines of F or a fixed-point of an elliptic element of order 2. (All the previous definitions such as conjugate, elliptic cycles, etc. apply to this extended set of vertices.)

The only non-trivial finite cyclic subgroups of $PSL(2, \mathbb{R})$ (even of $PSL(2, \mathbb{C})$) are those generated by elliptic elements (see §§5.2 and 5.7) and each elliptic element of $PSL(2, \mathbb{R})$ has a unique fixed-point in $\mathscr{U}$. The same remark therefore applies to elliptic elements of Γ. Also, if a point in $\mathscr{U}$ has a non-trivial stabiliser in Γ then this stabiliser is a finite cyclic subgroup of Γ by Theorem 5.7.3; by Lemma 5.2.3, this finite cyclic subgroup is a maximal finite cyclic subgroup.

Theorem 5.8.6. *There is a one-to-one correspondence between the elliptic cycles of F and the conjugacy classes of non-trivial maximal finite cyclic subgroups of* Γ. $\square$

Example. Let Γ be the modular group. The Dirichlet region F in Fig. 5.10 has vertices in $\mathscr{U}$ at $z_3 = (-1 + i\sqrt{3})/2$, $w_3 = (1 + i\sqrt{3})/2$ and i. These are stabilised by the cyclic subgroups generated respectively by $z \mapsto (-z-1)/z$, $z \mapsto (z-1)/z$, and $z \mapsto 1/z$. As $z \mapsto z + 1$ maps z_3 to w_3 these two vertices belong to the same elliptic cycle. As there are no other vertices of F whose angle is $\leqslant 2\pi/3$ these two vertices form an elliptic cycle. The point i is fixed by an elliptic element of order 2.

As no other point of the boundary of F is fixed by an elliptic element of order 2 (see Exercise 5Q), $\{i\}$ is an elliptic cycle consisting of just one vertex. By Theorem 5.8.6, the modular group has two conjugacy classes of maximal finite cyclic subgroups, one consisting of groups of order 2, the other consisting of groups of order 3.

Definition. The orders of the maximal finite subgroups of Γ are called the *periods* of Γ. Each period is repeated as many times as there are conjugacy classes of maximal finite subgroups of Γ of that order.

Thus the modular group has periods 2, 3. A triangle group obtained from a hyperbolic triangle with angles π/l, π/m, π/n as described in §5.6 has periods l, m, n.

Note. A parabolic element is often treated as being an elliptic element of infinite order. Then one allows infinite periods, the period ∞ occurring the same number of times as there are conjugacy classes of maximal parabolic subgroups. For example, it is easily calculated that in the modular group every parabolic element is conjugate to $z \mapsto z + n$ for some $n \in \mathbb{Z}$ so that the modular group has periods $2, 3, \infty$. Each parabolic cyclic subgroup of an arbitrary Fuchsian group Γ fixes one point of $\mathbb{R} \cup \{\infty\}$. In Beardon [1983] it is shown that there is a Dirichlet region for Γ in which this point is a 'vertex', that is two bounding H-lines of the Dirichlet region meet there. The angle at this 'vertex' is 0. For example, with this convention, the Dirichlet region for the modular group described in Theorem 5.8.4 has a vertex at ∞ whose angle is $\pi/\infty = 0$. Thus the modular group could be considered as a triangle group obtained from a hyperbolic triangle with angles $\pi/2$, $\pi/3$, π/∞.

We now consider congruence of sides. Let s be a side of a Dirichlet region F for a Fuchsian group Γ. If $T \in \Gamma \setminus \{I\}$ and $T(s)$ is a side of F then s and $T(s)$ are called *congruent sides*. Now $T(s)$ is also a side of $T(F)$ so that $T(s) \subseteq F \cap T(F)$. As $T(F)$ is a neighbouring face of F in the Dirichlet tessellation must have $T(s) = F \cap T(F)$ (see Exercise 5R). There cannot be more than two sides in a congruent set for suppose that $T_1(s)$ is also a side of F ($T_1 \in \Gamma$). Then $T_1(s) = F \cap T_1(F)$ and thus $s = T^{-1}(F) \cap F = T_1^{-1}(F) \cap F$ so that $T = T_1$. Thus the sides of a Dirichlet region fall into congruent pairs. If a side has a fixed-point of an elliptic element S of order two on it then S interchanges the two segments of the side of which are separated by this fixed-point and thus this side is congruent to itself. It can be shown that this is the only case where a side of F is congruent to itself (Lehner [1966] p. 37). Alternatively, we could regard these two segments as being distinct sides separated by the fixed-point.

Example. The two vertical sides of the fundamental region for the modular group found in Theorem 5.8.4 are congruent via the transformation $z \mapsto z + 1$. The side on the unit circle between z_3 and w_3 (see Fig. 5.10) is mapped to itself by the elliptic element $z \mapsto -1/z$ of order 2. Alternatively we can regard this as the union of two congruent sides, one from z_3 to i, the other from i to w_3.

Theorem 5.8.7. *Let $\{T_i\}$ be the subset of Γ consisting of those elements which pair the sides of some fixed Dirichlet region F. Then $\{T_i\}$ is a set of generators for Γ.*

Proof. Let Λ be the subgroup of Γ generated by the set $\{T_i\}$. We have to show that $\Lambda = \Gamma$. Suppose that $S_1 \in \Lambda$ and that $S_2(F)$ is a neighbouring face of $S_1(F)$. Then $S_1^{-1}S_2(F)$ is a neighbouring face of F. Hence $S_1^{-1}S_2 = T_k$ for some $T_k \in \{T_i\}$ and as $S_2 = S_1 T_k$ we must have $S_2 \in \Lambda$. If $S_3(F)$ intersects $S_1(F)$ in a vertex v then as there can only be finitely many faces with vertex v, by Theorem 5.8.5, the above argument can be used repeatedly to show that $S_3 \in \Lambda$. Hence, if we let $X = \bigcup_{S \in \Lambda} S(F)$, $Y = \bigcup_{S \in \Gamma \setminus \Lambda} S(F)$, then $X \cap Y = \varnothing$. Clearly, $X \cup Y = \mathcal{U}$, so if we show that X and Y are closed subsets of $\mathcal{U}$ then as $\mathscr{U}$ is connected and $X \neq \varnothing$ we must have $X = \mathscr{U}$ and $Y = \varnothing$. This shows that $\Lambda = \Gamma$ and the result is proved.

We now show that any union $\bigcup V_j(F)$ of faces of the tessellation is closed. For suppose that there is an infinite sequence (z_i) of points of $\bigcup V_j(F)$ which tends to some limit $l \in \mathscr{U}$. Then $l \in A(F)$ for some $A \in \Gamma$ and so by Theorem 5.8.5 there exists a neighbourhood N of l intersecting only finitely many of the $V_j(F)$.

Hence one face of this finite family, $V_m(F)$, say, must contain a subsequence of (z_i) tending to l and as $V_m(F)$ is closed, $l \in V_m(F) \subseteq \bigcup V_i(F)$. Thus $\bigcup V_i(F)$ is closed and in particular X and Y are closed.   □

Example.   Theorem 5.8.7 implies that the modular group is generated by $z \mapsto z + 1$ and $z \mapsto -1/z$. (In the next chapter defining relations for the modular group will also be obtained.)

A similar situation occurs in the case of lattices. The opposite sides of a fundamental parallelogram or hexagon are congruent in pairs and the transformations which pair them generate the lattice. By identifying the sides in each pair we obtain the quotient space. We now wish to consider the quotient-space $\mathscr{U}/\Gamma$ and we will show that we can also obtain it by identifying the congruent sides of a Dirichlet region.

5.9 The quotient-space $\mathscr{U}/\Gamma$

We constructed the quotient-space $\mathbb{C}/\Omega$ in §3.5 and in an analogous way we can construct the quotient-space $\mathscr{U}/\Gamma$. We let $[z]_\Gamma$ or just $[z]$ denote the Γ-orbit of z and $\Pi:\mathscr{U} \to \mathscr{U}/\Gamma$ the natural projection given by $\Pi(z) = [z]$. As before, a set $V \subseteq \mathscr{U}/\Gamma$ is said to be *open* if $\Pi^{-1}(V) = \{z \in \mathscr{U} \mid \Pi(z) \in V\}$ is open in $\mathscr{U}$; with this definition, Π is a continuous and open map.

Theorem 5.9.1. $\mathscr{U}/\Gamma$ *is a connected Riemann surface and* $\Pi:\mathscr{U} \to \mathscr{U}/\Gamma$ *is a holomorphic map.*

Proof.   We show that $\mathscr{U}/\Gamma$ is a connected Hausdorff space in exactly the same way as we showed that $\mathbb{C}/\Omega$ is a connected Hausdorff space in Theorem 4.11.1. There we used the fact that Ω is a discontinuous group of Euclidean isometries; here we use the result that Γ is a properly discontinuous group of hyperbolic isometries. In fact, if Γ contains no elliptic elements then we can adapt the proof of Theorem 4.11.1 in a straightforward way to complete the proof of the theorem. However, if Γ does contain elliptic elements then the projection $\Pi:\mathscr{U} \to \mathscr{U}/\Gamma$ is not one-to-one in a neighbourhood of an elliptic fixed-point and so we have to define our charts in a rather different way.

First we define a suitable open covering of $\mathcal{U}$. For each $q \in \mathcal{U}$ let $\delta(q)$ be the least hyperbolic distance from q to a point of the orbit of q, other than q itself, and let W_q be an open hyperbolic disc, centre q, radius $\delta(q)/2$. By Corollary 5.6.2, $\delta(q) > 0$ and by the same argument as in the proof of Theorem 5.6.3(i),

$$W_q \cap T(W_q) \neq \varnothing \text{ for } T \in \Gamma \text{ implies that } T(q) = q. \qquad (5.9.2)$$

Also, $\delta(S(q)) = \delta(q)$, for $S \in \Gamma$, so that if $r = S(q)$ then $W_r = S(W_q)$.

Note that in (5.9.2), $T = I$ unless q is an elliptic fixed-point of Γ, and that W_q contains no elliptic fixed-points of Γ other than q. Let Π_q be the restriction of Π to W_q. Then as in the proof of Theorem 4.11.1, Π_q is continuous and open.

Let $m(q)$ be the order of the stabiliser of q in Γ. Thus $m(q) = 1$ unless q is an elliptic fixed-point of Γ. Define

$$f_q(z) = \left(\frac{z - q}{z - \bar{q}} \right)^{m(q)}.$$

Now $z \mapsto (z - q)/(z - \bar{q})$ maps $\mathcal{U}$ onto $\mathcal{D}$, the open unit disc. (The calculation is the same as that in §4.17 where $q = i$.) Hence f_q maps $\mathcal{U}$ onto $\mathcal{D}$. As f_q, being an analytic function, is continuous and open, and as $f_q(q) = 0$ it follows that $f_q(W_q)$ is an open disc with centre 0 contained in $\mathcal{D}$.

By §5.2 and Theorem 5.7.3, every elliptic element V of Γ fixing q satisfies

$$\frac{V(z) - q}{V(z) - \bar{q}} = \alpha \left(\frac{z - q}{z - \bar{q}} \right),$$

where α is an $m(q)$th root of unity, and thus $f_q(z_1) = f_q(z_2)$ if and only if $z_2 = V_1(z_1)$ where V_1 is an elliptic element of Γ fixing q.

Now define

$$\Phi_q = f_q \circ \Pi_q^{-1},$$

as depicted in Fig. 5.12. Then $\Phi_q : \Pi_q(W_q) \to f_q(W_q) \subseteq \mathcal{D}$. We wish to show that Φ_q is a homeomorphism for all $q \in \mathcal{U}$. This is obvious if q is not an elliptic fixed-point, for then $\Pi_q^{-1} : \Pi_q(W_q) \to W_q$ is a homeomorphism and as

Fig. 5.12

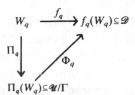

$m(q) = 1$, f_q is a homeomorphism. Now suppose that q is an elliptic fixed-point. Then (5.9.2) implies that $\Pi_q(z_1) = \Pi_q(z_2)$ if and only if $z_2 = V_1(z_1)$ for some $V_1 \in \Gamma$ which fixes q. This occurs if and only if $f_q(z_1) = f_q(z_2)$. Thus $\Phi_q = f_q \circ \Pi_q^{-1}$ is a well-defined function, in fact $\Phi_q[z] = f_q \circ \Pi_q^{-1}[z] = f_q(z)$. We now show that Φ_q is one-to-one. If $\Phi_q[z_1] = \Phi_q[z_2]$ then $f_q(z_1) = f_q(z_2)$ and so $z_2 = V_1(z_1)$ for some elliptic element $V_1 \in \Gamma$ which fixes q. Hence $[z_2] = [V_1(z_1)] = [z_1]$ and hence Φ_q is one-to-one. As both Π_q and f_q are continuous and open, $\Phi_q : \Pi_q(W_q) \to f_q(W_q)$ is a homeomorphism. Thus $\mathscr{U}/\Gamma$ is a surface and $\{\Pi_q(W_q), \Phi_q\}$ is an atlas defined on $\mathscr{U}/\Gamma$. We now show that this atlas is analytic.

Suppose that $\Pi_q(W_q) \cap \Pi_r(W_r) \neq \varnothing$ and consider the homeomorphism

$$\Phi_r \circ \Phi_q^{-1} : \Phi_q(\Pi_q(W_q) \cap \Pi_r(W_r)) \to \Phi_r(\Pi_q(W_q) \cap \Pi_r(W_r)).$$

We have

$$\Phi_r \circ \Phi_q^{-1} = (f_r \circ \Pi_r^{-1} \circ \Pi_q) \circ f_q^{-1}.$$

Now

$$(f_r \circ \Pi_r^{-1} \circ \Pi_q)(z) = (f_r \circ \Pi_r^{-1})[z] = f_r(T(z))$$

for some $T \in \Gamma$, and hence $f_r \circ \Pi_r^{-1} \circ \Pi_q : \mathscr{U} \to \mathscr{D}$ is analytic. Also, each branch of f_q^{-1} is analytic except at 0 when $m(q) > 1$, so that with these possible exceptions $\Phi_r \circ \Phi_q^{-1}$ is analytic. In the exceptional cases $0 \in \Phi_q(\Pi_q(W_q) \cap \Pi_r(W_r))$ and $m(q) > 1$. As q is the only point of W_q which f_q maps to 0, $[q] \in \Pi_q(W_q) \cap \Pi_r(W_r)$. Now W_r contains no elliptic fixed-points except possibly r and so $r = S(q)$ for some $S \in \Gamma$. By the way in which the H-discs W_q were chosen we have $W_r = S(W_q)$ and thus $\Pi_q(W_q) = \Pi_r(W_r)$ and

$$f_r \circ \Pi_r^{-1}[z] = (f_{S(q)} \circ \Pi_{S(q)}^{-1})[z] = f_{S(q)}(S(z))$$

$$= \left(\frac{S(z) - S(q)}{S(z) - S(\bar{q})} \right)^{m(q)} = k(q) \left(\frac{z - q}{z - \bar{q}} \right)^{m(q)}$$

$$= k(q) f_q(z) = k(q) f_q \circ \Pi_q^{-1}[z]$$

where $k(q)$ depends only on q. Hence $\Phi_r \circ \Phi_q^{-1}$ is analytic. Thus in all cases $\Phi_r \circ \Phi_q^{-1}$ is analytic and so $\{(\Pi_q(W_q), \Phi_q)\}$ is an analytic atlas.

Finally, as $\Phi_q \circ \Pi_q = f_q$ is analytic, $\Pi : \mathscr{U} \to \mathscr{U}/\Gamma$ is a holomorphic map. □

Thus quotient-spaces of Fuchsian groups are Riemann surfaces. Now, by Theorem 4.19.8 the universal covering space of a Riemann surface X not homeomorphic to the sphere, plane or torus is $\mathscr{U}$, and $X = \mathscr{U}/\Lambda$ where Λ is a properly discontinuous group of automorphisms of $\mathscr{U}$ acting without fixed-points (Λ is the group of covering transformations). Thus by Theorem 5.6.3

Λ is a Fuchsian group and as it acts without fixed-points, Λ contains no elliptic elements. If we include Fuchsian groups with elliptic elements then *every* Riemann surface can be represented as the quotient-space of $\mathscr{U}$ by a Fuchsian group. For example, we shall see that the quotient-space for a triangle group is a sphere and as every Riemann surface homeomorphic to the sphere is conformally equivalent to the Riemann sphere Σ (see Theorem 4.17.2) it follows that Σ is conformally equivalent to $\mathscr{U}/\Gamma$ where Γ is a triangle group.

The advantage of using Fuchsian groups without elliptic elements to represent Riemann surfaces is apparent from the following theorem which shows that there is a one-one correspondence between Riemann surfaces and conjugacy classes of Fuchsian groups without elliptic elements. It is analogous to Theorem 4.18.1 and the proof follows along the same lines.

Theorem 5.9.3. *Let* Λ_1, Λ_2 *be Fuchsian groups without elliptic elements. Then* $\mathscr{U}/\Lambda_1$ *and* $\mathscr{U}/\Lambda_2$ *are conformally equivalent if and only if there exists* $T \in PSL(2, \mathbb{R})$ *such that* $T\Lambda_1 T^{-1} = \Lambda_2$.

Proof. First of all as Λ_1, Λ_2 act without fixed-points $\mathscr{U}$ is a covering space of $\mathscr{U}/\Lambda_1$ and $\mathscr{U}/\Lambda_2$ (Exercise 4U), and hence as $\mathscr{U}$ is simply connected it is the universal covering space of $\mathscr{U}/\Lambda_1$ and $\mathscr{U}/\Lambda_2$. Thus if $g: \mathscr{U}/\Lambda_1 \to \mathscr{U}/\Lambda_2$ is a conformal homeomorphism then, as in the proof of Theorem 4.18.1, there exists an automorphism $\tilde{g}: \mathscr{U} \to \mathscr{U}$ such that the diagram in Fig. 5.13 commutes, where $\Pi_i: \mathscr{U} \to \mathscr{U}/\Lambda_i$ are the natural projection maps, $(i = 1, 2)$.

Fig. 5.13

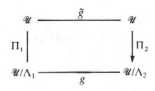

By Theorem 4.17.3, $\tilde{g} = T \in PSL(2, \mathbb{R})$ and so

$$g[z]_{\Lambda_1} = [T(z)]_{\Lambda_2}.$$

Now if $S \in \Lambda_1$, then $[z]_{\Lambda_1} = [S(z)]_{\Lambda_1}$ and so $[TS(z)]_{\Lambda_2} = [T(z)]_{\Lambda_2}$. It follows that there exists $V \in \Lambda_2$ such that $TS(z) = VT(z)$ for all $z \in \mathscr{U}$ and so $TST^{-1} = V \in \Lambda_2$. Therefore $T\Lambda_1 T^{-1} \subseteq \Lambda_2$ and a similar argument using $g^{-1}[z]_{\Lambda_2} = [T^{-1}(z)]_{\Lambda_1}$ shows that $T^{-1}\Lambda_2 T \subseteq \Lambda_1$, so that $T\Lambda_1 T^{-1} = \Lambda_2$.

Conversely, if $T\Lambda_1 T^{-1} = \Lambda_2$ then the map

$$[z]_{\Lambda_1} \mapsto [T(z)]_{\Lambda_2}$$

is a conformal homeomorphism of $\mathscr{U}/\Lambda_1$ onto $\mathscr{U}/\Lambda_2$. $\quad\square$

The theorem is false for arbitrary Fuchsian groups, for if we consider two triangle groups with different periods then they are not isomorphic and hence not conjugate in $PSL(2, \mathbb{R})$. However, the quotient-space of each group is the Riemann sphere as remarked above. The converse statement, that the quotient-spaces of conjugate Fuchsian groups are conformally equivalent Riemann surfaces, is true for arbitrary Fuchsian groups as the above proof applies.

Theorem 5.9.4. *If Λ is a Fuchsian group without elliptic elements then* $\mathrm{Aut}(\mathcal{U}/\Lambda) \cong N(\Lambda)/\Lambda$ *where $N(\Lambda)$ is the normaliser of Λ in $PSL(2, \mathbb{R})$.*

Proof. Let $t : \mathcal{U}/\Lambda \to \mathcal{U}/\Lambda$ be an automorphism of $\mathcal{U}/\Lambda$. Then $t[z]_\Lambda = [T(z)]_\Lambda$ for some $T \in PSL(2, \mathbb{R})$. By putting $\Lambda_1 = \Lambda_2 = \Lambda$ in the above theorem we see that $T\Lambda T^{-1} = \Lambda$ so that $T \in N(\Lambda)$. Conversely, if $T \in N(\Lambda)$ then the map $t : [z]_\Lambda \to [T(z)]_\Lambda$ is an automorphism of $\mathcal{U}/\Lambda$. It is now easy to see that $T \mapsto t$ gives a homomorphism of $N(\Lambda)$ onto $\mathrm{Aut}(\mathcal{U}/\Lambda)$ whose kernel is Λ. Hence by the first isomorphism theorem $\mathrm{Aut}(\mathcal{U}/\Lambda) \cong N(\Lambda)/\Lambda$. $\square$

Corollary 5.9.5. *If Λ is a non-cyclic Fuchsian group without elliptic elements then every group of automorphisms of $\mathcal{U}/\Lambda$ is isomorphic to Γ/Λ where Γ is some Fuchsian group such that $\Lambda \lhd \Gamma$.*

Proof. This follows directly from Theorem 5.9.4 in conjunction with Theorem 5.7.5. $\square$

These results will be used shortly to find information concerning groups of automorphisms of compact Riemann surfaces.

Let F be a Dirichlet region for a Fuchsian group Γ. As the edges of F are paired by elements of Γ it seems likely that we can obtain the quotient-space $\mathcal{U}/\Gamma$ by identifying the sides in each pair in the same way as we obtained a torus by identifying the opposite sides of a parallelogram. (Actually, the proof of the result that we can obtain $\mathbb{C}/\Omega$ by identifying the opposite sides of a fundamental parallelogram or hexagon for Ω is not completely obvious but it does follow by adapting the proof below.) As the only points of F which are identified under Γ are corresponding points of paired sides, the space we obtain by identifying paired sides is F/Γ.

Theorem 5.9.6. *If F is a Dirichlet region for Γ then F/Γ is homeomorphic to $\mathcal{U}/\Gamma$.*

Proof.   Let $i: F \to \mathcal{U}$ be the inclusion map, and $\psi: F \to F/\Gamma$, $\Pi: \mathcal{U} \to \mathcal{U}/\Gamma$ be the natural projection maps. We define $\theta: F/\Gamma \to \mathcal{U}/\Gamma$ to be the map which makes the diagram (Fig. 5.14) commute. That is, if $z \in F$ then $\theta(\psi(z)) = \Pi(z)$. If $\psi(z_1) = \psi(z_2)$ then $z_2 = S(z_1)$ where S pairs sides of F, so that as $S \in \Gamma$, $\Pi(z_1) = \Pi(z_2)$ and hence θ is well-defined and similarly, θ is bijective. Now if $V \subseteq U/\Gamma$ then

$$\psi^{-1}(\theta^{-1}(V)) = F \cap \Pi^{-1}(V),$$

so if V is open then as Π is continuous, $\Pi^{-1}(V)$ is open in U and so $F \cap \Pi^{-1}(V)$ is open in F. Thus $\psi^{-1}(\theta^{-1}(V))$ is open in F and so by definition of the quotient topology $\theta^{-1}(V)$ is open in F/Γ. Thus θ is continuous. We now show that θ is an open map.

Fig. 5.14

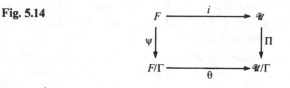

Let $A \subseteq F/\Gamma$ be open and let $\langle z \rangle = \psi(z) \in A$. By Theorem 5.8.5, F is locally finite so that

$$\psi^{-1}(\langle z \rangle) = \{z = T_0(z), T_1(z), T_2(z), \ldots, T_s(z)\},$$

a finite set. As $\psi^{-1}(A) = \tilde{A}$ is relatively open in F and $\psi^{-1}(\langle z \rangle)$ is finite, there is a hyperbolic disc B with centre z, such that

$T_j(B) \cap F \subseteq \tilde{A}$ and $T(B) \cap F \neq \varnothing$ implies that $T = T_j$ $(1 \leq j \leq s)$. (5.9.7)

(This is illustrated in Fig. 5.15; $\tilde{A}$ is the shaded region.)

Fig. 5.15

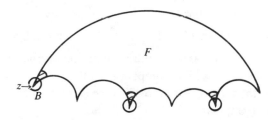

Now $\theta(\langle z \rangle) = \Pi(z) \in \Pi(B)$ which is open, so if we show that $\Pi(B) \subseteq \theta(A)$ then $\theta(A)$ is an open neighbourhood of $\theta(\langle z \rangle)$.

Let $[w] \in \Pi(B)$ where $w \in B$. Then there exists $T \in \Gamma$ such that $T(w) \in F$. Hence $T(B) \cap F \neq \varnothing$ so that by (5.9.7) $T = T_j$ $(1 \leq j \leq s)$. Therefore

$T_j(w) \in T_j(B)$ and $T_j(w) \in F$ and so by (5.9.7), $T_j(w) \in \tilde{A}$. Hence

$$[w] = \Pi(w) = \Pi(T_j(w)) \in \Pi(\tilde{A}) = \Pi i(\tilde{A}) = \theta(A). \quad \square$$

In the previous chapter we observed that a particularly important role was played by compact Riemann surfaces. We now prove two results about Fuchsian groups whose quotient-spaces are compact.

Theorem 5.9.8.   *Let F be a Dirichlet region for a Fuchsian group* Γ. *Then* $\mathcal{U}/\Gamma$ *is compact if and only if F is a compact subset of* $\mathcal{U}$.

Proof.   If F is compact then F/Γ, being a continuous image of F, is compact. Hence, by Theorem 5.9.6, $\mathcal{U}/\Gamma$ is compact. Now suppose that $\mathcal{U}/\Gamma$ is compact. By local finiteness (Theorem 5.8.5) it follows that if $[z] \in \mathcal{U}/\Gamma$ then there are at most a finite number of $z' \in F$ such that $\Pi(z') = [z]$. Now let $z_1, z_2, \ldots$ be an infinite sequence of distinct points of F. We will show that this sequence has a limit point in F so that F is compact (see §1.2). By local finiteness $[z_1], [z_2], \ldots$ is a sequence containing infinitely many points so that it must have a limit point $[l] \in \mathcal{U}/\Gamma$; $[l]$ has finitely many pre-images $l_1, \ldots, l_r$ in F and we show that at least one of them is a limit point of the sequence (z_j). For otherwise each l_i has a neighbourhood V_i in $\mathcal{U}$ containing only a finite number of points of the sequence and as $l_1, \ldots, l_r$ are in the same Γ-orbit, each V_i contains finitely many points of the form $S(z_j)$. Hence $\bigcap_{i=1}^{r} \Pi(V_i)$ is an open neighbourhood of $[l]$ containing only finitely many points of the sequence $([z_j])$, a contradiction. $\quad \square$

Theorem 5.9.9.   *If* $\mathcal{U}/\Gamma$ *is compact then* Γ *contains no parabolic elements.*

Proof.   Let F be a Dirichlet region for Γ. Then by Theorem 5.9.8, F is compact. By Corollary 5.6.2,

$$\eta(z) = \inf \{\rho(z, T(z)) | T \in \Gamma \backslash \{I\}, \ T \text{ not elliptic}\} > 0.$$

As for each $T \in \Gamma$, $\rho(z, T(z))$ is a continuous function of z, $\eta(z)$ is a continuous function of z. Therefore, as F is compact $\eta = \inf \{\eta(z) | z \in F\}$ is attained and $\eta > 0$. If $z \in \mathcal{U}$ then there exists $S \in \Gamma$ such that $w = S(z) \in F$. Hence if $T_0 \in \Gamma \backslash \{I\}$ is not elliptic,

$$\rho(z, T_0(z)) = \rho(S(z), S(T_0(z))) = \rho(w, ST_0S^{-1}(w)) \geqslant \eta,$$

and therefore

$$\inf \{\rho(z, T_0(z)) | z \in \mathcal{U}\} = \eta > 0.$$

Now suppose that Γ contains parabolic elements. As $\mathcal{U}/\Gamma$ is homeomorphic to $\mathcal{U}/\Gamma_1$ when Γ is conjugate to Γ_1 in $PSL(2, \mathbb{R})$ we may assume by §5.2 that $T_0(z)$ or $T_0^{-1}(z)$ is the transformation $z \mapsto z + 1$. However, by (5.4.4), $\rho(z, z + 1) \to 0$ as $\text{Im}(z) \to \infty$, a contradiction. $\square$

It is appropriate here to mention ways in which a Dirichlet region can be non-compact in $\mathscr{U}$. First of all, a Dirichlet region with an infinite number of sides is non-compact (see Exercise 5P). However, finite-sided Dirichlet regions can be non-compact in $\mathscr{U}$ in the following ways. Firstly, there might be a 'vertex' in $\mathbb{R} \cup \{\infty\}$. For example, it is known that if the group has a parabolic element T, then there is a Dirichlet region for the group which has a 'vertex' at the fixed-point x of T and such that the two sides whose end-point is at x are paired by T (see Beardon [1983] and see Fig. 5.16). Such a vertex is called *parabolic vertex* and the quotient-space has a 'puncture' at points corresponding to parabolic vertices. For example, the modular group has ∞ as the fixed-point of the parabolic element $z \mapsto z + 1$ and the two vertical sides of the Dirichlet region of Fig. 5.10 are paired by this element. The quotient-space is then a sphere with one puncture and so is homeomorphic to the plane. Secondly, the Dirichlet region might be bounded by a section of the real axis. For example, the Dirichlet region centred at i for the parabolic cyclic group generated by $z \mapsto z + 1$ is $\{z \in \mathscr{U} \mid -\frac{1}{2} \leqslant \text{Re}(z) \leqslant \frac{1}{2}\}$. This is bounded by the section $\{x \in \mathbb{R} \mid -\frac{1}{2} \leqslant x \leqslant \frac{1}{2}\}$ of the real axis; this Dirichlet region also has a vertex at ∞. The quotient-space is a sphere with a closed disc removed (corresponding to the section of the real axis) and a puncture (corresponding to ∞). This is homeomorphic to a cylinder. (A removed disc is topologically, but not conformally, equivalent to a puncture.) (See §6.10 for more details about the quotient-space for the modular group.)

Fig. 5.16

5.10 The hyperbolic area of a fundamental region

By using Theorem 5.5.1 we observe that the proof of the following Theorem is exactly analogous to that of Theorem 3.4.6.

Theorem 5.10.1. *Let F_1, F_2, be two fundamental regions for a Fuchsian group Γ. Suppose that the boundaries of F_1, F_2 have zero hyperbolic area. Then $\mu(F_1) = \mu(F_2)$.* $\square$

The importance of this theorem is that the hyperbolic area of a fundamental region is a numerical invariant of the group. The condition that the boundary has zero hyperbolic area will almost always be satisfied. For example, the boundary of a Dirichlet region is a countable union of H-lines and so has zero hyperbolic area.

It is possible for a fundamental region to have infinite hyperbolic area. For example, the Dirichlet region for the parabolic cyclic group generated by $z \mapsto z + 1$ mentioned at the end of the previous section has infinite hyperbolic area. The theorem is then interpreted as saying that if one fundamental region for the group has infinite hyperbolic area then so do all fundamental regions for the group (still assuming that the boundaries have zero hyperbolic area). A compact fundamental region (indeed, every compact subset of $\mathcal{U}$) has finite hyperbolic area. However non-compact fundamental regions may have finite hyperbolic area. For example, the Dirichlet region for the modular group in Fig. 5.10 is a hyperbolic triangle with angles $\pi/3$, $\pi/3$, 0 and so by the Gauss–Bonnet theorem (5.5.2) has hyperbolic area $\pi/3$.

The Gauss–Bonnet theorem will be used to compute the hyperbolic area of a wide class of fundamental regions. In order to apply it we need to know the sum of the angles at the vertices.

Theorem 5.10.2. *Let F be a Dirichlet region for a Fuchsian group* Γ. *Let* $\theta_1, \theta_2, \ldots, \theta_t$ *be the internal angles at a congruent set of vertices of F. Let m be the order of the stabiliser in* Γ *of one of these vertices. Then*

$$\theta_1 + \theta_2 + \cdots + \theta_t = 2\pi/m.$$

(*Remarks.*

1. As F is locally finite there are only finitely many vertices in a congruent set.
2. As the stabilisers of two points in a congruent set are conjugate subgroups of Γ they have the same order.)

Proof. Let $v_1, v_2, \ldots, v_t$ be the vertices belonging to the congruent set, the internal angle at v_i being θ_i. Let $H = \{I, S, S^2, \ldots, S^{m-1}\}$ be the stabiliser of v_1 in Γ. Then the regions $S^r(F)(0 \leqslant r \leqslant m-1)$ all have a vertex at v_1 whose angle is θ_1. Now suppose that $T_k(v_k) = v_1$ ($T_k \in \Gamma$). Then the set of all elements which map v_k to v_1 is just the coset HT_k which also has m elements so that all the regions $S^r T_k(F)$ have v_1 as a vertex, the angle there being θ_k. On the other hand, if a region $A(F)$ ($A \in \Gamma$) has a vertex v_1 then $A^{-1}(v_1) \in F$ and so

$A^{-1}(v_1) = v_i$ for some i $(1 \leqslant i \leqslant t)$. Thus $A \in HT_i$ so that the region $A(F)$ has been included in the above description. Thus these are mt regions surrounding the vertex v_1, and each angle θ_i is repeated m times; these regions are distinct for if $S^r T_k(F) = S^q T_l(F)$ then $S^r T_k = S^q T_l$ so that $T_k = T_l$ and $S^r = S^q$. Hence

$$m(\theta_1 + \theta_2 + \cdots + \theta_t) = 2\pi,$$

which gives the result. $\quad\square$

We now make the assumption that Γ is a Fuchsian group with $\mathcal{U}/\Gamma$ compact. Then by Theorem 5.9.8, a Dirichlet region F for Γ is compact and so has a finite number of sides (Exercise 5P). Therefore F has finitely many vertices and hence only finitely many elliptic cycles. By Theorem 5.8.6, Γ has a finite number of periods $m_1, \ldots, m_r$, say. If $\mathcal{U}/\Gamma$ has genus g then we say that Γ has signature $(g; m_1, \ldots, m_r)$.

Theorem 5.10.3.   *Let Γ have signature $(g; m_1, \ldots, m_r)$. If F is a fundamental region for Γ whose boundary has zero hyperbolic area then*

$$\mu(F) = 2\pi \left\{ (2g - 2) + \sum_{i=1}^{r} \left(1 - \frac{1}{m_i} \right) \right\}.$$

Proof.   By Theorem 5.10.1 we may assume that F is a Dirichlet region for Γ. By Theorem 5.8.6 we know that F has r elliptic cycles of vertices. (As described in §5.8 we include the interior point of a side fixed by an elliptic element of order 2 as a vertex whose angle is π, and then regard this side as being composed of two sides separated by this vertex.) By Theorem 5.10.2, the sum of the angles at all the elliptic vertices is $2\pi \sum_{i=1}^{r}(1/m_i)$. Suppose that there are s other conjugacy classes of vertices. None of these is fixed by an elliptic element so that by Theorem 5.10.2 the sum of the angles at these vertices is $2\pi s$. Hence the sum of the angles of F is

$$2\pi \left\{ \left(\sum_{i=1}^{r} \frac{1}{m_i} \right) + s \right\}.$$

The sides of F are paired by elements of Γ. If we identify these paired sides we obtain, by Theorem 5.9.6, a compact orientable surface of genus g. If there are n such pairs then the images under the projection from $\mathcal{U}$ to $\mathcal{U}/\Gamma$ of the vertices, edges and interior of F gives a decomposition of $\mathcal{U}/\Gamma$ into $(r + s)$ vertices, n edges and one simply connected face. Hence by the Euler–Poincaré formula (Theorem 4.16.2),

$$2 - 2g = (r + s) - n + 1. \tag{5.10.4}$$

By the Gauss–Bonnet theorem (5.5.2) we obtain

$$\mu(F) = (2n - 2)\pi - 2\pi \left\{ \sum_{i=1}^{r} \left(\frac{1}{m_i} \right) + s \right\}$$

$$= (4g - 4 + 2r + 2s)\pi - 2\pi \sum_{i=1}^{r} \frac{1}{m_i} - 2\pi s$$

$$= 2\pi \left\{ (2g - 2) + \sum_{i=1}^{r} \left(1 - \frac{1}{m_i} \right) \right\}. \quad \square$$

Equation (5.10.4) can prove useful for calculating the genus of $\mathcal{U}/\Gamma$. For example, let Γ be a triangle group obtained from a hyperbolic triangle with angles π/l, π/m, π/n as described in §5.6. If the reflections in the sides of the triangle are R_1, R_2, R_3 as indicated in Fig. 5.17 then the sides AB, AB' are paired by $R_1 R_3$ and the sides BC, $B'C$ are paired by $R_1 R_2$. Thus there are two conjugate pairs and so $n = 2$. ($ABCD'$ is a fundamental region for Γ.) $\{B, B'\}$ is an elliptic cycle, both these vertices being stabilised by cyclic groups of order m, $\{A\}$ is an elliptic cycle, A being stabilised by a cyclic group of order l, and $\{C\}$ is an elliptic cycle, C being stabilised by a cyclic group of order n. Therefore $r = 3$, $s = 0$ hence

$$2 - 2g = 3 - 2 + 1 = 2,$$

giving $g = 0$. (Alternatively, by 'glueing' AB to AB' and CB to CB' we see that we obtain a surface homeomorphic to the sphere.) Thus $\mathcal{U}/\Gamma$ has genus 0 and Γ has periods l, m, n. Therefore the signature of Γ is $(0; l, m, n)$. The hyperbolic area of a fundamental region for Γ is

$$2\pi \left\{ 1 - \frac{1}{l} - \frac{1}{m} - \frac{1}{n} \right\}.$$

As another example, consider a Fuchsian group Λ which is the group of covering transformations of a compact Riemann surface S of genus $g > 1$ (regarding $\mathcal{U}$ as a covering space of S). As Λ acts without fixed-points (§4.19) it has no elliptic elements and its signature is $(g; —)$, where the dash indicates no periods. By Theorem 5.9.9, Λ has no parabolic elements so that, besides the identity, it contains only hyperbolic elements.

We now sketch a proof of the existence of Fuchsian groups with given signature. This first appeared in Poincaré's original paper on Fuchsian

Fig. 5.17

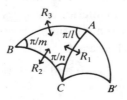

groups (Poincaré [1882]) but rigorous proofs have been given only comparatively recently (Maskit [1971]).

Theorem 5.10.5. *If $g \geqslant 0$, $m_i \geqslant 2$ are integers and if*

$$2g - 2 + \sum_{i=1}^{r} (1 - (1/m_i)) > 0,$$

then there exists a Fuchsian group with signature $(g; m_1, \ldots, m_r)$.

Indication of Proof. It is convenient to use the unit disc model of hyperbolic geometry (see Exercise 5H). Here the H-lines are circles and Euclidean lines orthogonal to the boundary of the unit disc $\mathcal{D}$. From the centre of $\mathcal{D}$ draw $(4g + r)$ radii at equal angles. Let $0 < t < 1$ and choose points at Euclidean distance t from the centre along each radius. If we join successive points by H-lines then we get a regular hyperbolic polygon $M(t)$. On the first r sides of $M(t)$ construct r external isosceles hyperbolic triangles such that the angles between the equal sides of the triangles are $2\pi/m_1, \ldots, 2\pi/m_r$. The union of $M(t)$ with these triangles is a starlike hyperbolic polygon $N(t)$ with $4g + 2r$ sides. Label these sides α_1, β_1, α_1', $\beta_1', \ldots, \alpha_g, \beta_g, \alpha_g', \beta_g', \xi_1, \xi_1', \ldots, \xi_r, \xi_r'$, and orient them as indicated in Fig. 5.18 (where $g = 2$, $r = 4$).

Now the hyperbolic area of $N(t)$ tends to 0 as t approaches 0, and by Corollary 5.5.6 tends to $(4g + 2r - 2)\pi - \sum_{i=1}^{r} 2\pi/m_i$ as t approaches 1. Hence for some t_0 between 0 and 1 the hyperbolic area of $N(t_0)$ is

$$2\pi \left\{ (2g - 2) + \sum_{i=1}^{r} \left(1 - \frac{1}{m_i} \right) \right\}.$$

Fig. 5.18

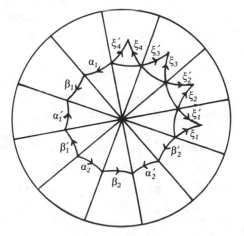

By construction α_i, α'_i have the same hyperbolic length as do β_j, β'_j and ξ_k, ξ'_k. Hence there exist hyperbolic isometries $A_i, B_j, X_k (i, j = 1, \ldots, g; k = 1, \ldots, r)$ such that

$$A_i(\alpha'_i) = \alpha_i, \ B_j(\beta'_j) = \beta_j, \ X_k(\xi'_k) = \xi_k.$$

We now compute the congruence classes of vertices (consult Fig. 5.18). Start by calling the end of α'_1 the vertex v_1; this is congruent to the end of α_1 which we call v_2, which in turn is congruent to the beginning of β'_1 which we call v_3, etc. In this way we see that the $(4g + r)$ vertices of the original polygon $M(t_0)$ form a congruent set. The other r vertices $w_1, \ldots, w_r$ form r congruent sets each with just one element.

As the area of $N(t_0)$ is

$$2\pi \left\{ (2g - 2) + \sum_{i=1}^{r} \left(1 - \frac{1}{m_i} \right) \right\}$$

the Gauss–Bonnet theorem 5.5.2 shows that the sum of the angles at the congruent set of vertices $v_1, v_2, \ldots, v_{4g+r}$ is 2π. Now let Γ be the group generated by

$$\{ A_i, B_j, X_k : i, j = 1, \ldots, g; k = 1, \ldots, r \}.$$

We wish to show that $N(t_0)$ is a fundamental region for Γ. If we compare Theorem 5.10.2 we see that we certainly need the sum of the angles at $v_1, \ldots, v_{4g+r}$ to be 2π and the angle at w_k to be $2\pi/m_k$. As we have these conditions it can be shown that the Γ-images of $N(t_0)$ cover $\mathcal{D}$ without overlap so that $N(t_0)$ is a fundamental region for Γ. (The same idea was used in §5.6 when we constructed triangle groups.) Thus Γ is a properly discontinuous group of hyperbolic isometries and if we transfer back to $\mathcal{U}$ we get a Fuchsian group.

The quotient-space $\mathcal{D}/\Gamma$ is decomposed into $(r + 1)$ vertices (corresponding to the $(r + 1)$ congruent sets of vertices), $(2g + r)$ edges and one simply connected face. Hence by the Euler–Poincaré formula (Theorem 4.16.2) its genus h satisfies

$$2 - 2h = (r + 1) - (2g + r) + 1 = 2 - 2g,$$

and thus $h = g$. There are r elliptic cycles namely $\{w_1\}, \ldots, \{w_r\}$ and the stabilisers have orders $m_1, \ldots, m_r$. Hence Γ has signature $(g; m_1, \ldots, m_r)$. □

The signature of a Fuchsian group also determines its algebraic structure. The presentation of a group with signature $(g; m_1, \ldots, m_r)$ is

$$\langle A_1, B_1, A_2, B_2, \ldots, A_g, B_g, X_1, \ldots, X_r | X_1^{m_1} = \cdots = X_r^{m_r}$$
$$= X_1 X_2 \ldots X_r A_1 B_1 A_1^{-1} B_1^{-1} \ldots A_g B_g A_g^{-1} B_g^{-1} = I \rangle. \quad (5.10.6)$$

Here, the X_i are elliptic and the A_k, B_k are hyperbolic. (For details see Lehner [1964]. Presentations are discussed in Appendix 2.) If

$$(2g - 2) + \sum_{i=1}^{r} \left(1 - \frac{1}{m_i}\right) \leqslant 0,$$

then there clearly does not exist a Fuchsian group with signature $(g; m_1, \ldots, m_r)$. (By a simple arithmetical calculation there are only a finite number of such signatures.) For example, there is no Fuchsian group of signature $(1; -)$. This gives an alternative proof that no Fuchsian group without elliptic elements represents a compact Riemann surface of genus 1 (see the Note in §5.7). However, it can be shown that if

$$2g - 2 + \sum_{i=1}^{r} \left(1 - \frac{1}{m_i}\right) = 0,$$

then there is a group of Euclidean isometries acting on $\mathbb{C}$ corresponding to this signature. For example, a lattice corresponds to the signature $(1; -)$, and we get the Euclidean triangle groups $(0; 2, 4, 4)$, $(0; 2, 3, 6)$, $(0; 3, 3, 3)$ from triangles with angles π/l, π/m, π/n where $1/l + 1/m + 1/n = 1$. If

$$(2g - 2) + \sum_{i=1}^{r} \left(1 - \frac{1}{m_i}\right) < 0$$

we get groups of isometries of the sphere including the spherical triangle groups treated in Theorem 2.13.5.

Theorem 5.10.7. *If F is a Dirichlet region for a Fuchsian group Γ with $\mathcal{U}/\Gamma$ compact then $\mu(F) \geqslant \pi/21$. If $\mu(F) = \pi/21$ then Γ is a triangle group with signature $(0; 2, 3, 7)$.*

Proof. This is by calculation using Theorem 5.10.3. If $g \geqslant 2$ then $\mu(F) \geqslant 4\pi$. If $g = 1$ then as $\mu(F) > 0$, Γ must have periods and then $\mu(F) \geqslant \pi$, the minimum being attained for a group with signature $(1; 2)$. If $g = 0$ then

$$\mu(F) = 2\pi\left(-2 + \sum_{i=1}^{r}\left(1 - \frac{1}{m_i}\right)\right).$$

As $1 - (1/m_i) \geqslant \frac{1}{2}$, $\mu(F) \geqslant 2\pi(-2 + (r/2)) = \pi(r - 4)$ so that if $r \geqslant 5$ then $\mu(F) \geqslant \pi$, the minimum being attained for a group with signature $(0; 2, 2, 2, 2, 2)$. If $r = 4$ and $m_1 = m_2 = m_3 = m_4 = 2$ then $\mu(F) = 0$ which is inadmissable. The minimum value of $\mu(F)$ (for $r = 4$) corresponds to a group of signature $(0; 2, 2, 2, 3)$ giving $\mu(F) \geqslant \pi/3$. The only cases left to consider are $g = 0, r \leqslant 3$. If $g = 0, r \leqslant 2$ then $\mu(F) < 0$ and so we only consider $g = 0, r = 3$.

Therefore Γ has signature $(0; m_1, m_2, m_3)$ and

$$\mu(F) = 2\pi\left(1 - \frac{1}{m_1} - \frac{1}{m_2} - \frac{1}{m_3}\right).$$

We may assume that $m_1 \leqslant m_2 \leqslant m_3$. If $m_1 \geqslant 4$ then $\mu(F) \geqslant \pi/2$. If $m_1 = 3$ then $(0; 3, 3, 3)$ gives $\mu(F) = 0$ and the minimum is attained for $(0; 3, 3, 4)$ giving $\mu(F) \geqslant \pi/6$. If $m_1 = 2$ then $m_2 > 2$ and if $m_2 \geqslant 4$ then $\mu(F) \geqslant \pi/10$, the minimum corresponding to $(0; 2, 4, 5)$. If $m_2 = 3$ then $(0; 2, 3, 6)$ gives $\mu(F) = 0$ and the minimum is attained for $(0; 2, 3, 7)$, giving $\mu(F) \geqslant \pi/21$. ☐

As we have been dealing, for simplicity, with groups Γ for which $\mathcal{U}/\Gamma$ is compact, we have not considered groups with parabolic elements (see Theorem 5.9.9). It can be shown (Lehner [1966]) that if Γ has a Dirichlet region F with finite hyperbolic area then F has a finite number of sides and so by Theorem 5.8.7, Γ is finitely generated. Suppose that Γ has r conjugacy classes of elliptic cyclic subgroups of orders $m_1, \ldots, m_r$, s conjugacy classes of parabolic cyclic subgroups and $\mathcal{U}/\Gamma$ has genus g. Then we say that Γ has signature

$$(g; m_1, \ldots, m_r; s). \tag{5.10.8}$$

By a similar proof to Theorem 5.10.3 we can show that

$$\mu(F) = 2\pi\left\{(2g - 2) + \sum_{i=1}^{r}\left(1 - \frac{1}{m_i}\right) + s\right\},$$

and as in Theorem 5.10.7 we can show that if $s > 0$ then $\mu(F) \geqslant \pi/3$, this minimum being attained for the modular group, which has signature $(0; 2, 3; 1)$. (Thus in Theorem 5.10.7 the hypothesis that $\mathcal{U}/\Gamma$ is compact is not necessary.)

If $\mu(F) > 0$ then we can show, by a similar method of proof to Theorem 5.10.5, that a group Γ with signature (5.10.8) exists. (We need s of the isosceles triangles to have vertices on the boundary of the disc, the angle at these vertices being 0.) The algebraic structure of the group Γ is determined by its signature, a group with signature (5.10.8) having the presentation

$$\langle A_1, B_1, \ldots, A_g, B_g, X_1, \ldots, X_r, P_1, \ldots, P_s | X_1^{m_1} = \cdots = X_r^{m_r}$$
$$= P_1 \ldots P_s X_1 \ldots X_r A_1 B_1 A_1^{-1} B_1^{-1} \ldots A_g B_g A_g^{-1} B_g^{-1} = I \rangle.$$

Theorem 5.10.9.   *Let Γ be a Fuchsian group and Λ a subgroup of index n. If*

$$\Gamma = \Lambda T_1 \cup \Lambda T_2 \cup \ldots \cup \Lambda T_n$$

is a decomposition of Γ into Λ-cosets and if F is a fundamental region for Γ

then

(i) $F_1 = T_1(F) \cup T_2(F) \cup \ldots \cup T_n(F)$ *is a fundamental region for* Λ,
(ii) *if* $\mu(F)$ *is finite and the H-area of the boundary of F is zero then*
$$\mu(F_1)/\mu(F) = n.$$

Proof. (i) Let $z \in \mathcal{U}$. Then as F is a fundamental region for Γ, there exists $w \in F$ such that $z = T(w)$, $T \in \Gamma$. Now $T = ST_i$ for $S \in \Lambda$ and $1 \leqslant i \leqslant n$. Therefore

$$z = ST_i(w) = S(T_i(w)).$$

As $T_i(w) \in F_1$, z is in the Λ-orbit of some point of F_1. Hence the union of the Λ-images of F_1 is $\mathcal{U}$.

Now suppose that $z \in \mathring{F}_1$ (the interior of F_1) and that $S(z) \in \mathring{F}_1$, for $S \in \Lambda$. We need to prove that $S = I$. Let $\varepsilon > 0$ be so small that $B_\varepsilon(z)$ (the open H-disc, centre z, radius ε) is contained in $\mathring{F}_1$. Then $B_\varepsilon(z)$ has non-trivial intersection with precisely k of the images of $\mathring{F}$ under $T_1, \ldots, T_n$, where $1 \leqslant k \leqslant n$; suppose that these images are $T_{i_1}(\mathring{F}), \ldots, T_{i_k}(\mathring{F})$. Let $B_\varepsilon(S(z)) = S(B_\varepsilon(z))$ have non-empty intersection with $T_j(\mathring{F})$ say, $(1 \leqslant j \leqslant n)$. It follows that $B_\varepsilon(z)$ has non-empty intersection with $S^{-1}T_j(\mathring{F})$ so that $S^{-1}T_j = T_{i_l}$ where $1 \leqslant l \leqslant k$. Therefore

$$\Lambda T_j = \Lambda S^{-1}T_j = \Lambda T_{i_l},$$

so that $T_j = T_{i_l}$ and $S = I$. Hence $\mathring{F}_1$ contains precisely one point of each Λ-orbit.

(ii) This follows immediately as $\mu(T(F)) = \mu(F)$, for all $T \in PSL(2, \mathbb{R})$, and $\mu(T_i(F) \cap T_j(F)) = 0$ for $i \neq j$. $\quad\square$

5.11 Automorphisms of compact Riemann surfaces

In §2.1 we considered the automorphism group of the Riemann sphere, the unique Riemann surface of genus 0, and in §4.18 we considered the automorphism groups of compact Riemann surfaces of genus 1. In these cases we found that the automorphism groups were infinite; by contrast we show that groups of automorphisms of compact Riemann surfaces of genus $g \geqslant 2$ are necessarily finite.

Theorem 5.11.1. *Let S be a compact Riemann surface of genus $g \geqslant 2$. Then* $|\text{Aut } S| \leqslant 84(g - 1)$.

Proof. Let $S = \mathcal{U}/\Lambda$ where Λ is a Fuchsian group without elliptic elements. (The existence of such a group follows from Theorem 4.19.8 and was discussed before Theorem 5.10.5.) As Λ has signature $(g; -)$ the hyperbolic area of a Dirichlet region for Λ is $2\pi(2g - 2)$. By Corollary 5.9.5, and Exercise 5T, Aut $S \cong \Gamma/\Lambda$ for some Fuchsian group Γ containing Λ as a normal subgroup. Now $[z]_\Lambda \mapsto [z]_\Gamma$ is a well-defined continuous map from $\mathcal{U}/\Lambda$ onto $\mathcal{U}/\Gamma$ and as $\mathcal{U}/\Lambda$ is compact it follows that $\mathcal{U}/\Gamma$ is compact. Hence a Dirichlet region F for Γ is compact in $\mathcal{U}$ by Theorem 5.9.8 and so $\mu(F)$ is finite; also by Theorem 5.10.7, $\mu(F) \geqslant \pi/21$. It follows from Theorem 5.10.9 that

$$|\text{Aut } S| = |\Gamma/\Lambda| = \frac{\mu(F_1)}{\mu(F)} = \frac{2\pi(2g - 2)}{\mu(F)} \leqslant 84(g - 1),$$

where F_1 is a fundamental region for Λ. □

The finiteness of the automorphism group of a compact Riemann surface of genus $g \geqslant 2$ was first proved by Schwarz in 1878 and the bound given in Theorem 5.11.1 was proved by Hurwitz in 1893.

We now investigate briefly the question of when the bound of Theorem 5.11.1 is attained. A group of $84(g - 1)$ automorphisms of a compact Riemann surface of genus $g \geqslant 2$ is called a *Hurwitz group*.

Theorem 5.11.2. *A finite group H is a Hurwitz group if and only if H is non-trivial and has two generators x, y obeying the relations*

$$x^2 = y^3 = (xy)^7 = 1.$$

Proof. If H is a Hurwitz group then by Corollary 5.9.5 and Theorem 5.10.7, $H \cong \Gamma/\Lambda$ where Γ is a triangle group of signature $(0; 2, 3, 7)$ and Λ has signature $(g; -)$ for some integer $g \geqslant 2$. Now Γ has two generators X, Y, obeying the relations $X^2 = Y^3 = (XY)^7 = I$, so that if $\theta : \Gamma \to H$ is the canonical homomorphism then $x = \theta(X)$, $y = \theta(Y)$ generate H and obey the relations $x^2 = y^3 = (xy)^7 = 1$. Also H is non-trivial as its order is divisible by 84.

Conversely, let H be a non-trivial finite group with two generators x, y obeying the relations $x^2 = y^3 = (xy)^7 = 1$. Let Γ be a triangle group with signature $(0; 2, 3, 7)$ and presentation $\langle X, Y | X^2 = Y^3 = (XY)^7 = I \rangle$. By the results of Appendix 2 there is a homomorphism θ from Γ onto H such that $\theta(X) = x$ and $\theta(Y) = y$. Let Λ be the kernel of θ. We show that Λ contains no elliptic elements. By Theorem 5.8.6 every elliptic element of Γ is conjugate to a power of X, Y or XY, so if Λ contains elliptic elements then as $\Lambda \trianglelefteq \Gamma$ and as 2, 3 and 7 are prime it must contain X, Y or XY.

Suppose that Λ contains X. Then $x = 1$ and hence $y^3 = y^7 = 1$ so that $y = 1$ and H is trivial. Similarly, if Λ contains Y or XY then H is trivial. Thus Λ contains no elliptic elements. Also, by Theorem 5.9.9, Γ contains no parabolic elements and hence Λ contains no parabolic elements. As Λ has finite index in Γ, it has a compact fundamental region by Theorem 5.10.9, and so Λ must have signature $(g; —)$ for some integer g which must be greater than or equal to 2 by the discussion before Theorem 5.10.7. Hence Γ/Λ is a group of automorphisms of $\mathscr{U}/\Lambda$, a compact Riemann surface of genus $g \geq 2$, and by Theorem 5.10.9,

$$|\Gamma/\Lambda| = \frac{2\pi(2g - 2)}{\pi/21} = 84(g - 1). \qquad \square$$

Theorem 5.11.3. *Let H be a Hurwitz group and let H_1 be a non-trivial homomorphic image of H. Then H_1 is a Hurwitz group.*

Proof. Let x, y generate H and obey the relations $x^2 = y^3 = (xy)^7 = 1$. Let ϕ be the homomorphism from H onto H_1. If $\phi(x) = x_1$, $\phi(y) = y_1$ then $x_1^2 = y_1^3 = (x_1 y_1)^7 = 1$. $\square$

Corollary 5.11.4. *A Hurwitz group of smallest order is a simple group.* $\square$

Theorem 5.11.5.

(i) *There is no Hurwitz group of order 84.*

(ii) *If $\mathbb{Z}_7$ denotes the field with seven elements then $PSL(2, \mathbb{Z}_7)$ is a Hurwitz group of order 168.*

(There is a unique field with seven elements, namely the integers $\{0, 1, 2, 3, 4, 5, 6\}$ under addition and multiplication mod (7). For the definition of $PSL(2, F)$, where F is a field, see §2.2.)

Proof. (i) By Corollary 5.11.4, a Hurwitz group of order 84 would be simple. We now use the Sylow theorems (Rose [1978]) to show that there is no simple group of order 84. A group of order 84 must contain $(7k + 1)$ Sylow 7-subgroups where $(7k + 1)$ divides 84. The only possibility is that $k = 0$ so that the group must contain a unique Sylow 7-subgroups which must then be normal.

(ii) $PSL(2, \mathbb{Z}_7) = SL(2, \mathbb{Z}_7)/\{\pm I\}$ so that $PSL(2, \mathbb{Z}_7)$ is a homomorphic image of $SL(2, \mathbb{Z}_7)$. Let $A, B \in SL(2, \mathbb{Z}_7)$ be defined by

$$A = \begin{pmatrix} 0 & 1 \\ -1 & 0 \end{pmatrix} \qquad B = \begin{pmatrix} 0 & -1 \\ 1 & 1 \end{pmatrix}.$$

Then

$$AB = \begin{pmatrix} 1 & 1 \\ 0 & 1 \end{pmatrix}$$

and

$$A^2 = B^3 = -I, \quad (AB)^7 = I.$$

Let x, y be the images of A, B respectively under the homomorphism from $SL(2, \mathbb{Z}_7)$ to $PSL(2, \mathbb{Z}_7)$. Then

$$x^2 = y^3 = (xy)^7 = 1,$$

so we just need to show that x, y generate $PSL(2, \mathbb{Z}_7)$. Direct calculation shows that if $C = ABA^{-1}B^{-1}$ then $C^2 = \begin{pmatrix} 5 & 3 \\ 3 & 2 \end{pmatrix}$ and $C^4 = -I$. Thus if u is the image of C in $PSL(2, \mathbb{Z}_7)$ then u has order 4. Thus x, y generate a group whose order is divisible by $3 \times 4 \times 7 = 84$. It cannot be 84 by part (i) and a simple calculation (repeated in §6.9) shows that $|PSL(2, \mathbb{Z}_7)| = 168$. Hence x, y generate $PSL(2, \mathbb{Z}_7)$. □

(Corollary 5.11.4 and Theorem 5.11.5 imply that $PSL(2, \mathbb{Z}_7)$ is simple. This is a special case of the theorem that $PSL(2, F)$ is simple for any field F with more than three elements, see Dickson [1958].)

Thus the Hurwitz bound is not attained when $g = 2$ but is attained when $g = 3$. It can be shown that it is attained for infinitely many values of g and not attained for infinitely many values of g (Exercises 5W, X, Y). The precise values of g for which the Hurwitz bound is attained are unknown; the first four values are $g = 3, 7, 14, 17$.

5.12 Automorphic functions and uniformisation

We mention, without proof, a few results about automorphic functions and their relationship to algebraic curves. *Automorphic functions* are meromorphic functions defined on $\mathcal{U}$ which are invariant under transformations of a Fuchsian group. Thus they are related to elliptic functions which are meromorphic functions defined on $\mathbb{C}$ invariant under transformations of a lattice.

Let Γ be a Fuchsian group; for simplicity we shall assume that $\mathcal{U}/\Gamma$ is compact. A function f which is meromorphic on $\mathcal{U}$ is called Γ-*automorphic* if $f(T(z)) = f(z)$ for all $T \in \Gamma$. As with elliptic functions a major problem is to show that non-constant automorphic functions exist. This can be achieved by writing down quotients of appropriate infinite series. It is then found that automorphic functions possess many properties analogous to those of elliptic functions. For example, the Γ-automorphic functions clearly form a field which we denote by $\mathscr{F}(\Gamma)$. If f, $g \in \mathscr{F}(\Gamma)$ then there is a polynomial $\Phi(x, y)$ such that $\Phi(f(z), g(z)) \equiv 0$ (see Theorem 3.11.2).

Now suppose that Γ has no elliptic elements. Then as with elliptic functions (§4.12) there is an isomorphism between $\mathscr{F}(\Gamma)$ and the field of meromorphic functions on the Riemann surface $\mathscr{U}/\Gamma$; if f is Γ-automorphic and if we define $f : \mathscr{U}/\Gamma \to \Sigma$ by $f[z]_\Gamma = f(z)$ then f is meromorphic on $\mathscr{U}/\Gamma$ and $f \leftrightarrow f$ gives this isomorphism.

Now suppose that $A(x, y) = 0$ is an irreducible algebraic function with Riemann surface S of genus greater than one. Then by §§4.13 and 4.14 there exist meromorphic functions ϕ, ψ defined on S such that $\phi(s) = x$, $\psi(s) = y$. By Theorem 4.19.8, $S = \mathscr{U}/\Gamma$ for some Fuchsian group Γ without elliptic elements and then ϕ, ψ give rise to Γ-automorphic functions $\bar{\phi}$, $\bar{\psi}$ such that $\bar{\phi}(z) = \phi[z]_\Gamma$, $\bar{\psi}(z) = \psi[z]_\Gamma$. Hence $A(\bar{\phi}(z), \bar{\psi}(z)) \equiv 0$. Thus from the algebraic function $A(x, y) = 0$ which expresses y as a many-valued function of x we have found single-valued functions $\bar{\phi}$, $\bar{\psi}$ such that $x = \bar{\phi}(z)$, $y = \bar{\psi}(z)$.

The process of representing many-valued functions in terms of single-valued functions in this way is called *uniformisation*. Theorem 4.17.2 is often called the uniformisation theorem because it is a crucial step in establishing the above process.

There are simpler examples of uniformisation not involving Fuchsian groups. There are two well-known examples of the uniformisation of $x^2 + y^2 = 1$, namely $x = \sin t$, $y = \cos t$ and $x = (2t)/(1 + t^2)$, $y = (1 - t^2)/(1 + t^2)$. (These are in terms of simply periodic functions and rational functions respectively. In fact any irreducible algebraic function whose Riemann surface has genus 0 can be uniformised by rational functions as these are the meromorphic functions on the sphere as shown in Theorem 1.4.1.) In Chapter 3 we saw how we could uniformise some cubic curves $y^2 = 4x^3 - g_2 x - g_3$ with elliptic functions. In the next chapter we shall see that all such curves with $g_2^3 \neq 27 g_3^2$ can be so uniformised. Interestingly, the proof involves functions which are automorphic with respect to the modular group.

EXERCISES

5A. Show that every transformation in $PSL(2, \mathbb{R})$ is a composition of transformations of the form $z \mapsto \lambda z$ ($\lambda \in \mathbb{R} \setminus \{0\}$), $z \mapsto z + \mu$ ($\mu \in \mathbb{R}$) and $z \mapsto -1/z$ (see §2.3).

5B. Prove that $T : z \mapsto z + 1$ and $T^{-1} : z \mapsto z - 1$ are not conjugate in $PSL(2, \mathbb{R})$.

5C.   Show that the transformations in $PSL(2, \mathbb{R})$ together with the transformations

$$z \mapsto \frac{a\bar{z} + b}{c\bar{z} + d}, \quad a, b, c, d \in \mathbb{R}, \quad ad - bc = -1 \qquad (*)$$

form a group $\mathscr{L}$ which contains $PSL(2, \mathbb{R})$ with index 2. Prove that every transformation of $\mathscr{L}$ is an angle-preserving homeomorphism of $\mathscr{U}$.

5D.   Show that if $a + d \neq 0$ then the transformation (*) has two fixed-points in $\mathbb{R} \cup \{\infty\}$, and if $a + d = 0$ then the fixed-point set in $\mathscr{U}$ of the transformation (*) is either a semi-circle centred on $\mathbb{R}$ or a line perpendicular to $\mathbb{R}$. (In the terminology of §5.3 these are H-lines. In the hyperbolic geometry of $\mathscr{U}$ the transformations (*) are orientation-reversing isometries; those with $a + d = 0$ are hyperbolic reflections and those with $a + d \neq 0$ are hyperbolic glide-reflections.)

5E.   Divide the transformations (*) into conjugacy classes in $\mathscr{L}$ and compute the centraliser of an element of each conjugacy class.

5F.   If $z = (ai + b)/(ci + d)$, $a, b, c, d \in \mathbb{R}$, $ad - bc = 1$, prove that

$$\cosh \rho(i, z) = \tfrac{1}{2}(a^2 + b^2 + c^2 + d^2),$$

where ρ is the hyperbolic metric in $\mathscr{U}$.

5G.   Let T be a hyperbolic element with fixed-points $a, b \in \mathbb{R} \cup \{\infty\}$ and let Q be the H-line joining a and b. If $p \in Q$ show that $\rho(p, T(p))$ is independent of p. (Hint: map Q onto the imaginary axis and use Theorem 5.3.5.)

5H.   Using the result that $z \mapsto (z - i)/(z + i)$ maps $\mathscr{U}$ bijectively onto the unit disc $\mathscr{D}$ (§4.17), prove that $\mathscr{D}$ carries an induced hyperbolic metric given by

$$ds = \frac{2|dw|}{1 - |w|^2}$$

and show that the geodesics of this metric are segments of Euclidean circles or Euclidean lines perpendicular to the unit circle.

5I.   If ρ_1 denotes the hyperbolic distance in the unit disc prove that

$$\sinh^2 \left[\tfrac{1}{2} \rho_1(z, w) \right] = \frac{|z - w|^2}{(1 - |z|^2)(1 - |w|^2)}.$$

5J.   Show that a hyperbolic circle in $\mathscr{D}$ with centre at 0 is a Euclidean circle with centre at 0. Find a formula relating the hyperbolic radius and the Euclidean radius.

5K.   (i)   Show that every hyperbolic element of $PSL(2, \mathbb{R})$ is a product of two parabolic elements of $PSL(2, \mathbb{R})$ (see Exercise 2N).

     (ii) Still using 2N show that every elliptic element of $PSL(2, \mathbb{C})$ which lies in Aut $\mathscr{D}$ is a product of two parabolic elements in Aut $\mathscr{D}$.

     (iii) Hence show that $PSL(2, \mathbb{R})$ is generated by parabolic elements and deduce that $PSL(2, \mathbb{R})$ is a simple group.

5L.   Consider a hyperbolic triangle T which contains a hyperbolic triangle T_1 in its interior. Compute the H-area of the (non-simply connected) hexagon consisting of those points exterior to T_1 and interior to T in terms of the angles of the polygon. Hence show that the Gauss–Bonnet formula given in Corollary 5.5.6 is false for non-simply connected polygons.

5M. Let $m: PSL(2, \mathbb{R}) \times PSL(2, \mathbb{R}) \to PSL(2, \mathbb{R})$ be defined by $m(S, T) = ST$ and let $i: PSL(2, \mathbb{R}) \to PSL(2, \mathbb{R})$ be defined by $i(S) = S^{-1}$. Prove that m and i are both continuous (so that $PSL(2, \mathbb{R})$ is a topological group – see §3.2).

5N. Show that every transformation in $PSL(2, \mathbb{R})$ can be written uniquely in the form TR where R is an elliptic element fixing i and $T(z) = az + b$ $(a, b \in \mathbb{R}, a > 0)$. Deduce that as a topological space $PSL(2, \mathbb{R})$ is homeomorphic to $\mathbb{R}^2 \times S^1$, where S^1 is a circle.

5P. Show that the vertices of a Dirichlet region F are *isolated*, that is every vertex of F has a neighbourhood containing no other vertices of F. (Hint: use local finiteness.) Deduce that a compact Dirichlet region has a finite number of vertices.

5Q. Show that i is the only point of the region F in Fig. 5.10 that is fixed by an involution of the modular group.

5R. Let F be a Dirichlet region for a Fuchsian group Γ and let s be a side of F. If $T \in \Gamma$ and $T(s)$ is a side of F prove that

$$F \cap T(F) = T(s).$$

5S. If F is a Dirichlet region for Γ centred at p prove that $T(F)$ is a Dirichlet region for $T\Gamma T^{-1}$ centred at $T(p)$.

5T. (i) Let Λ be the cyclic group generated by $z \mapsto \lambda z \, (\lambda > 1)$. Find the Dirichlet region for Λ centred at i.

(ii) Let Λ be the cyclic group generated by $z \mapsto z + 1$. Find the Dirichlet region for Λ centred at i.

(iii) Prove that if Λ is a cyclic Fuchsian group then $\mathscr{U}/\Lambda$ is not compact (use Theorem 5.9.8).

5U. Let $p = g - 1 > 84$ be a prime number. Show that there is no compact Riemann surface of genus g admitting $84(g - 1)$ automorphisms. (Hint: use Theorem 5.11.3 to show that a Hurwitz group of order $84p$ must be simple and then use the Sylow theorems to show that a Sylow p-subgroup must be normal.)

5V. Let $p \geqslant 5$ be a prime number. Construct an epimorphism θ from a triangle group with signature $(0; p, p, p)$ onto C_p, the cyclic group of order p, such that the kernel of θ has no elliptic elements. Deduce that C_p acts as a group of automorphisms of a Riemann surface of genus $(p - 1)/2$.

The following problems show that for infinitely many values of g there is a compact Riemann surface of genus g admitting $84(g - 1)$ automorphisms. The proof follows an idea of Macbeath [1961]. The first of these problems is purely group-theoretic.

5W. (i) Let G be a group. A subgroup K of G is called *characteristic* if $\alpha(K) = K$ for all group automorphisms $\alpha: G \to G$. Prove that if $K \leqslant N \leqslant G$ and N is normal is G and K is characteristic in N then K is normal in G. (Hint: consider the automorphism of N defined by $n \mapsto yny^{-1}$, $y \in G$.)

(ii) Let G^m be the subgroup of G generated by the mth powers of elements of G. Let $[G, G]$ denote the commutator subgroup of G, that is the subgroup of G generated by elements of the form $aba^{-1}b^{-1}$, $(a, b \in G)$. Prove that G^m, $[G, G]$ and $G^m[G, G]$ are characteristic subgroups of G.

(It follows from results in Appendix 2 that if a presentation for G is

$$gp\langle a_1, a_2, \ldots, a_n | R_i(a_j) = 1 \rangle,$$

then a presentation for $G/G^m[G, G] = \mathbf{G}$ is

$$gp\langle \mathbf{a}_1, \mathbf{a}_2, \ldots, \mathbf{a}_n | R_i(\mathbf{a}_j) = 1, \mathbf{a}_s^m = 1, \mathbf{a}_t\mathbf{a}_u = \mathbf{a}_u\mathbf{a}_t (1 \leqslant s, t, u \leqslant n) \rangle,$$

where $\mathbf{a}_i$ is the image of a_i under the canonical homomorphism from G to $\mathbf{G}$.)

5X. If Λ is a Fuchsian group of signature $(g; -)$ use the presentation (5.10.6) to show that $\Lambda/\Lambda^m[\Lambda, \Lambda]$ is a finite abelian group of order m^{2g}.

5Y. Let Γ be a triangle group with signature $(0; 2, 3, 7)$ and let Λ be a normal subgroup of signature $(g; -)$. (Theorem 5.11.5 shows that at least one such normal subgroup exists when $g = 3$, its index being $84(3 - 1) = 168$.) Prove that for each positive integer m there is a normal subgroup of Γ of index $84(g - 1).m^{2g}$ and hence show that for each such m there is a compact Riemann surface of genus

$$g' = m^{2g}(g - 1) + 1,$$

which admits $84(g' - 1)$ automorphisms.

6

The modular group

The modular group $\Gamma = PSL(2, \mathbb{Z})$ is the most widely studied of all Fuchsian groups, and several books have been written about Γ, its action on $\mathcal{U}$, and the associated modular functions (meromorphic functions invariant under Γ). The importance of Γ lies in its many connections with other branches of mathematics, and especially with number theory; indeed, interest in Γ first arose out of the investigations of Gauss into number-theoretic properties of quadratic forms $ax^2 + bxy + cy^2$ $(a, b, c \in \mathbb{Z})$. We will concentrate on those aspects of Γ which are related to topics considered in earlier chapters; for example, we will apply Γ to the problem (partially dealt with in §4.18) of classifying the compact Riemann surfaces of genus 1.

In §6.1 we construct a bijection between the set of all conformal equivalence classes of tori $\mathbb{C}/\Omega$ (equivalently, the similarity classes of lattices $\Omega \subset \mathbb{C}$) and the orbits of Γ on $\mathcal{U}$. In order to apply this to the Riemann surface S of $\sqrt{p(z)}$ (considered in §4.9), where p is a cubic polynomial, we introduce in §6.2 the discriminant Δ, the non-vanishing of which is equivalent to p having distinct roots, that is, to S having genus 1. We use Δ in §6.3 to obtain an analytic function $J : \mathcal{U} \to \mathbb{C}$ invariant under the action of Γ; detailed study in §6.4 of the analytic properties of J enables us to show in §6.5 that J maps $\mathcal{U}$ onto $\mathbb{C}$ and that the level sets $J^{-1}(c)$, $c \in \mathbb{C}$, are the orbits of Γ. This gives a bijection between conformal equivalence classes of tori and points $c \in \mathbb{C}$, and from this we obtain a proof (independent of the uniformisation theorem) that if p has distinct roots then $S \cong \mathbb{C}/\Omega$ for some lattice Ω. In §6.6 we give one of the classical applications of the J-function, the proof of Picard's theorem on entire functions, and in §6.7 we show how J is related to the cross-ratio function λ considered in §2.5. In §6.8 we use the action of Γ on $\mathcal{U}$ to obtain a presentation for Γ in terms of generators and relations, and we use this in §6.9 to study the homomorphic images of Γ, many of which have appeared earlier in this book in different contexts. This involves studying certain normal subgroups N of Γ, and in §6.10 we examine the corresponding quotient surfaces $\mathcal{U}/N$.

6.1 Lattices, tori and moduli

We saw in §4.18 that if Ω and Ω' are lattices in $\mathbb{C}$, then the tori $\mathbb{C}/\Omega$ and $\mathbb{C}/\Omega'$ are conformally equivalent if and only if Ω and Ω' are similar, that is, $\Omega' = \mu\Omega$ for some $\mu \in \mathbb{C} \setminus \{0\}$. If $\{\omega_1, \omega_2\}$ and $\{\omega_1', \omega_2'\}$ are bases for Ω and Ω', then by Theorem 3.4.2 this is equivalent to

$$\left.\begin{aligned} \omega_2' &= \mu(a\omega_2 + b\omega_1), \\ \omega_1' &= \mu(c\omega_2 + d\omega_1), \end{aligned}\right\} \tag{6.1.1}$$

where $a, b, c, d \in \mathbb{Z}$ and $ad - bc = \pm 1$.

Since ω_1 and ω_2 are linearly independent over $\mathbb{R}$ we have $\mathrm{Im}\,(\omega_2/\omega_1) \neq 0$; interchanging ω_1 and ω_2 if necessary, we may assume that $\mathrm{Im}\,(\omega_2/\omega_1) > 0$. We define the *modulus* of the basis $\{\omega_1, \omega_2\}$ to be

$$\tau = \omega_2/\omega_1,$$

where the numbering is such that

$$\mathrm{Im}\,(\tau) > 0.$$

Each lattice Ω determines a set of moduli, the moduli of its various bases, and since $\mu\omega_2/\mu\omega_1 = \omega_2/\omega_1$, similar lattices determine the same sets of moduli. Putting $\tau = \omega_2/\omega_1$ and $\tau' = \omega_2'/\omega_1'$ (the moduli of the above bases for Ω and Ω'), we see from (6.1.1) that Ω and Ω' are similar if and only if

$$\tau' = \frac{a\tau + b}{c\tau + d}, \tag{6.1.2}$$

where $a, b, c, d \in \mathbb{Z}$, and $ad - bc = \pm 1$. Now both τ and τ', being moduli, lie in the upper half-plane

$$\mathscr{U} = \{z \in \mathbb{C} \,|\, \mathrm{Im}\,(z) > 0\};$$

as shown in §2.8, if $ad - bc = -1$ then the Möbius transformation $T{:}z \mapsto (az + b)/(cz + d)$, which is in $PGL(2, \mathbb{R}) \setminus PSL(2, \mathbb{R})$, maps $\mathscr{U}$ onto the lower half-plane, so we must therefore have $ad - bc = 1$. Conversely, if $a, b, c, d \in \mathbb{Z}$ and $ad - bc = 1$, then (6.1.1) gives a basis $\{\omega_1', \omega_2'\}$ for a lattice Ω' similar to Ω.

As we saw in §5.6, the Möbius transformations

$$T{:}z \mapsto \frac{az + b}{cz + d} \quad (a, b, c, d \in \mathbb{Z}, ad - bc = 1) \tag{6.1.3}$$

form a discrete subgroup of $PSL(2, \mathbb{R})$, the *modular group* $PSL(2, \mathbb{Z})$, which we shall denote throughout this chapter by Γ. Summarising the above argument, we have:

Theorem 6.1.4. *If $\Omega = \Omega(\omega_1, \omega_2)$ and $\Omega' = \Omega(\omega_1', \omega_2')$ are lattices in $\mathbb{C}$,* *with moduli* $\tau = \omega_2/\omega_1$ *and* $\tau' = \omega_2'/\omega_1'$ *(so that* $\tau, \tau' \in \mathcal{U}$*), then the following are equivalent:*

  (i) *the tori* $\mathbb{C}/\Omega$ *and* $\mathbb{C}/\Omega'$ *are conformally equivalent;*
 (ii) *the lattices* Ω *and* Ω' *are similar;*
(iii) $\tau' = T(\tau)$ *for some* $T \in \Gamma$. $\square$

In §5.7 we observed that every compact Riemann surface of genus 1 is conformally equivalent to $\mathbb{C}/\Omega$ for some lattice Ω, so Theorem 6.1.4 gives a bijection between the conformal equivalence classes of such Riemann surfaces and the points on the quotient-space $\mathcal{U}/\Gamma$. Thus we can think of $\mathcal{U}/\Gamma$ as representing the set of all complex structures which can be imposed on a surface of genus 1. Now by Theorem 5.9.1, $\mathcal{U}/\Gamma$ is itself a Riemann surface! If we consider the Dirichlet region (see Fig. 6.1) $F = \{z \in \mathcal{U} \,||\, z| \geqslant 1$ and $|\mathrm{Re}\,(z)| \leqslant \frac{1}{2}\}$ for Γ, given by Theorem 5.8.4, then Theorem 5.9.6 shows that $\mathcal{U}/\Gamma$ is homeomorphic to F/Γ, and by using the transformations $z \mapsto z + 1$ and $z \mapsto -1/z$ to identify congruent sides of F we see that F/Γ (and hence $\mathcal{U}/\Gamma$) is homeomorphic to the plane $\mathbb{C}$ (later in this chapter we shall introduce an analytic function $J : \mathcal{U} \to \mathbb{C}$ which induces a conformal equivalence between $\mathcal{U}/\Gamma$ and $\mathbb{C}$).

Fig. 6.1

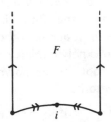

More generally, for each integer $g \geqslant 0$, let R_g be the set of all conformal equivalence classes of compact Riemann surfaces of genus g; this set is called the *Riemann space*, or *space of moduli* of genus g. We have seen that R_1 can be identified in a natural way with $\mathbb{C}$, so that Riemann surfaces of genus 1 are parametrised by one complex parameter (or, equivalently, two real parameters). Riemann observed that, for any given $g > 1$, Riemann surfaces of genus g are parametrised in some sense by $6g - 6$ real parameters, and this inspired attempts to realise R_g as a $(6g - 6)$-dimensional space. In the 1930s the theory of quasi-conformal mappings led to the discovery of Teichmüller space T_g (O. Teichmüller [1913–41]); this is a metric space, homeomorphic (for $g > 1$) to $\mathbb{R}^{6g-6}$, admitting a discontinuous group Γ_g of isometries such that $R_g = T_g/\Gamma_g$. For $g = 1$, T_1 is the upper half-plane $\mathcal{U}$ (homeomorphic to $\mathbb{R}^2$), the metric is the hyperbolic metric, Γ_1 is the modular group Γ, and (as we have seen) R_1 can be identified with $\mathcal{U}/\Gamma$. For $g = 0$, R_0 consists of a single point since Theorem 4.17.2 implies that any Riemann surface of genus 0 is conformally equivalent to Σ.

For an account of this theory, see Abikoff [1980], or Bers' survey article [1972].

6.2 The discriminant of a cubic polynomial

In §4.9 (iv) we saw that if $p(z)$ is any cubic polynomial with distinct roots, then the Riemann surface S of $\sqrt{p(z)}$ has genus 1. Our aim is to produce a lattice Ω such that $S \cong \mathbb{C}/\Omega$, or, equivalently, to parametrise S by means of elliptic functions. In this section we give a necessary and sufficient condition for p to have distinct roots.

In §3.10 we saw that Weierstrass' elliptic function $\wp$ satisfies a differential equation $\wp' = \sqrt{p(\wp)}$, where p is a cubic polynomial of the form

$$p(z) = 4z^3 - c_2 z - c_3 \quad (c_2, c_3 \in \mathbb{C}); \tag{6.2.1}$$

any polynomial of the form (6.2.1) is said to be in *Weierstrass normal form*. By means of a substitution $\theta: z \mapsto az + b$ $(a, b \in \mathbb{C}, a \neq 0)$, any cubic polynomial may be brought into this form; now $\theta: \mathbb{C} \to \mathbb{C}$ is a bijection, preserving the multiplicities of roots, so without loss of generality we can restrict our attention to cubic polynomials p in Weierstrass normal form.

If e_1, e_2 and e_3 are the roots of the polynomial p in (6.2.1), then we define the *discriminant* of p to be

$$\Delta_p = 16(e_1 - e_2)^2 (e_2 - e_3)^2 (e_3 - e_1)^2; \tag{6.2.2}$$

clearly these roots are distinct if and only if $\Delta_p \neq 0$.

Theorem 6.2.3. $\Delta_p = c_2^3 - 27c_3^2$.

Proof. Putting

$$p(z) = 4(z - e_1)(z - e_2)(z - e_3), \tag{6.2.4}$$

and equating coefficients between this and (6.2.1), we have

$$\left. \begin{aligned} e_1 + e_2 + e_3 &= 0, \\ e_1 e_2 + e_2 e_3 + e_3 e_1 &= -\frac{c_2}{4}, \\ e_1 e_2 e_3 &= \frac{c_3}{4}. \end{aligned} \right\} \tag{6.2.5}$$

The remaining symmetric functions of the roots may be obtained from (6.2.5), for example

$$e_1^2 + e_2^2 + e_3^2 = (e_1 + e_2 + e_3)^2 - 2(e_1 e_2 + e_2 e_3 + e_3 e_1) = \frac{c_2}{2},$$

and

$$e_1^2 e_2^2 + e_2^2 e_3^2 + e_3^2 e_1^2 = (e_1 e_2 + e_2 e_3 + e_3 e_1)^2 - 2e_1 e_2 e_3 (e_1 + e_2 + e_3) = \frac{c_2^2}{16}.$$

Now differentiating (6.2.1) and (6.2.4) at $z = e_1$, we have
$$4(e_1 - e_2)(e_1 - e_3) = p'(e_1)$$
$$= 12e_1^2 - c_2,$$

with similar expressions for $p'(e_2)$ and $p'(e_3)$. Hence
$$\Delta_p = -\tfrac{1}{4} p'(e_1) p'(e_2) p'(e_3)$$

$$= -\tfrac{1}{4} \prod_i (12e_i^2 - c_2)$$

$$= -\tfrac{1}{4}(1728(e_1 e_2 e_3)^2 - 144 c_2 (e_1^2 e_2^2 + e_2^2 e_3^2 + e_3^2 e_1^2)$$
$$+ 12 c_2^2 (e_1^2 + e_2^2 + e_3^2) - c_2^3)$$

$$= -\tfrac{1}{4}(108 c_3^2 - 9 c_2^3 + 6 c_2^3 - c_2^3)$$

$$= c_2^3 - 27 c_3^2. \quad \square$$

Corollary 6.2.6. *p has distinct roots if and only if* $c_2^3 - 27 c_3^2 \neq 0$. $\quad \square$

One can give a direct proof of Corollary 6.2.6, without introducing Δ_p, by eliminating z between the equations $p(z) = 0$ and $p'(z) = 0$, thus giving a necessary and sufficient condition for p and p' to have a common root. We have chosen the above proof since the discriminant is an interesting function in its own right.

6.3 The modular function *J*

Theorem 6.1.4 suggests that we can obtain information about lattices and tori by studying the action of the modular group Γ on the upper half-plane $\mathcal{U}$. With this in mind we shall construct a function $J : \mathcal{U} \to \mathbb{C}$ with the property that $J(\tau') = J(\tau)$ if and only if $\tau' = T(\tau)$ for some $T \in \Gamma$; thus J distinguishes between different similarity classes of lattices and hence between different conformal equivalence classes of tori.

By Theorem 3.10.5, the Weierstrass function $\wp$ associated with a lattice Ω satisfies $\wp' = \sqrt{p(\wp)}$, where p is a cubic polynomial in Weierstrass normal form

$$p(z) = 4z^3 - g_2 z - g_3, \tag{6.3.1}$$

with

$$g_2 = g_2(\Omega) = 60 \sum_{\omega \in \Omega}{}' \omega^{-4},$$

and

$$g_3 = g_3(\Omega) = 140 \sum_{\omega \in \Omega}{}' \omega^{-6}.$$

If we write $\Delta(\Omega)$ for the discriminant Δ_p of p, then by Theorem 6.2.3 we have

$$\Delta(\Omega) = g_2(\Omega)^3 - 27 g_3(\Omega)^2. \tag{6.3.2}$$

Theorem 3.10.9 implies that p has distinct roots, so $\Delta(\Omega) \neq 0$ by Corollary 6.2.6, and hence we may define the *modular function* $J(\Omega)$ by

$$\left.\begin{aligned} J(\Omega) &= \frac{g_2(\Omega)^3}{\Delta(\Omega)} \\[2mm] &= \frac{g_2(\Omega)^3}{g_2(\Omega)^3 - 27 g_3(\Omega)^2}. \end{aligned}\right\} \tag{6.3.3}$$

For a similar lattice $\mu\Omega$ $(\mu \neq 0)$ we have

$$g_2(\mu\Omega) = 60 \sum_{\omega \in \Omega}{}' (\mu\omega)^{-4} = \mu^{-4} g_2(\Omega)$$

and

$$g_3(\mu\Omega) = 140 \sum_{\omega \in \Omega}{}' (\mu\omega)^{-6} = \mu^{-6} g_3(\Omega),$$

so that

$$\Delta(\mu\Omega) = \mu^{-12} \Delta(\Omega).$$

It follows that

$$J(\mu\Omega) = J(\Omega) \tag{6.3.4}$$

for all $\mu \in \mathbb{C} \setminus \{0\}$, so that similar lattices determine the same value of J (the converse is also true, as we shall prove in §6.5).

We can regard g_2, g_3, Δ and J as functions of $\tau \in \mathscr{U}$ by evaluating them on the lattice $\Omega = \Omega(1, \tau)$ which has τ as one of its moduli. Thus

$$\left.\begin{aligned} g_2(\tau) &= 60 \sum_{m,n}{}' (m + n\tau)^{-4} \\[2mm] g_3(\tau) &= 140 \sum_{m,n}{}' (m + n\tau)^{-6}, \end{aligned}\right\} \tag{6.3.5}$$

where $\sum_{m,n}'$ denotes summation over all $(m,n) \in \mathbb{Z} \times \mathbb{Z}$ except $(0,0)$; then

$$\Delta(\tau) = g_2(\tau)^3 - 27 g_3(\tau)^2 \tag{6.3.6}$$

and

$$J(\tau) = \frac{g_2(\tau)^3}{\Delta(\tau)}. \tag{6.3.7}$$

If $\tau' = T(\tau)$ for some $T \in \Gamma$, then by Theorem 6.1.4 the lattices $\Omega = \Omega(1, \tau)$ and $\Omega' = \Omega(1, \tau')$ are similar, and hence $J(\tau') = J(\tau)$ by (6.3.4). This proves

Theorem 6.3.8. $J(T(\tau)) = J(\tau)$ for all $\tau \in \mathcal{U}$ and $T \in \Gamma$. $\square$

Thus $J(\tau)$ is invariant under the action of the modular group Γ; we now show that the functions $g_2(\tau)$, $g_3(\tau)$ and $\Delta(\tau)$ come close to sharing this property. If $T : \tau \mapsto (a\tau + b)/(c\tau + d)$ is an element of Γ, then

$$g_2(T(\tau)) = 60 \sum_{m,n}' \left(m + n \frac{(a\tau + b)}{(c\tau + d)} \right)^{-4}$$

$$= 60(c\tau + d)^{-4} \sum_{m,n}' (m(c\tau + d) + n(a\tau + b))^{-4}$$

$$= 60(c\tau + d)^{-4} \sum_{m,n}' ((md + nb) + (mc + na)\tau)^{-4};$$

since $ad - bc = 1$, the transformation $(m, n) \mapsto (md + nb, mc + na)$ merely permutes the elements of the indexing set $(\mathbb{Z} \times \mathbb{Z}) \backslash \{(0,0)\}$, by Theorem 3.4.2, so that by absolute convergence (Theorem 3.9.2) we have

$$g_2(T(\tau)) = 60(c\tau + d)^{-4} \sum_{m,n}' (m + n\tau)^{-4}$$

$$= (c\tau + d)^{-4} g_2(\tau). \tag{6.3.9}$$

Similarly,

$$g_3(T(\tau)) = (c\tau + d)^{-6} g_3(\tau), \tag{6.3.10}$$

and hence

$$\Delta(T(\tau)) = (c\tau + d)^{-12} \Delta(\tau), \tag{6.3.11}$$

from which we immediately obtain an alternative proof of Theorem 6.3.8. In the special case where $a = b = d = 1$ and $c = 0$, we have $T(\tau) = \tau + 1$, giving

Theorem 6.3.12. *The functions $g_2(\tau)$, $g_3(\tau)$, $\Delta(\tau)$ and $J(\tau)$ are periodic with respect to $\mathbb{Z}$.* $\square$

It is also useful to determine the effect on these functions of the orientation-reversing transformations of $\mathcal{U}$ of the form

$$T(\tau) = \frac{a\bar{\tau} + b}{c\bar{\tau} + d} \quad (a, b, c, d \in \mathbb{Z}, \; ad - bc = -1). \tag{6.3.13}$$

Calculations similar to those above give

$$\left.\begin{aligned}
g_2(T(\tau)) &= (c\bar{\tau} + d)^{-4}\overline{g_2(\tau)}, \\
g_3(T(\tau)) &= (c\bar{\tau} + d)^{-6}\overline{g_3(\tau)}, \\
\Delta(T(\tau)) &= (c\bar{\tau} + d)^{-12}\overline{\Delta(\tau)}, \\
J(T(\tau)) &= \overline{J(\tau)}.
\end{aligned}\right\} \tag{6.3.14}$$

6.4   Analytic properties of g_2, g_3, Δ and J

The main result of this section is:

Theorem 6.4.1.   *The functions g_2, g_3, Δ and $J: \mathcal{U} \to \mathbb{C}$ are analytic on $\mathcal{U}$.*

Proof.   Given any $\tau_0 \in \mathcal{U}$, let $\delta = \frac{1}{2}\text{Im}(\tau_0)$ (so that $\delta > 0$) and let $K(\tau_0)$ be the compact disc $\{\tau \in \mathcal{U} \,||\, \tau - \tau_0| \leqslant \delta\}$. Now the functions $(m + n\tau)^{-4}$ and $(m + n\tau)^{-6}$ are analytic on $\mathcal{U}$ for all $(m,n) \in (\mathbb{Z} \times \mathbb{Z}) \setminus \{(0,0)\}$, so if we can show that the series (6.3.5) defining $g_2(\tau)$ and $g_3(\tau)$ are normally convergent on each $K(\tau_0)$, $\tau_0 \in \mathcal{U}$, then by §3.7 it will follow that these two functions are analytic on $\mathcal{U}$.

For all $m, n \in \mathbb{Z}$ with $n \neq 0$ we have $-m/n \in \mathbb{R}$ and hence

$$\left|\frac{m}{n} + \tau_0\right| \geqslant \text{Im}(\tau_0) = 2\delta$$

(see Fig. 6.2); therefore for all $m, n \in \mathbb{Z}$ (including $n = 0$) and $\tau \in K(\tau_0)$ we have

$$\begin{aligned}
|(m + n\tau) - (m + n\tau_0)| &= |n||\tau - \tau_0| \\
&\leqslant |n|\delta \\
&\leqslant \tfrac{1}{2}|m + n\tau_0|,
\end{aligned}$$

Fig. 6.2

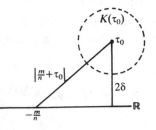

and so the triangle inequality gives

$$\begin{aligned}
|m + n\tau| &\geqslant |m + n\tau_0| - |(m + n\tau) - (m + n\tau_0)| \\
&\geqslant \tfrac{1}{2}|m + n\tau_0|.
\end{aligned}$$

Thus for any $r > 0$ we have

$$|m + n\tau|^{-2r} \leqslant 2^{2r}|m + n\tau_0|^{-2r}$$

for all $\tau \in K(\tau_0)$ and $(m, n) \neq (0, 0)$. By Theorem 3.9.2 the series $\sum'_{m,n} |m + n\tau_0|^{-2r}$ converges for each $r > 1$, so $\sum'_{m,n}(m + n\tau)^{-2r}$ is normally convergent on $K(\tau_0)$. Putting $r = 2$ and $r = 3$, we see that $g_2(\tau)$ and $g_3(\tau)$ are analytic on $\mathcal{U}$.

It immediately follows from (6.3.6) that $\Delta(\tau)$ is analytic on $\mathcal{U}$, and since $\Delta(\tau) \neq 0$ on $\mathcal{U}$ by Theorem 3.10.9 and Corollary 6.2.6, it follows from (6.3.7) that $J(\tau)$ is analytic on $\mathcal{U}$. $\square$

Our next task is to discover the behaviour of the functions $g_2(\tau)$, $g_3(\tau)$, $\Delta(\tau)$ and $J(\tau)$ as $\mathrm{Im}(\tau) \to +\infty$; by Theorem 6.3.12 all four functions are periodic, so by Theorem 3.3.1 we can express them as analytic functions of $q = e^{2\pi i \tau}(0 < |q| < 1)$, and we then see how they behave as $q \to 0$. First we need the following result:

Lemma 6.4.2. $\sum_{m=1}^{\infty} m^{-4} = \pi^4/90$ and $\sum_{m=1}^{\infty} m^{-6} = \pi^6/945$.

Proof. We saw in §3.8 and Exercise 3J that

$$\sum_{m=-\infty}^{\infty} (z - m)^{-2} = \pi^2 \operatorname{cosec}^2 \pi z. \tag{6.4.3}$$

By squaring and then inverting the series

$$\sin \pi z = \pi z - \frac{\pi^3 z^3}{3!} + \frac{\pi^5 z^5}{5!} - \cdots$$

we obtain the Laurent series

$$\pi^2 \operatorname{cosec}^2 \pi z = \frac{1}{z^2} + \frac{\pi^2}{3} + \frac{\pi^4 z^2}{15} + \frac{2\pi^6 z^4}{189} + \cdots, \tag{6.4.4}$$

valid for $|z| < 1$ since $\pi^2 \operatorname{cosec}^2 \pi z - (1/z^2)$ is analytic on the unit disc. By differentiating twice in (6.4.3) and (6.4.4) we have

$$6 \sum_{m=-\infty}^{\infty} (z - m)^{-4} = \frac{6}{z^4} + \frac{2\pi^4}{15} + \frac{8\pi^6 z^2}{63} + \cdots,$$

and differentiating twice again we have

$$120 \sum_{m=-\infty}^{\infty} (z - m)^{-6} = \frac{120}{z^6} + \frac{16\pi^6}{63} + \cdots,$$

both valid for $|z| < 1$. Cancelling the principal parts at $z = 0$ with the summands corresponding to $m = 0$, and then putting $z = 0$, we have

$$6\sum_m{}' (-m)^{-4} = \frac{2\pi^4}{15}$$

and

$$120\sum_m{}' (-m)^{-6} = \frac{16\pi^6}{63}$$

(where $\sum_m'$ denotes summation over all $m \in \mathbb{Z} \setminus \{0\}$), and hence the result. $\square$

(This method allows us to evaluate the Riemann zeta-function $\sum_{m=1}^{\infty} m^{-s}$ at each even integer $s \geqslant 2$; for example, (6.4.3) and (6.4.4) immediately give $\sum_{m=1}^{\infty} m^{-2} = \pi^2/6$.)

We can obtain the Fourier series for the periodic function $\pi^2 \operatorname{cosec}^2 \pi z$ from that of $\sin \pi z$: if we put $\zeta = e^{2\pi i z}$, then since

$$\sin \pi z = \frac{e^{\pi i z} - e^{-\pi i z}}{2i}$$

we have

$$\pi^2 \operatorname{cosec}^2 \pi z = \frac{-4\pi^2}{\zeta - 2 + \zeta^{-1}}$$

$$= \frac{-4\pi^2 \zeta}{(1 - \zeta)^2}$$

$$= -4\pi^2 \sum_{r=1}^{\infty} r\zeta^r,$$

the Fourier series for $\pi^2 \operatorname{cosec}^2 \pi z$, valid for $|\zeta| < 1$, that is, for $\operatorname{Im}(z) > 0$. By (6.4.3) we therefore have

$$\sum_m (m + z)^{-2} = \sum_m (z - m)^{-2}$$

$$= \pi^2 \operatorname{cosec}^2 \pi z$$

$$= -4\pi^2 \sum_{r=1}^{\infty} r\zeta^r$$

(where $\sum_m$ denotes summation over all $m \in \mathbb{Z}$). Differentiating this twice with respect to z, and using $d\zeta/dz = 2\pi i \zeta$, we have

$$6\sum_m (m + z)^{-4} = 16\pi^4 \sum_{r=1}^{\infty} r^3 \zeta^r,$$

and differentiating twice again,

$$120\sum_m (m + z)^{-6} = -64\pi^6 \sum_{r=1}^{\infty} r^5 \zeta^r,$$

both valid for $\mathrm{Im}(z) > 0$. We now put $z = n\tau$ where n is any positive integer and $\tau \in \mathcal{U}$; we have $\zeta = q^n$ where $q = e^{2\pi i \tau}$, so

$$\sum_m (m + n\tau)^{-4} = \tfrac{8}{3}\pi^4 \sum_{r=1}^\infty r^3 q^{nr}$$

and

$$\sum_m (m + n\tau)^{-6} = -\tfrac{8}{15}\pi^6 \sum_{r=1}^\infty r^5 q^{nr}.$$

For $n < 0$ we have

$$\sum_m (m + n\tau)^{-4} = \sum_m (-m - n\tau)^{-4}$$

$$= \sum_m (m - n\tau)^{-4}$$

$$= \tfrac{8}{3}\pi^4 \sum_{r=1}^\infty r^3 q^{-nr},$$

and similarly,

$$\sum_m (m + n\tau)^{-6} = -\tfrac{8}{15}\pi^6 \sum_{r=1}^\infty r^5 q^{-nr},$$

while for $n = 0$, Lemma 6.4.2 gives

$$\sideset{}{'}\sum_m (m + n\tau)^{-4} = 2 \sum_{m=1}^\infty m^{-4}$$

$$= \frac{\pi^4}{45},$$

and similarly,

$$\sideset{}{'}\sum_m (m + n\tau)^{-6} = \frac{2\pi^6}{945}.$$

As shown in the proof of Theorem 6.4.1, $\sideset{}{'}\sum_{m,n}(m + n\tau)^{-4}$ is normally convergent on compact subsets of $\mathcal{U}$, so it is absolutely convergent on $\mathcal{U}$; we may therefore combine the cases $n > 0$, $n = 0$ and $n < 0$ to obtain

$$\sideset{}{'}\sum_{m,n} (m + n\tau)^{-4} = \frac{\pi^4}{45} + \frac{16\pi^4}{3} \sum_{n=1}^\infty \sum_{r=1}^\infty r^3 q^{nr}, \tag{6.4.5}$$

valid for $0 < |q| < 1$. Each power series $\sum_{r=1}^\infty r^3 q^{nr}$, having radius of convergence 1, is absolutely convergent for $|q| < 1$. Moreover, $\sum_{n=1}^\infty \sum_{r=1}^\infty |r^3 q^{nr}|$ converges for $|q| < 1$: there is no problem when $q = 0$, and if $|q| = q_0$ satisfies $0 < q_0 < 1$ then

$$\sum_{n=1}^\infty \sum_{r=1}^\infty |r^3 q^{nr}| = \sum_{n=1}^\infty \sum_{r=1}^\infty r^3 q_0^{nr}$$

converges by (6.4.5) since, putting $q_0 = e^{2\pi i \tau_0}$ $(\tau_0 \in \mathcal{U})$, $\sum'_{m,n}(m + n\tau_0)^{-4}$ converges. Thus the double series in (6.4.5) is absolutely convergent, so we may rearrange it and collect the powers of q: if we define

$$\sigma_3(k) = \sum_{r|k} r^3$$

for each positive integer k (summing over all positive divisors r of k), then (6.4.5) gives

$$\sum_{m,n}{}'(m + n\tau)^{-4} = \frac{\pi^4}{45} + \frac{16\pi^4}{3} \sum_{k=1}^{\infty} \sigma_3(k)q^k$$

$$= \frac{\pi^4}{45} + \frac{16\pi^4}{3}(q + 9q^2 + 28q^3 + 73q^4 + \ldots),$$

and similarly, if we define

$$\sigma_5(k) = \sum_{r|k} r^5,$$

then

$$\sum_{m,n}{}'(m + n\tau)^{-6} = \frac{2\pi^6}{945} - \frac{16\pi^6}{15} \sum_{k=1}^{\infty} \sigma_5(k)q^k$$

$$= \frac{2\pi^6}{945} - \frac{16\pi^6}{15}(q + 33q^2 + 244q^3 + 1057q^4 + \ldots).$$

By (6.3.5) we therefore have

$$g_2(\tau) = 60 \sum_{m,n}{}'(m + n\tau)^{-4}$$

$$= \pi^4(\tfrac{4}{3} + 320q + 2880q^2 + 8960q^3 + \ldots) \qquad (6.4.6)$$

and

$$g_3(\tau) = 140 \sum_{m,n}{}'(m + n\tau)^{-6}$$

$$= \pi^6(\tfrac{8}{27} - \tfrac{448}{3}q - 4928q^2 - \ldots), \qquad (6.4.7)$$

so (6.3.6) gives

$$\Delta(\tau) = g_2(\tau)^3 - 27g_3(\tau)^2$$

$$= \pi^{12}(4096q - 98304q^2 + \ldots), \qquad (6.4.8)$$

and then (6.3.7) gives

$$J(\tau) = \frac{g_2(\tau)^3}{\Delta(\tau)}$$

$$= \frac{1}{1728}\left(\frac{1}{q} + 744 + 196884q + \ldots\right). \qquad (6.4.9)$$

These expansions, valid near $q = 0$, can be used to define g_2, g_3, Δ and J as functions of q at $q = 0$ (that is, at $\tau = \infty$): notice that Δ and J have a simple zero and a simple pole respectively.

One can show (see Exercise 6B) that (6.4.8) can be written in the form

$$\Delta(\tau) = (2\pi)^{12} \sum_{n=1}^{\infty} a_n q^n,$$

with the coefficients a_n all *integers*. The function a_n, of considerable importance in number theory, is *Ramanujan's tau-function*, usually denoted by $\tau(n)$ (a notation we have avoided, having used τ already). Similarly the coefficients $c(n)$ in

$$J(\tau) = \frac{1}{1728} \left(\frac{1}{q} + \sum_{n=0}^{\infty} c(n) q^n \right)$$

are all integers, of great interest to number-theorists; recent work of J. H. Conway, S. Norton and J. G. Thompson has shown that these coefficients arise in a rather mysterious way in connection with a finite simple group known as the Fischer–Griess monster (see Conway–Norton[1979]). For connections with number theory, see Apostol [1976], which includes a proof of the remarkable infinite product representation

$$\Delta(\tau) = (2\pi)^{12} q \prod_{n=1}^{\infty} (1 - q^n)^{24}.$$

6.5 The Riemann surface of $\sqrt{p(z)}$, p a cubic polynomial

Our main aim in this chapter is to show that if $p(z)$ is a cubic polynomial in $\mathbb{C}[z]$, with distinct roots, then there is a lattice Ω such that the Riemann surface of $\sqrt{p(z)}$ is conformally equivalent to $\mathbb{C}/\Omega$. The main step in the argument is to show that J maps $\mathcal{U}$ onto $\mathbb{C}$, a non-trivial result requiring virtually all the information we have so far acquired concerning J and Γ.

Lemma 6.5.1.

(i) *If* $2 \operatorname{Re}(\tau) \in \mathbb{Z}$ *then* $g_2(\tau)$, $g_3(\tau)$, $\Delta(\tau)$ *and* $J(\tau)$ *are all real.*

(ii) *If* $|\tau| = 1$ *then* $g_2(\tau) = \tau^4 \overline{g_2(\tau)}$, $g_3(\tau) = \tau^6 \overline{g_3(\tau)}$, $\Delta(\tau) = \tau^{12} \overline{\Delta(\tau)}$ *and* $J(\tau) = \overline{J(\tau)}$.

Proof.

(i) If $2 \operatorname{Re}(\tau) = n \in \mathbb{Z}$ then τ is fixed by the reflection $T: \tau \mapsto n - \bar{\tau}$, which is of type (6.3.13) with $a = -1, b = n, c = 0$ and $d = 1$, so the result follows from (6.3.14).

(ii) If $|\tau| = 1$ then τ is fixed by the inversion $T:\tau \mapsto 1/\bar{\tau}$ in the unit circle; putting $a = d = 0$ and $b = c = 1$ in (6.3.14) we have

$$g_2(\tau) = g_2(1/\bar{\tau})$$
$$= (\bar{\tau})^{-4}\overline{g_2(\tau)}$$
$$= \tau^4\overline{g_2(\tau)},$$

and similarly for the other three functions. $\quad\square$

We proved in Theorem 5.8.4 that Γ has a fundamental region $F = \{\tau \in \mathcal{U} \mid |\tau| \geqslant 1 \text{ and } |\text{Re}(\tau)| \leqslant \frac{1}{2}\}$. We immediately have:

Corollary 6.5.2. *$J(\tau)$ is real whenever τ is on the imaginary axis or on the boundary ∂F of F.* $\quad\square$

Corollary 6.5.3. *$g_2(\rho) = g_3(i) = J(\rho) = 0$ and $J(i) = 1$, where $\rho = e^{2\pi i/3}$.*

Proof. Part (i) of Lemma 6.5.1 shows that g_2 and g_3 both take real values at i and ρ, while part (ii) shows that $g_2(\rho) = \rho \overline{g_2(\rho)}$ and $g_3(i) = -\overline{g_3(i)}$. Thus $g_2(\rho) = g_3(i) = 0$, so (6.3.7) gives $J(\rho) = 0$ and $J(i) = 1$. $\quad\square$

Let $L = L_1 \cup L_2 \cup L_3$, where

$L_1 = \{\tau \in \mathcal{U} \mid |\tau| \geqslant 1 \text{ and } \text{Re}(\tau) = -\frac{1}{2}\}$,
$L_2 = \{\tau \in \mathcal{U} \mid |\tau| = 1 \text{ and } -\frac{1}{2} \leqslant \text{Re}(\tau) \leqslant 0\}$,
$L_3 = \{\tau \in \mathcal{U} \mid |\tau| \geqslant 1 \text{ and } \text{Re}(\tau) = 0\}$, as illustrated in Fig. 6.3.

Fig. 6.3

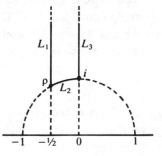

By Corollary 6.5.2 we have $J(L) \subseteq \mathbb{R}$, but in fact we can prove equality:

Theorem 6.5.4. *J maps L onto $\mathbb{R}$.*

Proof. If $\tau \in L_3$ then we have $\tau = iy$ with $y \geqslant 1$. Then $q = e^{2\pi i \tau} = e^{-2\pi y}$, and as $y \to +\infty$ we have $q \to 0$ through positive real values, so (6.4.9) gives $J(\tau) = (q^{-1} + 744 + \ldots)/1728 \to +\infty$. Similarly, on L_1 we have $\tau = -\frac{1}{2} + iy$ and $q = -e^{2\pi y}$, so as $y \to +\infty$ we have $q \to 0$ through negative real values, and hence $J(\tau) \to -\infty$. Thus, as a real-valued function on L, J is unbounded above and below. Being analytic on $\mathscr{U}$ (by Theorem 6.4.1), J maps L continuously to $\mathbb{R}$; since L is connected, it follows that $J(L)$ is connected. Now it is easily seen that a connected subset of $\mathbb{R}$, unbounded above and below, must be $\mathbb{R}$ itself, so $J(L) = \mathbb{R}$. $\quad\square$

By Theorem 6.3.8, J is constant on each orbit of Γ in $\mathscr{U}$. The next two results explain why we have devoted so much attention to J:

Theorem 6.5.5. *For each $c \in \mathbb{C}$ there is exactly one orbit of Γ in $\mathscr{U}$ on which J takes the value c.*

Combining this with Theorem 6.1.4, we immediately have:

Corollary 6.5.6. *If lattices Ω and Ω' have moduli $\tau, \tau' \in \mathscr{U}$, then the tori $\mathbb{C}/\Omega$ and $\mathbb{C}/\Omega'$ are conformally equivalent if and only if $J(\tau) = J(\tau')$.* $\quad\square$

(By Theorem 6.5.5, J induces a homeomorphism $\mathscr{U}/\Gamma \to \mathbb{C}$; this confirms our claim at the end of §6.1 that the Riemann space R_1 of genus 1 is homeomorphic to the plane $\mathbb{C}$, and it is not hard to show that these two surfaces are conformally equivalent.)

Proof of Theorem 6.5.5. By §5.8, each orbit of Γ meets the fundamental region F either at a unique point in the interior $\overset{\circ}{F}$, or else at one or two equivalent points on ∂F.

First suppose that $c \in \mathbb{C} \backslash \mathbb{R}$; since $J(\partial F) \subseteq \mathbb{R}$ by Corollary 6.5.2, it is sufficient to show that there is a unique solution of $J(\tau) = c$ in $\overset{\circ}{F}$. By Theorem 6.4.1 and Corollary 6.5.3, J is analytic and not identically equal to c, so the function

$$g(\tau) = \frac{J'(\tau)}{J(\tau) - c}$$

is meromorphic on $\mathscr{U}$; by an argument analogous to that used in the proof of Theorem 3.6.4, $a \in \mathscr{U}$ is a solution of $J(\tau) = c$ with multiplicity k if and only if $g(\tau)$ has a pole with residue k at a. We can use (6.4.9) to express $g(\tau)$ as a function of $q = e^{2\pi i \tau}$, meromorphic at $q = 0$ since $J(\tau)$ is; hence $g(\tau)$ is analytic for sufficiently small non-zero $|q|$, that is, provided $\text{Im}(\tau)$

is sufficiently large, say $\text{Im}(\tau) \geqslant K$ for some $K > 1$. Thus the poles of $g(\tau)$ in F all lie within the interior of $G = \{\tau \in F \mid \text{Im}(\tau) \leqslant K\}$, illustrated in Fig. 6.4, so the sum of the residues of $g(\tau)$ in F (and hence the number of solutions, counting multiplicities, of $J(\tau) = c$ in F) is equal to

$$\frac{1}{2\pi i} \int_{\partial G} g(\tau) d\tau, \tag{6.5.7}$$

where the boundary ∂G has the positive orientation.

Fig. 6.4

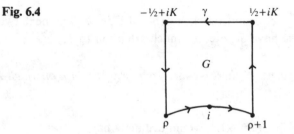

Now the sides $\text{Re}(\tau) = -\frac{1}{2}$ and $\text{Re}(\tau) = \frac{1}{2}$ of G are equivalent under the transformation $\tau \mapsto \tau + 1$ of Γ, so $J(\tau)$ and hence $g(\tau)$ take the same values at equivalent points on these sides; hence the integrals of $g(\tau)$ along these sides cancel in (6.5.7), and similarly the integral along the unit circle from ρ to i cancels with the integral from i to $\rho + 1$, using the transformation $\tau \mapsto -1/\tau$. Hence

$$\int_{\partial G} g(\tau) d\tau = \int_{\gamma} g(\tau) d\tau,$$

where γ is the side $\text{Im}(\tau) = K$ of G, oriented from $\frac{1}{2} + iK$ to $-\frac{1}{2} + iK$. Away from the poles of $g(\tau)$, each branch of the logarithm function satisfies

$$\frac{d}{d\tau}(\log(J(\tau) - c)) = \frac{J'(\tau)}{J(\tau) - c} = g(\tau),$$

so

$$\int_{\gamma} g(\tau) d\tau = [\log(J(\tau) - c)]_{\gamma},$$

the change in the value of $\log(J(\tau) - c)$ arising from analytic continuation

Fig. 6.5

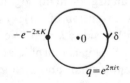

along γ. As τ follows γ, q winds once (in the negative direction) around the circle δ given by $|q| = e^{-2\pi K}$, starting and finishing at $-e^{-2\pi K}$ (see Fig. 6.5). By (6.4.9), $q(J(\tau) - c)$ is analytic and non-zero for $0 \leqslant |q| \leqslant e^{-2\pi K}$, and, since this set is simply connected, the monodromy theorem (4.5.3) implies that

$$[\log q(J(\tau) - c)]_\gamma = 0,$$

so

$$\begin{aligned}
[\log (J(\tau) - c)]_\gamma &= [\log q(J(\tau) - c) - \log q]_\gamma \\
&= [-\log q]_\gamma \\
&= 2\pi i.
\end{aligned}$$

Hence (6.5.7) shows that the number of solutions of $J(\tau) = c$ in F is equal to $(1/2\pi i) \cdot 2\pi i = 1$, as required.

Finally, suppose that $c \in \mathbb{R}$. By Theorem 6.5.4 there is at least one orbit of Γ on which J takes the value c. If there were more than one such orbit, there would be two inequivalent solutions τ_1, τ_2 of $J(\tau) = c$, so by choosing $c' \in \mathbb{C} \backslash \mathbb{R}$ sufficiently close to c we would have two inequivalent solutions τ_1' and τ_2' of $J(\tau) = c'$, close to τ_1 and τ_2 respectively. We have already shown that this is impossible, so the orbit is unique. $\square$

Corollary 6.5.8. *If $c_2, c_3 \in \mathbb{C}$ satisfy $c_2^3 - 27c_3^2 \neq 0$, then there is a lattice $\Omega \subset \mathbb{C}$ with $g_k(\Omega) = c_k$ for $k = 2, 3$.*

Proof. First suppose that $c_2 = 0$, so that $c_3 \neq 0$. By Corollary 6.5.3, $g_2(\rho) = 0$ and hence $g_3(\rho) \neq 0$ since $g_2(\tau)^3 - 27g_3(\tau)^2 = \Delta(\tau)$ does not vanish on $\mathscr{U}$. We can therefore choose $\mu \in \mathbb{C} \backslash \{0\}$ such that $\mu^{-6} g_3(\rho) = c_3$, so putting

$$\begin{aligned}
\Omega &= \mu\Omega(1, \rho) \\
&= \Omega(\mu, \mu\rho)
\end{aligned}$$

we have $g_2(\Omega) = \mu^{-4} g_2(\rho) = 0 = c_2$ and $g_3(\Omega) = \mu^{-6} g_3(\rho) = c_3$, as required.

Similarly, if $c_3 = 0$ then $c_2 \neq 0$. We have $g_3(i) = 0 \neq g_2(i)$, so we can choose $\mu \in \mathbb{C} \backslash \{0\}$ satisfying $\mu^{-4} g_2(i) = c_2$, and then $\Omega = \Omega(\mu, \mu i)$ satisfies $g_2(\Omega) = \mu^{-4} g_2(i) = c_2$ and $g_3(\Omega) = 0 = c_3$.

Finally we consider the general case, where $c_2 \neq 0 \neq c_3$. By Theorem 6.5.5 there exists $\tau \in \mathscr{U}$ such that

$$J(\tau) = \frac{c_2^3}{c_2^3 - 27c_3^2}. \tag{6.5.9}$$

For any $\mu \in \mathbb{C} \backslash \{0\}$ the lattice $\Omega = \Omega(\mu, \mu\tau)$ satisfies $g_2(\Omega) = \mu^{-4} g_2(\tau)$ and

$g_3(\Omega) = \mu^{-6}g_3(\tau)$, both non-zero since

$$\left.\begin{array}{c} \dfrac{g_2(\Omega)^3}{g_2(\Omega)^3 - 27g_3(\Omega)^2} = J(\Omega) \\[2mm] = J(\tau) \end{array}\right\} \qquad (6.5.10)$$

does not take the value 0 or 1, by (6.5.9) and the fact that $c_2 \neq 0 \neq c_3$. We can therefore choose $\mu \neq 0$ so that

$$\mu^2 = \frac{c_2}{c_3} \cdot \frac{g_3(\tau)}{g_2(\tau)},$$

and hence

$$\frac{g_2(\Omega)}{g_3(\Omega)} = \frac{\mu^{-4}g_2(\tau)}{\mu^{-6}g_3(\tau)}$$

$$= \frac{c_2}{c_3}.$$

Thus $g_k(\Omega) = \lambda c_k$ $(k = 2, 3)$ for some $\lambda \neq 0$, so substituting in (6.5.10) and using (6.5.9) we have

$$\frac{c_2^3}{c_2^3 - 27c_3^2} = J(\Omega)$$

$$= \frac{\lambda^3 c_2^3}{\lambda^3 c_2^3 - 27\lambda^2 c_3^2}$$

$$= \frac{c_2^3}{c_2^3 - 27\lambda^{-1}c_3^2}.$$

Hence $\lambda = 1$ and so $g_k(\Omega) = c_k$ $(k = 2, 3)$, as required. $\square$

We can now achieve our main aim by proving:

Theorem 6.5.11. *If $p(z)$ is a cubic polynomial in $\mathbb{C}[z]$ with distinct roots, then the Riemann surface S of $w = \sqrt{p(z)}$ is conformally equivalent to $\mathbb{C}/\Omega$ for some lattice $\Omega \subset \mathbb{C}$.*

Proof. Since the transformations $\theta : z \mapsto az + b$ $(a, b \in \mathbb{C}, a \neq 0)$ are automorphisms of $\mathbb{C}$, they leave the complex structure of S unchanged, so by using such a transformation we may assume that $p(z)$ is in Weierstrass normal form

$$p(z) = 4z^3 - c_2 z - c_3.$$

Since $p(z)$ has distinct roots, $c_2^3 - 27c_3^2 \neq 0$ by Corollary 6.2.6; hence

Corollary 6.5.8 implies that there is a lattice $\Omega \subset \mathbb{C}$ with $g_k(\Omega) = c_k$ $(k = 2, 3)$, and now Theorem 4.18.3 gives $\mathbb{C}/\Omega \cong S$. □

By quoting (without proof) the very deep uniformisation theorem (4.17.2), we were able to prove in §5.7 a generalisation of Theorem 6.5.11, namely the result that *every* compact Riemann surface of genus 1 is conformally equivalent to $\mathbb{C}/\Omega$ for some lattice Ω; Theorem 6.5.11 is independent of the uniformisation theorem, as is Exercise 6Q where we prove the same result for the Riemann surface of $w = \sqrt{p(z)}, p(z)$ a polynomial of degree 4 with distinct roots (by §4.9(v) we know that this surface has genus 1).

6.6 The mapping $J: \mathscr{U} \to \mathbb{C}$

In this section we see how J induces an infinite-sheeted branched covering of $\mathbb{C}$ by $\mathscr{U}$, and from this we deduce an important theorem of C. E. Picard (1856–1941) concerning entire functions. First we need to know which images of F under Γ are adjacent to F.

Consider the elements

$$\left.\begin{array}{l} X: \tau \mapsto -1/\tau, \\ Y: \tau \mapsto -1/(\tau + 1), \\ Z: \tau \mapsto \tau + 1 \end{array}\right\} \tag{6.6.1}$$

of Γ. Of these, X and Y are elliptic, fixing i and $\rho = e^{2\pi i/3}$, while Z is parabolic, fixing ∞. They satisfy

$$\left.\begin{array}{l} X^2 = Y^3 = 1, \\ XY = Z. \end{array}\right\} \tag{6.6.2}$$

By applying these transformations successively to F, we obtain that part of the tessellation of $\mathscr{U}$ shown in Fig. 6.6.

We see that F meets $Z(F)$, $Z^{-1}(F)$ and $X(F)$ across its edges $\text{Re}(\tau) = \frac{1}{2}$, $\text{Re}(\tau) = -\frac{1}{2}$, and $|\tau| = 1$, and that each point $\tau \in \partial F$ is contained in exactly

Fig. 6.6

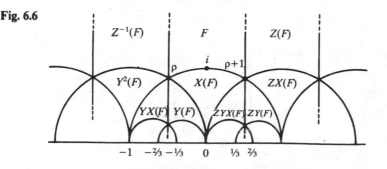

two images of F under Γ (including F itself), except that ρ and $\rho + 1$ (vertices with internal angle $2\pi/6$) are each contained in six images of F.

Theorem 6.6.3. *$J: \mathcal{U} \to \mathbb{C}$ is a branched covering map, with infinitely many sheets, and with branch-points of order 1 and 2 on the orbits $J^{-1}(1)$ and $J^{-1}(0)$ of Γ containing i and ρ respectively.*

Proof. Since J is an analytic function from $\mathcal{U}$ onto $\mathbb{C}$, §1.5 shows that away from the branch-points (the zeros of J') J is an unbranched covering map; there are infinitely many sheets since, for each $c \in \mathbb{C}$, $J^{-1}(c)$ is infinite (being an orbit of Γ).

A point $a \in \mathcal{U}$ is a branch-point of order $k - 1$ (with $k > 1$) if and only if J is locally k-to-one near a; applying an element of Γ to a, we may assume that $a \in F$. Now if $a \in \mathring{F}$ then there is a neighbourhood N of a in $\mathring{F}$; since no two elements of N are equivalent under Γ, J is one-to-one on N by the uniqueness result in Theorem 6.5.5, and so a is not a branch-point.

A similar argument applies if $a \in \partial F \setminus \{i, \rho, \rho + 1\}$. Suppose, for example, that $\mathrm{Re}\,(a) = \frac{1}{2}$ and $|a| > 1$. Let N be an open disc, centred at a, of radius less than $\min(|a| - 1, \frac{1}{2})$, so N is contained in the interior of $F \cup Z(F)$ (see Fig. 6.7). It follows that J is one-to-one on N, for if $\tau, \tau' \in N$ are equivalent

Fig. 6.7

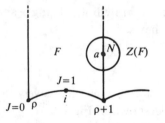

Fig. 6.8

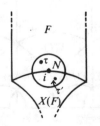

under Γ then (relabelling if necessary) we must have $\tau \in F$ and $\tau' = \tau + 1 \in Z(F)$, impossible since N has diameter less than 1. Similar arguments may be applied to the other sides of F, so the only possible branch-points in F are at the vertices i, ρ and $\rho + 1$.

We can choose arbitrarily small neighbourhoods N of i in the interior of $F \cup X(F)$, such that for each $\tau \in N \setminus \{i\}$ there is a unique equivalent point $\tau' = X(\tau) = -1/\tau \in N$ (see Fig. 6.8). Thus J is locally two-to-one near i, so $J'(i) = 0 \neq J''(i)$ and i is a branch-point of order 1, as are all equivalent points $T(i) \in J^{-1}(1)$.

Similarly, if τ is sufficiently close to ρ, as in Fig. 6.9, then there are equivalent points $\tau' = Y(\tau)$ and $\tau'' = Y^2(\tau)$ near ρ, and no others (for example, if $\tau \to \rho$ within $\overset{\circ}{F}$ then the element $\tilde{\tau}$ of $Z^{-1}(F)$ equivalent to τ satisfies $\tilde{\tau} = \tau - 1 \to \rho - 1$, so $\tilde{\tau}$ is *not* close to ρ). Thus J is locally three-to-one near ρ, so $J'(\rho) = J''(\rho) = 0 \neq J'''(\rho)$ and ρ is a branch-point of order 2 (as are all its equivalent points in $J^{-1}(0)$, such as $\rho + 1$). $\square$

Fig. 6.9

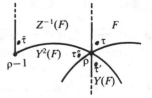

It follows from Theorems 6.5.4 and 6.5.5 that if

$$A = \{\tau \in \overset{\circ}{F} \mid \mathrm{Re}\,(\tau) < 0\}$$

and

$$B = \{\tau \in \overset{\circ}{F} \mid \mathrm{Re}\,(\tau) > 0\},$$

then J maps $A \cup B$ homeomorphically onto $\mathbb{C} \setminus \mathbb{R}$, so A and B must be

Fig. 6.10

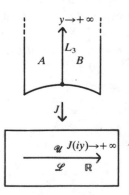

mapped to the two connected components $\mathcal{U}$ and $\mathcal{L}$ of $\mathbb{C}\backslash\mathbb{R}$, illustrated in Fig. 6.10; either $J(A) = \mathcal{U}$ and $J(B) = \mathcal{L}$, or vice-versa. Now we saw in the proof of Theorem 6.5.4 that if $\tau = iy\in L_3$, then $J(\tau) \to +\infty$ as $y \to +\infty$. For $y > 1$ we have $J'(\tau) \neq 0$ by Theorem 6.6.3, so J is directly conformal and hence A and B (on the left and right of L_3 as y increases) are mapped to the regions on the left and right of $\mathbb{R}$ as $J(\tau)$ increases, that is,

$$J(A) = \mathcal{U} \quad \text{and} \quad J(B) = \mathcal{L}.$$

We now give an important application of the J-function. A function $f:\mathbb{C}\to\mathbb{C}$ *omits a value* a if $a\in\mathbb{C}\backslash f(\mathbb{C})$; we say that f is *entire* if f is analytic on $\mathbb{C}$ (equivalently, f is represented by a power series with infinite radius of convergence). Now Liouville's theorem states that a bounded entire function must be constant, but by using the monodromy theorem and the properties of J we can prove a much stronger result, Picard's theorem:

Theorem 6.6.4. *If f is an entire function omitting more than one value, then f is constant.*

(Notice that e^z is entire and omits just 0, while a polynomial of positive degree omits no values, by the fundamental theorem of algebra.)

Proof. Suppose that f omits two distinct values a and b; then the entire function $f^* = (f-a)/(b-a)$ omits 0 and 1, and f is constant if and only if f^* is, so by replacing f by f^* we may assume that f omits 0 and 1, that is, $f(\mathbb{C}) \subseteq \mathbb{C}\backslash\{0,1\}$.

By Theorem 6.6.3, $J:\mathcal{U}\to\mathbb{C}$ restricts to an unbranched covering map

$$J:\mathcal{U}\backslash J^{-1}(\{0,1\})\to\mathbb{C}\backslash\{0,1\},$$

depicted in the commutative diagram, Fig. 6.11. Thus $\mathbb{C}\backslash\{0,1\}$ is a union of elementary neighbourhoods U such that $J^{-1}(U)$ is a disjoint union of sets V mapped homeomorphically onto U by J (see §4.19). Using the inverse homeomorphisms $g = J^{-1}:U\to V$ we obtain a set $\mathcal{G}$ of analytic germs $[g\circ f]_z(z\in\mathbb{C})$ representing the local branches of the many-valued function $J^{-1}\circ f:\mathbb{C}\to\mathcal{U}\backslash J^{-1}(\{0,1\})$. Now $\mathcal{G}$ is sufficient for analytic

Fig. 6.11

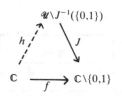

continuation within the simply connected region $\mathbb{C}$, so by Corollary 4.13.2 (which is essentially the monodromy theorem) any germ in $\mathcal{G}$ extends to a single-valued analytic function $h:\mathbb{C} \to \mathcal{U} \backslash J^{-1}(\{0,1\})$, locally equal to a branch of $J^{-1} \circ f$, so that $J \circ h = f$.

Since h is entire and maps $\mathbb{C}$ into $\mathcal{U}$, the function $h^* = (h - i)/(h + i)$ is entire and satisfies $|h^*(z)| < 1$ for all $z \in \mathbb{C}$. Liouville's theorem now implies that h^* is constant, so h is constant and hence so is $f = J \circ h$. $\square$

6.7 The λ-function

Let $\Omega = \Omega(\omega_1, \omega_2)$ be a lattice with modulus $\tau = \omega_2/\omega_1 \in \mathcal{U}$, and let ω_3 denote $\omega_1 + \omega_2$. Then as shown in §3.15 the Weierstrass elliptic function $\wp(z) = \wp(z, \Omega)$ induces a 2-sheeted covering map $\hat{\wp} : \mathbb{C}/\Omega \to \Sigma$ with branch-points at $[0], [\frac{1}{2}\omega_j]$ lying above the points $\infty, e_j (j = 1, 2, 3)$ of Σ; the points e_j are the roots of the cubic polynomial p associated with $\wp$ (see (6.2.1)), and are given by

$$e_j = \wp(\tfrac{1}{2}\omega_j)$$
$$= 4\omega_j^{-2} + \sum_{\omega \in \Omega}{}' ((\tfrac{1}{2}\omega_j - \omega)^{-2} - \omega^{-2}).$$

Taking these four points in the order ∞, e_2, e_3, e_1 we have the cross-ratio

$$\lambda = (\infty, e_2; e_3, e_1)$$
$$= \lim_{z \to \infty} \frac{(z - e_2)(e_3 - e_1)}{(e_2 - e_3)(e_1 - z)}$$
$$= \frac{e_3 - e_1}{e_3 - e_2}, \tag{6.7.1}$$

and if we replace ω_1 and ω_2 by the corresponding basis $\mu\omega_1, \mu\omega_2$ of a similar lattice $\mu\Omega (\mu \neq 0)$ then each e_j is replaced by $\wp(\tfrac{1}{2}\mu\omega_j, \mu\Omega) = \mu^{-2}e_j$, so λ is unchanged. Thus we may regard λ as a function $\lambda(\tau)$ of the modulus τ, since two bases with the same modulus generate similar lattices and hence determine the same value of λ.

Suppose, on the other hand, that we replace $\{\omega_1, \omega_2\}$ with another basis

$$\omega_2' = a\omega_2 + b\omega_1,$$
$$\omega_1' = c\omega_2 + d\omega_1$$

for the same lattice Ω; thus $a, b, c, d \in \mathbb{Z}$ with $ad - bc = \pm 1$ by Theorem 3.4.2, and by transposing ω_1' and ω_2' if necessary (so that $\tau' = \omega_2'/\omega_1' \in \mathcal{U}$) we see that $ad - bc = 1$. The branch-points of $\hat{\wp}$ now lie above ∞ and $e_j' = \wp(\tfrac{1}{2}\omega_j')(j = 1, 2, 3)$, and since these points depend only on Ω (and not on the particular basis for Ω), we must have $\{e_1', e_2', e_3'\} = \{e_1, e_2, e_3\}$.

Thus each matrix $A = \begin{pmatrix} a & b \\ c & d \end{pmatrix} \in SL(2, \mathbb{Z})$ induces a permutation $\alpha = \psi(A)$ of $\{e_1, e_2, e_3\}$ given by $\alpha(e_j) = e'_j$, so that by taking the four points in the new order ∞, e'_2, e'_3, e'_1 we obtain a possibly different cross-ratio $\lambda_\alpha = (\infty, e'_2; e'_3, e'_1) = (e'_3 - e'_1)/(e'_3 - e'_2)$.

Now $e'_1 = \wp(\tfrac{1}{2}\omega'_1) = \wp(\tfrac{1}{2}(c\omega_2 + d\omega_1))$ and similarly $e'_2 = \wp(\tfrac{1}{2}(a\omega_2 + b\omega_1))$; it follows (since $\wp$ is periodic) that if $A, B \in SL(2, \mathbb{Z})$ are congruent mod (2) in the sense that their corresponding entries are congruent mod (2), then they induce the same permutation $\psi(A) = \psi(B)$ of $\{e_1, e_2, e_3\}$. Every element of $SL(2, \mathbb{Z})$ is congruent mod (2) to precisely one of

$$A = \begin{pmatrix} 1 & 0 \\ 0 & 1 \end{pmatrix}, \begin{pmatrix} 0 & 1 \\ -1 & 0 \end{pmatrix}, \begin{pmatrix} 1 & 1 \\ 0 & 1 \end{pmatrix}, \begin{pmatrix} 1 & 0 \\ 1 & 1 \end{pmatrix},$$

$$\begin{pmatrix} 1 & 1 \\ -1 & 0 \end{pmatrix} \quad \text{and} \quad \begin{pmatrix} 0 & -1 \\ 1 & 1 \end{pmatrix},$$

and it is easily seen that the corresponding permutations are

$$\psi(A) = 1, (e_1 e_2), (e_2 e_3), (e_1 e_3), (e_1 e_2 e_3) \quad \text{and} \quad (e_1 e_3 e_2).$$

For example, if $A = \begin{pmatrix} 1 & 1 \\ -1 & 0 \end{pmatrix}$ then

$$\omega'_2 = \omega_2 + \omega_1 = \omega_3,$$
$$\omega'_1 = -\omega_2,$$

and

$$\omega'_3 = \omega'_1 + \omega'_2 = \omega_1,$$

so that

$$e'_1 = \wp(\tfrac{1}{2}\omega'_1) = \wp(-\tfrac{1}{2}\omega_2) = \wp(\tfrac{1}{2}\omega_2) = e_2,$$
$$e'_2 = \wp(\tfrac{1}{2}\omega'_2) = \wp(\tfrac{1}{2}\omega_3) = e_3,$$

and

$$e'_3 = \wp(\tfrac{1}{2}\omega'_3) = \wp(\tfrac{1}{2}\omega_1) = e_1,$$

giving

$$\psi(A) = (e_1 e_2 e_3).$$

Thus ψ maps $SL(2, \mathbb{Z})$ *onto* the group S_3 of all permutations of $\{e_1, e_2, e_3\}$, and one easily sees that $\psi(AB) = \psi(A)\psi(B)$ for all $A, B \in SL(2, \mathbb{Z})$ (composing permutations from left to right in this case), so that ψ is a group-homomorphism. The kernel K of ψ is the normal subgroup of index 6 in $SL(2, \mathbb{Z})$ consisting of the matrices $A \equiv I \bmod(2)$, and we have $SL(2, \mathbb{Z})/K \cong S_3$.

The values $\lambda_\alpha = (e'_3 - e'_1)/(e'_3 - e'_2)$ corresponding to the six permutations $\alpha = \psi(A) \in S_3$ are, as shown in §2.5, given by

$$\lambda_\alpha = \lambda, \frac{1}{\lambda}, 1 - \lambda, \frac{\lambda}{\lambda - 1}, \frac{\lambda - 1}{\lambda} \quad \text{and} \quad \frac{1}{1 - \lambda}$$

respectively. For instance, in the above example $\alpha = (e_1 e_2 e_3)$, so that

$$\lambda_\alpha = \frac{e_1 - e_2}{e_1 - e_3}$$

$$= \left(\frac{e_3 - e_1}{e_3 - e_2} - 1 \right) \left(\frac{e_3 - e_2}{e_3 - e_1} \right)$$

$$= \frac{\lambda - 1}{\lambda}.$$

It follows that the function

$$\Phi = \prod_{\alpha \in S_3} (\lambda_\alpha + 1)$$

$$= (\lambda + 1)(1 - \lambda + 1)\left(\frac{\lambda}{\lambda - 1} + 1 \right)\left(\frac{1}{\lambda} + 1 \right)\left(\frac{\lambda - 1}{\lambda} + 1 \right)\left(\frac{1}{1 - \lambda} + 1 \right)$$

$$= -\frac{(\lambda + 1)^2 (2 - \lambda)^2 (2\lambda - 1)^2}{\lambda^2 (1 - \lambda)^2} \tag{6.7.2}$$

is independent of the choice of basis of Ω, since a change of basis (by a matrix $A \in SL(2, \mathbb{Z})$) merely permutes the six factors $\lambda_\alpha + 1$. Since similar lattices have similar bases, Φ depends only on the similarity class of Ω, so we may regard Φ as a function $\Phi(\tau)$ of the modulus $\tau \in \mathcal{U}$, invariant under the modular group $\Gamma = PSL(2, \mathbb{Z})$. The subgroup of Γ leaving $\lambda(\tau)$ invariant is the group $\Gamma(2)$ consisting of all Möbius transformations corresponding to matrices $A \in K = \ker(\psi)$, that is, satisfying $A \equiv I \bmod (2)$. Since K contains the kernel $\{\pm I\}$ of the natural map $SL(2, \mathbb{Z}) \to \Gamma = PSL(2, \mathbb{Z})$, we have

$$\Gamma/\Gamma(2) \cong SL(2, \mathbb{Z})/K \cong S_3.$$

We can calculate Φ as follows. We have

$$\lambda = \frac{e_3 - e_1}{e_3 - e_2},$$

so

$$\lambda + 1 = \frac{2e_3 - e_1 - e_2}{e_3 - e_2}$$

$$= \frac{3e_3}{e_3 - e_2},$$

using the relation $e_1 + e_2 + e_3 = 0$ (see (6.2.5)). Similarly

$$2 - \lambda = \frac{-3e_2}{e_3 - e_2},$$

$$2\lambda - 1 = \frac{-3e_1}{e_3 - e_2},$$

and

$$\lambda - 1 = \frac{e_2 - e_1}{e_3 - e_2},$$

so substituting in (6.7.2), and writing Δ, g_2 and J for $\Delta(\tau)$, etc., we have

$$\Phi = \frac{-9e_2^2.9e_1^2.9e_3^2}{(e_2 - e_1)^2(e_1 - e_3)^2(e_3 - e_2)^2}$$

$$= \frac{-16.9^3(e_1 e_2 e_3)^2}{\Delta} \qquad \text{(by 6.2.2)}$$

$$= -\frac{3^6 g_3^2}{\Delta} \qquad \text{(by 6.2.5)}$$

$$= 27(1 - J). \qquad \text{(by 6.3.3)}$$

It follows that

$$J = 1 - \frac{\Phi}{27}$$

$$= \frac{27\lambda^2(\lambda - 1)^2 - (\lambda + 1)^2(2 - \lambda)^2(2\lambda - 1)^2}{27\lambda^2(\lambda - 1)^2}$$

$$= \frac{4(1 - \lambda + \lambda^2)^3}{27\lambda^2(\lambda - 1)^2}.$$

6.8 A presentation for Γ

(The reader who is unfamiliar with group presentations should consult Appendix 2.)

We know from §5.8 that Γ is generated by the transformations

$$\left.\begin{array}{l} X: \tau \mapsto -1/\tau \\ \\ Z: \tau \mapsto \tau + 1, \end{array}\right\} \qquad (6.8.1)$$

and

which pair the sides of the fundamental region F. Putting

$$Y = XZ: \tau \mapsto -1/(\tau + 1)$$

we see that X and Y generate Γ (since $Z = XY$) and that they satisfy the relations

$$X^2 = Y^3 = 1. \tag{6.8.2}$$

We shall show that these are defining relations for Γ, so that Γ has a presentation

$$\Gamma = \langle X, Y | X^2 = Y^3 = 1 \rangle. \tag{6.8.3}$$

Any relation in Γ can be written in the form

$$W(X, Y) = 1, \tag{6.8.4}$$

where W is a word in X and Y. We shall show how, given any word $W(X, Y)$, we can test whether or not $W(X, Y) = 1$ in Γ: we use the relations (6.8.2) to reduce W to a simpler word $W'(X, Y)$, equal to W as an element of Γ, and then by considering the action of W' on $\mathscr{U}$, we see whether or not W' is the identity transformation.

We write

$$W = g_1 g_2 \cdots g_k,$$

where each $g_r (1 \leqslant r \leqslant k)$ is a power of X or of Y, and then we successively use the following two operations to simplify W:

(i) if consecutive terms g_r and g_{r+1} are powers of the *same* generator X or Y, then we amalgamate them, replacing $g_r g_{r+1}$ by a single power of that generator;

(ii) we use the relations (6.8.2) to reduce all powers of X and Y to $X^i (i = 0, 1)$ and $Y^i (i = 0, 1, -1)$, and then delete any powers with $i = 0$ (since they represent the identity element).

If we apply (i) and (ii) successively to W, then after a finite number of steps the process must stop (since the length k of W decreases), at which point either

(a) we have reduced W to a *non-trivial reduced word*

$$W' = h_1 h_2 \ldots h_l$$

of length $l > 0$, where each h_r is X, Y or Y^{-1}, and no consecutive terms h_r, h_{r+1} are powers of the same generator, or

(b) we have deleted *all* terms in W, producing the identity element $W' = 1$ (which we may regard as the *empty*, or *trivial reduced word* of length $l = 0$). Notice that in either case, $W' = W$ in Γ since steps (i) and (ii) do not change elements of Γ. For instance, $W = X \cdot Y^{-2} \cdot X \cdot X^3 \cdot Y \cdot X$ (with $k = 6$) gives $W = X Y^{-2} X^4 Y X = X Y X^0 Y X = X Y^2 X = X Y^{-1} X$, so $W' = X Y^{-1} X$, a simpler expression for the same element of Γ.

We shall show that if W' is a non-trivial reduced word then $W' \neq 1$ in Γ. It follows that if $W = 1$ in Γ then since $W = W'$ we have $W' = 1$ and hence we are in case (b), so that the reduction of W to $W' = 1$ enables us to deduce (6.8.4) from (6.8.2), as required.

Theorem 6.8.5. *If $W'(X, Y)$ is any non-trivial reduced word in X and Y, then $W'(X, Y) \neq 1$ in Γ.*

Proof. Let

$$A = \{\tau \in \mathcal{U} \mid \mathrm{Re}(\tau) < 0\},$$
$$B = \{\tau \in \mathcal{U} \mid |\tau + 1| > |\tau| \quad \text{and} \quad |\tau + 1| > 1\},$$

and

$$C = A \cap B,$$

as illustrated in Fig. 6.12. Clearly $\mathcal{U} = A \cup B$ and $C \neq \varnothing$.

Fig. 6.12

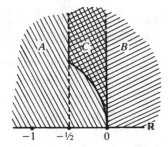

Since

$$X(A) = \{-1/\tau \mid \tau \in \mathcal{U} \quad \text{and} \quad \mathrm{Re}(\tau) < 0\}$$
$$= \{\tau \in \mathcal{U} \mid \mathrm{Re}(\tau) > 0\},$$

we have

$$X(A) \cap A = \varnothing.$$

If $\tau \in B$ then $|Y(\tau) + 1| = |1 - 1/(\tau + 1)| = |\tau/(\tau + 1)| < 1$, so $Y(\tau) \notin B$; thus $Y(B) \cap B = \varnothing$, and by applying Y^{-1} to this we have $Y^{-1}(B) \cap B = \varnothing$. Thus

$$X(A) \subseteq \mathcal{U} \setminus A \subseteq B \tag{6.8.6}$$

and

$$Y^i(B) \subseteq \mathcal{U} \setminus B \subseteq A \quad (i = \pm 1). \tag{6.8.7}$$

Now let $W' = h_1 h_2 \ldots h_l$ be a non-trivial reduced word in X and Y. Suppose first that $h_1 = h_l = X$, so that l is odd, $h_r = X$ for all odd r, and $h_r = Y^{i_r}(i_r = \pm 1)$ for all even r. Then

$$
\begin{aligned}
W'(C) &= h_1 h_2 \ldots h_{l-1} h_l(C) \\
&= X Y^{i_2} \ldots X Y^{i_l - 1} X(C) \\
&\subseteq X Y^{i_2} \ldots X Y^{i_l - 1} X(A) && \text{(since } C \subseteq A) \\
&\subseteq X Y^{i_2} \ldots X Y^{i_l - 1}(B) && \text{(by 6.8.6)} \\
&\subseteq X Y^{i_2} \ldots X(A) && \text{(by 6.8.7)} \\
&\quad \ldots \\
&\subseteq X(A) \\
&\subseteq \mathcal{U} \backslash A \\
&\subseteq \mathcal{U} \backslash C,
\end{aligned}
$$

so $W'(C) \cap C = \emptyset$ and hence $W' \neq 1$. Similar arguments apply in the other three cases, when either or both of h_1 and h_l are powers of Y. $\square$

From this and the preceding argument we immediately have:

Corollary 6.8.6. Γ *has presentation* $\langle X, Y | X^2 = Y^3 = 1 \rangle$. $\square$

This shows that Γ is an example of a free product. If $G = \langle \mathcal{X} | \mathcal{R} \rangle$ and $H = \langle \mathcal{Y} | \mathcal{S} \rangle$ are groups, where $\mathcal{X}$ and $\mathcal{Y}$ are disjoint sets of generators and $\mathcal{R}$ and $\mathcal{S}$ are sets of relations involving $\mathcal{X}$ and $\mathcal{Y}$ respectively, then the *free product* $G * H$ is the group with presentation $\langle \mathcal{X} \cup \mathcal{Y} | \mathcal{R} \cup \mathcal{S} \rangle$ (it can be shown that $G * H$ is independent of the chosen presentations for G and H). In our case, if we take

$$
G = \langle X | X^2 = 1 \rangle \cong C_2
$$

and

$$
H = \langle Y | Y^3 = 1 \rangle \cong C_3,
$$

then we have $\Gamma = G * H \cong C_2 * C_3$. For the general theory of presentations, including free products, see Johnson [1980], Lyndon & Schupp [1977], or Magnus, Karass & Solitar [1966].

6.9 Homomorphic images of Γ

Among the homomorphic images of Γ are a number of interesting groups, including many of those we have considered earlier. The next result follows directly from Theorem A.13.

Lemma 6.9.1. *Let G be any group. Then the following are equivalent:*

(i) *there is an epimorphism* $0 : \Gamma \to G$;

(ii) *there is a normal subgroup* $\Lambda \trianglelefteq \Gamma$ *with* $\Gamma / \Lambda \cong G$;

(*iii*) *G is generated by elements* x, y *satisfying* $x^2 = y^3 = 1$ (*and possibly other relations*). □

We now give some examples of homomorphic images of Γ.

(i) In §5.6 we introduced the *triangle groups* with signature $(0; l, m, n)$, generated by elements x, y satisfying $x^l = y^m = (xy)^n = 1$ (in fact one can show that these are defining relations). It follows from Lemma 6.9.1 that if G_n denotes the triangle group with signature $(0; 2, 3, n)$ then there is an epimorphism $\theta_n : \Gamma \to G_n$ given by $X \mapsto x$, $Y \mapsto y$; equivalently, $G_n \cong \Gamma / \Lambda$ where $\Lambda = \ker(\theta_n)$ is the smallest normal subgroup of Γ containing $(XY)^n = Z^n$. By considering the action of G_n on Σ (for $n \leqslant 5$), $\mathbb{C}$ (for $n = 6$) and $\mathcal{U}$ (for $n \geqslant 7$), one can show that G_n is finite for $n \leqslant 5$ and infinite for $n \geqslant 6$. The finite groups (which we have already met in §2.13) are

$$G_2 \cong D_3, G_3 \cong A_4, G_4 \cong S_4, G_5 \cong A_5,$$

the rotation groups of a triangle, tetrahedron, octahedron, and icosahedron inscribed in Σ; we can take x and y to be rotations of Σ (through π and $2\pi/3$) about the mid-points of an edge and an adjacent face, so that xy rotates Σ about a vertex.

(ii) The *Hurwitz groups* (introduced in §5.11) are groups of $84(g-1)$ automorphisms of a Riemann surface of genus $g > 1$, and by Theorem 5.11.2 they coincide with the non-trivial finite homomorphic images $H = \psi(G_7)$ of the triangle group G_7 with signature $(0; 2, 3, 7)$. Putting $\phi = \psi \circ \theta_7$ (see Fig. 6.13), where θ_7 is as defined in (i), we see that the Hurwitz groups are the non-trivial finite homomorphic images $\phi(\Gamma)$ of Γ such that $Z^7 \in \ker(\phi)$.

Fig. 6.13

(iii) For each integer $n \geqslant 2$, let $\mathbb{Z}_n$ denote the ring of integers $\mathrm{mod}\,(n)$; then the 2×2 unimodular matrices with coefficients in $\mathbb{Z}_n$ form a group $SL(2, \mathbb{Z}_n)$ in which the matrices $\pm I$ form a normal subgroup (which is trivial if $n = 2$). The natural ring-epimorphism $\mathbb{Z} \to \mathbb{Z}_n$, $a \mapsto [a]$, induces (in the obvious way) a group-homomorphism $SL(2, \mathbb{Z}) \to SL(2, \mathbb{Z}_n)$, and this in turn induces a group-homomorphism ϕ_n from $\Gamma = PSL(2, \mathbb{Z}) = SL(2, \mathbb{Z})/\{\pm I\}$ to $PSL(2, \mathbb{Z}_n) = SL(2, \mathbb{Z}_n)/\{\pm I\}$. The kernel $\Gamma(n)$ of ϕ_n, consisting of those transformations $\tau \mapsto (a\tau + b)/(c\tau + d)$ in Γ for which $a \equiv d \equiv \pm 1 \,\mathrm{mod}\,(n)$ and $b \equiv c \equiv 0 \,\mathrm{mod}\,(n)$, is called the *principal congruence subgroup of level n.* Any subgroup Λ of Γ, containing some $\Gamma(n)$, can be defined by finitely many congruences $\mod(n)$ between the coefficients a, b, c, d of the elements of Λ; we therefore call Λ a *congruence subgroup*, and the least such n is the *level* of Λ.

In §6.7 we constructed an epimorphism $\psi: SL(2, \mathbb{Z}) \to S_3$, such that $K = \ker(\psi)$ is the largest subgroup of Γ leaving invariant the function λ, that is,

$$K = \left\{ \begin{pmatrix} a & b \\ c & d \end{pmatrix} \in SL(2, \mathbb{Z}) \mid a \equiv d \equiv 1 \bmod(2) \quad \text{and} \quad b \equiv c \equiv 0 \bmod(2) \right\}.$$

Since $\{\pm I\} \leqslant K$, the matrices in K are precisely those corresponding to the transformations in $\Gamma(2) \leqslant \Gamma$, and we have

$$S_3 \cong SL(2, \mathbb{Z})/K \cong \Gamma/\Gamma(2) \leqslant PSL(2, \mathbb{Z}_2). \tag{6.9.2}$$

We shall now show that ϕ_n maps Γ *onto* $PSL(2, \mathbb{Z}_n)$, so it follows that the final inclusion in (6.9.2) is an isomorphism.

Theorem 6.9.3. *The map* $\phi_n: \Gamma \to PSL(2, \mathbb{Z}_n)$ *is an epimorphism.*

Proof. It is sufficient to show that the reduction $\mod(n)$ maps $SL(2, \mathbb{Z})$ onto $SL(2, \mathbb{Z}_n)$, that is, that if the coefficients of some element $A \in SL(2, \mathbb{Z}_n)$ are represented by integers a, b, c, d (so that $ad - bc \equiv 1 \bmod(n)$), then there exist integers a', b', c', d', congruent to $a, b, c, d \bmod(n)$, with $a'd' - b'c' = 1$ (for then $A' = \begin{pmatrix} a' & b' \\ c' & d' \end{pmatrix} \in SL(2, \mathbb{Z})$ is mapped to A).

We have $ad - bc = 1 + kn$ ($k \in \mathbb{Z}$) so that $(c, d, n) = 1$. For each $\lambda \in \mathbb{Z}$ we define $c_\lambda = c + \lambda n$, $d_\lambda = d + \lambda n$, $h_\lambda = (c_\lambda, d_\lambda)$. We first show that there exists $\lambda \in \mathbb{Z}$ such that $h_\lambda = 1$. As $c_\lambda - d_\lambda = c - d$, $h_\lambda | c - d$ (where '|' reads 'divides'), and thus for all $\lambda \in \mathbb{Z}$, h_λ takes only finitely many values so we can let $h = \text{l.c.m.}\{h_\lambda | \lambda \in \mathbb{Z}\}$. Now $(h_\lambda, n) | (c_\lambda, d_\lambda, n)$ so that $(h_\lambda, n) | (c, d, n)$ and hence $(h_\lambda, n) = 1$, for all $\lambda \in \mathbb{Z}$ and thus $(h, n) = 1$. Hence there exists $u \in \mathbb{Z}$ such that $nu \equiv -c \bmod h$. Letting $\lambda' = u + 1$ we find that $c_{\lambda'} = c + (u+1)n \equiv n \bmod h$. As $h_{\lambda'} | c_{\lambda'}$, $h_{\lambda'} | n$ and $(h, n) = 1$ we deduce that $h_{\lambda'} = 1$. Write $c' = c + \lambda'n$, $d' = d + \lambda'n$, $ad' - bc' = 1 + ln$ ($l \in \mathbb{Z}$). Choose $r, s \in \mathbb{Z}$ such that $rd' - sc' = -l$ and let $a' = a + rn$, $b' = b + sn$. Then $a'd' - b'c' = (a+rn)d' - (b+sn)c' = 1$. $\square$

Corollary 6.9.4. $\Gamma/\Gamma(n) \cong PSL(2, \mathbb{Z}_n)$. $\square$

We now calculate the order of $PSL(2, \mathbb{Z}_n)$ in the case where n is a prime p. First we determine the order of $GL(2, \mathbb{Z}_p)$. The elements of this group are the matrices $\begin{pmatrix} a & b \\ c & d \end{pmatrix}$, with $a, b, c, d \in \mathbb{Z}_p$, such that the row-vectors (a, b)

and (c,d) are linearly independent. There are $p^2 - 1$ choices for (a,b), excluding $(0,0)$, and for each (a,b) there are $p^2 - p$ choices for (c,d), excluding the p multiples of (a,b), so we have $|GL(2,\mathbb{Z}_p)| = (p^2 - 1)(p^2 - p)$. Now $SL(2,\mathbb{Z}_p)$ is the kernel of the epimorphism $\det:GL(2,\mathbb{Z}_p) \to \mathbb{Z}_p\backslash\{0\}$, so

$$|SL(2,\mathbb{Z}_p)| = |GL(2,\mathbb{Z}_p)|/|\mathbb{Z}_p\backslash\{0\}|$$
$$= p(p^2 - 1).$$

Since $PSL(2,\mathbb{Z}_p) = SL(2,\mathbb{Z}_p)/\{\pm I\}$ we have

$$|PSL(2,\mathbb{Z}_p)| = \begin{cases} \dfrac{p(p^2 - 1)}{2} & \text{if } p \text{ is an odd prime,} \\ 6 & \text{if } p = 2. \end{cases} \tag{6.9.5}$$

(For the general formula for $|PSL(2,\mathbb{Z}_n)|$, see Exercise 6L.)

If $\theta_n:\Gamma \to G_n$ is as in Example (i), then since $Z^n:\tau \mapsto \tau + n$ lies in $\Gamma(n)$ we have $\ker\theta_n \leqslant \Gamma(n)$, so there is an epimorphism $\psi_n:G_n \to PSL(2,\mathbb{Z}_n)$ satisfying $\psi_n \circ \theta_n = \phi_n$ (see Fig. 6.14). For $n \leqslant 5$ it can be shown that $\ker\theta_n = \Gamma(n)$, so that ψ_n is an isomorphism; however, if $n \geqslant 6$ then G_n is infinite whereas $PSL(2,\mathbb{Z}_n)$ is finite, so ψ_n is not an isomorphism. Taking $n = 7$ we see that $PSL(2,\mathbb{Z}_7)$ is a Hurwitz group, and by (6.9.5) it has order 168 (as we claimed in §5.11), so $PSL(2,\mathbb{Z}_7) \cong \mathrm{Aut}(S)$ for some compact Riemann surface S of genus 3.

Fig. 6.14

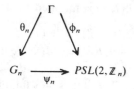

6.10 Quotient-surfaces for subgroups of Γ

In this section we show that if Λ is a subgroup of finite index in Γ then $\mathcal{U}/\Lambda$ (a non-compact Riemann surface) can be compactified by adding finitely many points; more precisely, there is a compact Riemann surface S such that $\mathcal{U}/\Lambda \cong S\backslash P$ for some finite subset $P \subset S$. We shall consider the genus and automorphisms of S in the simplest case, when Λ is normal in Γ; important examples include the principal congruence subgroups $\Lambda = \Gamma(n)$.

As an elementary example, consider the case where $\Lambda = \Gamma$. We saw in §6.5 that $J:\mathcal{U} \to \mathbb{C}$ induces a conformal equivalence between $\mathcal{U}/\Lambda$ and $\mathbb{C} = \Sigma\backslash\{\infty\}$, so we can take $S = \Sigma$ and $P = \{\infty\}$, the genus being 0. Without knowledge of J (not relevant in more complicated cases) we could argue that if F is the Dirichlet region for $\Lambda = \Gamma$ (given in Theorem 5.8.4 and §6.6) then $\mathcal{U}/\Lambda = F/\Lambda$ fails to be compact only 'near ∞', that is, for $\tau \in \mathcal{U}$ with large Im(τ). We therefore define, for each $r > 0$,

$$\mathcal{U}_r = \{\tau \in \mathcal{U} \,|\, \text{Im}\,(\tau) > r\}$$

and

$$\bar{\mathcal{U}}_r = \{\tau \in \mathcal{U} \,|\, \text{Im}\,(\tau) \geqslant r\},$$

the closure of $\mathcal{U}_r$ in $\mathcal{U}$ (though not in Σ). By considering the tessellation of $\mathcal{U}$ by images of F (see Fig. 6.6) we see that $\mathcal{U}_1$ is mapped into $\mathcal{U} \setminus \mathcal{U}_1$ by every element of Λ except for the translations $Z^n : \tau \mapsto \tau + n$, $n \in \mathbb{Z}$; hence $\tau, \tau' \in \mathcal{U}_1$ are in the same Λ-orbit if and only if $\tau' - \tau \in \mathbb{Z}$. Now the function $\varepsilon : \tau \mapsto e^{2\pi i \tau}$ maps $\mathcal{U}_1$ conformally onto the punctured disc

$$W = \{w \in \mathbb{C} \,|\, 0 < |w| < e^{-2\pi}\},$$

with $\varepsilon(\tau) = \varepsilon(\tau')$ if and only if $\tau' - \tau \in \mathbb{Z}$, so that the natural projection $p : \mathcal{U} \to \mathcal{U}/\Lambda$ induces a conformal equivalence $\varepsilon \circ p^{-1} : p(\mathcal{U}_1) \to W$. As τ approaches ∞ in $\mathcal{U}_1$, $w = \varepsilon(\tau)$ approaches 0 in W; we therefore add a single point to $p(\mathcal{U}_1)$ (denoted by $[\infty]_\Lambda$ since it corresponds to the orbit of Λ containing ∞), we define $(\varepsilon \circ p^{-1})([\infty]_\Lambda) = 0$, and we use $\varepsilon \circ p^{-1}$ as a chart at $[\infty]_\Lambda$, making $p(\mathcal{U}_1) \cup \{[\infty]_\Lambda\}$ a Riemann surface conformally equivalent to the open disc $W \cup \{0\}$. Now $p(\bar{\mathcal{U}}_2) \cup \{[\infty]_\Lambda\}$ is compact (being homeomorphic to a closed disc), as is $p(\mathcal{U} \setminus \mathcal{U}_2) = p(F \setminus \mathcal{U}_2)$ (since p is continuous and $F \setminus \mathcal{U}_2$ compact), so $(\mathcal{U}/\Lambda) \cup \{[\infty]_\Lambda\} = p(\mathcal{U}) \cup \{[\infty]_\Lambda\}$, being the union of these two subsets, is also compact. We shall show in Lemma 6.10.1 that $[\infty]_\Lambda = \mathbb{Q} \cup \{\infty\}$ for $\Lambda = \Gamma$, so if we define

$$\bar{\mathcal{U}} = \mathcal{U} \cup \mathbb{Q} \cup \{\infty\}$$

(not a surface!) then $J : \bar{\mathcal{U}}/\Gamma = (\mathcal{U}/\Gamma) \cup \{[\infty]_\Gamma\} \to \Sigma$ is a conformal equivalence, mapping $\bar{\mathcal{U}}/\Gamma$ onto $\mathbb{C}$ and the single point $[\infty]_\Gamma$ to ∞.

For convenience, let us denote $\mathbb{Q} \cup \{\infty\}$ by $\hat{\mathbb{Q}}$. The above argument suggests that when we consider $\mathcal{U}/\Lambda$ for arbitrary subgroups $\Lambda \leqslant \Gamma$, we must consider the orbits of Λ on $\hat{\mathbb{Q}}$ (such as $[\infty]$ above) and those parabolic transformations (such as Z above) which fix points in these orbits. This we now do.

If C is any subgroup of $PSL(2, \mathbb{R})$, then by Theorem 5.7.4, C is abelian if and only if it is cyclic, in which case, by Theorem 5.2.5, all non-identity elements of C have the same fixed-point set and are of the same type: elliptic, parabolic or hyperbolic. The *parabolic subgroups* of a Fuchsian group Λ are defined to be those non-identity cyclic subgroups $C \leqslant \Lambda$ which consist of parabolic elements (together with the identity) and which are maximal with respect to this property. The *parabolic class number s* of Λ is the number of conjugacy classes of parabolic subgroups of Λ. Here, we are interested in subgroups of Γ, so *from now on we will assume that* $\Lambda \leqslant \Gamma$.

Lemma 6.10.1. *Γ acts transitively on $\hat{\mathbb{Q}} = \mathbb{Q} \cup \{\infty\}$.*

Proof. Clearly $\hat{\mathbb{Q}}$ is invariant under Γ. If $r = a/c \in \mathbb{Q}$, with a, $c \in \mathbb{Z}$ and $(a, c) = 1$, then by the Euclidean algorithm there exist b, $d \in \mathbb{Z}$ with $ad - bc = 1$, so the transformation $\tau \mapsto (a\tau + b)/(c\tau + d)$, an element of Γ, maps ∞ to r, as required. $\square$

Corollary 6.10.2. *The parabolic subgroups of Λ are the non-trivial stabilisers Λ_r $(r \in \hat{\mathbb{Q}})$.*

Proof. Each parabolic subgroup $C \leqslant \Lambda$ has a unique fixed-point $r \in \mathbb{R} \cup \{\infty\}$, so $C \leqslant \Lambda_r$; solving $cr^2 + (d - a)r - b = 0$ ($a, b, c, d \in \mathbb{Z}, ad - bc = 1$, $|a + d| = 2$) we see that $r = (a - d)/2c \in \hat{\mathbb{Q}}$.

We now show that each stabiliser Λ_r $(r \in \hat{\mathbb{Q}})$ is either trivial or a parabolic subgroup. Conjugating Λ by an element of Γ mapping r to ∞ (possible by Lemma 6.10.1), we may assume that $r = \infty$, so that Λ_r is contained in Γ_∞, the subgroup generated by $Z : \tau \mapsto \tau + 1$. Thus Λ_r is cyclic and consists of parabolic elements together with the identity. If Λ_r is non-trivial then it must be parabolic, since any larger cyclic subgroup of Λ would also fix r (by Theorem 5.2.5) and would therefore be contained in Λ_r. Conversely, we have seen that every parabolic subgroup C is contained in some Λ_r $(r \in \hat{\mathbb{Q}})$; by the maximality of C we must have $C = \Lambda_r$, non-trivial by definition of parabolic subgroups. $\square$

Corollary 6.10.3. *The parabolic class number of Λ is the number of orbits of Λ on $\hat{\mathbb{Q}}$ for which the stabilisers Λ_r are non-trivial.*

Proof. Elements $r, r' \in \hat{\mathbb{Q}}$, with non-trivial stabilisers, lie in the same orbit of Λ if and only if Λ_r and $\Lambda_{r'}$ are conjugate in Λ, so the result follows from Corollary 6.10.2. $\square$

For example, Γ has parabolic class number $s = 1$, the parabolic subgroups of Γ being the conjugates of the subgroup C generated by Z; equivalently, Γ has a single orbit on $\hat{\mathbb{Q}}$ and the stabilisers are all non-trivial (being the conjugates of C). On the other hand, $\Lambda = C$ has infinitely many orbits on $\hat{\mathbb{Q}}$, but only the orbit $\{\infty\}$ has non-trivial stabiliser, so again $s = 1$, C

being its unique parabolic subgroup. The cyclic subgroups generated by elliptic or hyperbolic elements have $s = 0$, since they contain no parabolic elements.

We now assume, for the rest of this section, that Λ *is a subgroup of finite index* N *in* Γ; this will enable us to remove the 'non-triviality' conditions in Corollaries 6.10.2 and 6.10.3. First we need the following result.

Lemma 6.10.4. *If* A *and* B *are subgroups of a group* G, *and* $C = A \cap B$, *then* $|B{:}C| \leqslant |G{:}A|$.

Proof. To each coset bC of C in B we associate the coset bA of A in G; this is independent of the choice of coset representative, since if $b_1 C = b_2 C$, then $b_1^{-1} b_2 \in C \leqslant A$ implies $b_1 A = b_2 A$. Distinct cosets bC correspond to distinct cosets bA, for if $b_1 A = b_2 A$ (with $b_1, b_2 \in B$) then $b_1^{-1} b_2 \in A \cap B = C$ giving $b_1 C = b_2 C$. Thus there are at least as many cosets of A in G as there are of C in B. $\square$

(Notice that this Lemma is valid even when the indices may be infinite.) We now have the results we need when considering $\mathcal{U}/\Lambda$:

Corollary 6.10.5. *The parabolic subgroups of* Λ *are the stabilisers* Λ_r ($r \in \hat{\mathbb{Q}}$), *and the parabolic class number of* Λ *is the number of orbits of* Λ *on* $\hat{\mathbb{Q}}$.

Proof. By Corollaries 6.10.2 and 6.10.3, it is sufficient to prove that Λ_r is non-trivial for each $r \in \hat{\mathbb{Q}}$. Putting $G = \Gamma$, $A = \Lambda$ and $B = \Gamma_r$ in Lemma 6.10.4, we have $C = A \cap B = \Lambda_r$ (see Fig. 6.15), and hence $|\Gamma_r{:}\Lambda_r| \leqslant |\Gamma{:}\Lambda|$, which is finite. Thus Λ_r has finite index in the infinite group Γ_r (generated by some conjugate of Z), so Λ_r is non-trivial (in fact, infinite!). $\square$

Fig. 6.15

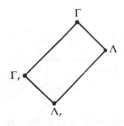

Corollary 6.10.6. *The parabolic class number* s *of* Λ *satisfies* $1 \leqslant s \leqslant N$, *where* N *is the index* $|\Gamma{:}\Lambda|$; *in particular,* s *is finite.*

Proof. Since s is the number of orbits of Λ on $\hat{\mathbb{Q}}$, we have $s \geqslant 1$. Let

$\Lambda T_1, \ldots, \Lambda T_N$ be the cosets of Λ in Γ. For any $r \in \hat{\mathbb{Q}}$ we have $r = T(\infty)$ for some $T \in \Gamma$, by Lemma 6.10.1, and putting $T = ST_i$ for some $S \in \Lambda$ and $i = 1, \ldots, N$ we see that $r = ST_i(\infty)$ lies in the Λ-orbit containing $T_i(\infty)$. Thus Λ has at most N orbits, so $s \leqslant N$. $\square$

We now show how to make $\mathcal{U}/\Lambda$ into a compact surface by adding s points, one for each orbit of Λ on $\hat{\mathbb{Q}}$. Suppose that $r_1, \ldots, r_s$ are elements of $\hat{\mathbb{Q}}$, one chosen from each orbit. For each $r = r_i$, Λ_r is a parabolic subgroup of Λ, and if $T \in \Gamma$ maps ∞ to r then $T^{-1}\Lambda_r T$ is a parabolic subgroup of $T^{-1}\Lambda T$, with fixed-point $\infty = T^{-1}(r)$, and therefore generated by $Z^l : \tau \mapsto \tau + l$ for some positive integer $l = l_i$ (we call r a *cusp* for Λ, and l its *amplitude*, or *width*); thus $T^{-1}\Lambda_r T$ is the unique subgroup of index l in $T^{-1}\Gamma_r T = \langle Z \rangle$, as is Λ_r in $\Gamma_r = \langle TZT^{-1} \rangle$. As before (in the case $\Lambda = \Gamma$) two points in $\mathcal{U}_1$ are in the same orbit of $T^{-1}\Lambda T$ if and only if they are equivalent under $T^{-1}\Lambda_r T$, that is, under some power of Z^l. Now $\varepsilon_l : \tau \mapsto e^{2\pi i \tau / l}$ maps $\mathcal{U}_1$ onto the punctured disc $W_l = \{w \in \mathbb{C} \mid 0 < |w| < e^{-2\pi/l}\}$. (If $l = 1$, then $\varepsilon_l = \varepsilon$ and $W_l = W$ as considered at the start of this section.) Hence $\varepsilon_l \circ T^{-1}$ maps $T(\mathcal{U}_1)$ onto W_l as in Fig. 6.16, and we have $(\varepsilon_l \circ T^{-1})(\tau) = (\varepsilon_l \circ T^{-1})(\tau')$ if and only if $T^{-1}(\tau') = T^{-1}(\tau) + lm \, (m \in \mathbb{Z})$, that is, if and only if $T^{-1}(\tau') = Z^{lm} T^{-1}(\tau)$. As remarked above, this is equivalent to $T^{-1}(\tau)$ and $T^{-1}(\tau')$ being in the same orbit of $T^{-1}\Lambda T$, that is, to τ and τ' being in the same Λ-orbit, so it follows that if $p : \mathcal{U} \to \mathcal{U}/\Lambda$ is the natural projection then $\varepsilon_l \circ T^{-1} \circ p^{-1}$ is a conformal equivalence from $(p \circ T)(\mathcal{U}_1)$ onto W_l. We therefore 'fill in' the puncture in $(p \circ T)(\mathcal{U}_1)$ by adjoining a single point, denoted $[r]_\Lambda$ since it corresponds to the Λ-orbit containing r; as before, we use $\varepsilon_l \circ T^{-1} \circ p^{-1}$ as a chart at $[r]_\Lambda$. This construction is independent of the choice of $r = r_i$, so doing it once for each orbit we obtain a surface $\bar{\mathcal{U}}/\Lambda = (\mathcal{U}/\Lambda) \cup (\hat{\mathbb{Q}}/\Lambda) = (\mathcal{U}/\Lambda) \cup \{[r_1]_\Lambda, \ldots, [r_s]_\Lambda\}$.

Fig. 6.16

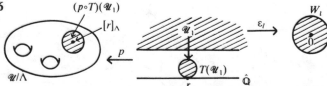

To show that $\bar{\mathcal{U}}/\Lambda$ is compact, it is sufficient to show that it is covered by finitely many compact subsets. First observe that each puncture $r = r_i$ $(i = 1, \ldots, s)$ is contained in a subset $C_i = (p \circ T)(\bar{\mathcal{U}}_2) \cup \{[r]_\Lambda\}$ which is compact, being homeomorphic under $\varepsilon_l \circ T^{-1} \circ p^{-1}$ to a closed disc $\{w \in \mathbb{C} \mid |w| \leqslant e^{-4\pi/l}\}$. Secondly, if $\Gamma = \Lambda T_1 \cup \ldots \cup \Lambda T_N$ and F is the fundamental region for Γ given in §6.1, then $G = T_1(F) \cup \ldots \cup T_N(F)$ is a fundamental region for Λ in $\mathcal{U}$. Now $F \backslash \mathcal{U}_2$ is compact (being closed and bounded), and hence so are the sets $T_j(F \backslash \mathcal{U}_2)$ for $j = 1, \ldots, N$; since the projection $p : \mathcal{U} \to \mathcal{U}/\Lambda$ is continuous, it follows that the sets $D_j = (p \circ T_j)(F \backslash \mathcal{U}_2)$ are all compact. It is easily seen that $\overline{\mathcal{U}}/\Lambda = C_1 \cup \ldots \cup C_s \cup D_1 \cup \ldots \cup D_N$, so that $\mathcal{U}/\Lambda$ is compact; by Theorem 4.16.1 it has a genus, which we will also refer to as the *genus* of Λ.

As an example, we consider $\Lambda = \Gamma(2)$, the principal congruence subgroup of index $N = 6$ and level 2 described in §6.7 and §6.9; we shall merely outline the details, since the reader should have no difficulty in reconstructing them. Firstly, Λ has three orbits on $\hat{\mathbb{Q}}$, namely

$[0]_\Lambda = \{p/q \,|\, p, q \in \mathbb{Z}, \ p \text{ is even and } q \text{ is odd}\}$,

$[1]_\Lambda = \{p/q \,|\, p, q \in \mathbb{Z}, \ p \text{ and } q \text{ are odd}\}$,

$[\infty]_\Lambda = \{p/q \,|\, p, q \in \mathbb{Z}, \ p \text{ is odd and } q \text{ is even}\} \cup \{\infty\}$,

represented by cusps $r_1 = 0$, $r_2 = 1$, $r_3 = \infty$ respectively. Thus Λ has parabolic class number $s = 3$, and in each case we have cusp-width $l_i = 2$: for example, Λ_∞ is the subgroup of index 2 in Γ_∞ generated by Z^2. As coset representatives for Λ in Γ we can take the elements

$$T_j = I, X, Z, ZY, Z^2 Y^2, ZYX \quad (j = 1, \ldots, 6),$$

corresponding to the matrices

$$\begin{pmatrix} 1 & 0 \\ 0 & 1 \end{pmatrix}, \begin{pmatrix} 0 & 1 \\ -1 & 0 \end{pmatrix}, \begin{pmatrix} 1 & 1 \\ 0 & 1 \end{pmatrix}, \begin{pmatrix} 1 & 0 \\ 1 & 1 \end{pmatrix}, \begin{pmatrix} -1 & 1 \\ -1 & 0 \end{pmatrix}, \begin{pmatrix} 0 & 1 \\ -1 & 1 \end{pmatrix}$$

in $SL(2, \mathbb{Z})$. These are congruent mod (2) to the six matrices given in §6.7; we have used these instead since they give a simpler fundamental region $G = \bigcup_{j=1}^6 T_j(F)$ for Λ, as illustrated in Fig. 6.17 where the index j denotes the region $T_j(F)$ $(j = 1, \ldots, 6)$. Notice that G has cusps at $r = 0, 1$ and (when we project $\mathcal{U}$ stereographically onto the sphere $\Sigma = S^2$) at ∞, each cusp formed from $l_i = 2$ regions $T_j(F)$; this is the motivation for using the

Fig. 6.17

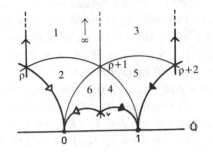

terms 'cusp' and 'width' for r_i and l_i. The arrows indicate how the sides of G are paired by elements of Λ: for example, the two vertical sides are equivalent under Z^2. When we form the quotient-surface $\mathcal{U}/\Lambda = G/\Lambda$ by identifying paired sides as indicated, we obtain a sphere with three punctures, corresponding to the orbits of Λ on $\hat{\mathbb{Q}}$ containing 0, 1 and ∞. (Notice that the three vertices $\rho, \rho + 1$ and $v = (\rho + 1)/(\rho + 2) = \frac{1}{2} + i/(2\sqrt{3})$ of G are all equivalent under Λ, so they are mapped to a *single* point in $\mathcal{U}/\Lambda$.) This corresponds to the fact that the cross-ratio function λ, invariant under $\Lambda = \Gamma(2)$, maps $\mathcal{U}$ onto $\Sigma \backslash \{0, 1, \infty\}$. If we compactify $\mathcal{U}/\Lambda$ as above, by adjoining three points $[0]_\Lambda$, $[1]_\Lambda$ and $[\infty]_\Lambda$ to fill the punctures, then the quotient-surface $\overline{\mathcal{U}}/\Lambda$ is a sphere, corresponding to the fact that λ maps $\overline{\mathcal{U}} = \mathcal{U} \cup \hat{\mathbb{Q}}$ onto Σ. This is illustrated in Fig. 6.18.

Fig. 6.18

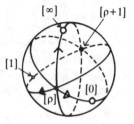

Any subgroup Λ of finite index in Γ has a finite number of cusp-widths $l_1, \ldots, l_s$, one for each orbit on $\hat{\mathbb{Q}}$. We define the *level* l of Λ to be the least common multiple of $l_1, \ldots, l_s$; we shall see later how this is related to the concept of the level of a congruence subgroup, defined in §6.9.

Lemma 6.10.7. *The integers k, which have the property that $P^k \in \Lambda$ for all parabolic elements $P \in \Gamma$, are the multiples of l (which is therefore their greatest common divisor).*

Proof. Any parabolic $P \in \Gamma$ fixes a unique $r \in \hat{\mathbb{Q}}$, so P^l lies in the unique subgroup $(\Gamma_r)^l$ of index l (consisting of the lth powers) in the stabiliser Γ_r. Since the cusp-width l_i of r divides l, P^l lies in $(\Gamma_r)^{l_i}$, the unique subgroup of index l_i in Γ_r; now $|\Gamma_r:\Lambda_r| = l_i$, so $P^l \in \Lambda_r \leqslant \Lambda$. Thus l has the stated property, and hence so do all its multiples.

Conversely, choose cusps $r_1, \ldots, r_s$, one from each Λ-orbit; then each Λ_{r_i} is generated by some conjugate $T_i^{-1} Z^{l_i} T_i$ of a power of Z. If $P^k \in \Lambda$ for all parabolics P, then putting $P = T_i^{-1} Z T_i$ we have $T_i^{-1} Z^k T_i \in \Lambda$ for all i; since $T_i^{-1} Z T_i$ fixes r_i, so does $T_i^{-1} Z^k T_i$, which therefore lies in Λ_{r_i}.

It follows that $T_i^{-1}Z^k T_i$ is a power of $T_i^{-1}Z^{l_i}T_i$, so each l_i divides k and hence so does l, their least common multiple. □

It follows easily (see Exercise 6C) that l is the order of the permutation induced by Z in the action of Γ on the cosets of Λ. *From now on we will assume that Λ is normal in* Γ, since the theory is then a little simpler; as before, $|Γ:Λ|$ is assumed to be finite.

Corollary 6.10.8. *If Λ is a normal subgroup of finite index N in* Γ, *then*

(i) *all cusp-widths l_i for Λ are equal to the level l;*
(ii) *l is the order of $ZΛ$ in the quotient group $Γ/Λ$;*
(iii) *$ls = N$, where s is the parabolic class number of Λ.*

Proof. (i) Let $r_1, r_2 \in \hat{\mathbb{Q}}$ have cusp-widths $l_1 = |Γ_{r_1}:Λ_{r_1}|$, $l_2 = |Γ_{r_2}:Λ_{r_2}|$, and choose $T \in Γ$ so that $T(r_1) = r_2$, and hence $T^{-1}Γ_{r_1}T = Γ_{r_2}$. Since Λ is normal we have $T^{-1}ΛT = Λ$, so $Λ_{r_1} = Λ \cap Γ_{r_1} = (T^{-1}ΛT) \cap (T^{-1}Γ_{r_2}T) = T^{-1}(Λ \cap Γ_{r_2})T = T^{-1}Λ_{r_2}T$. Thus conjugation by T induces an isomorphism from $Γ_{r_2}$ to $Γ_{r_1}$, with $Λ_{r_2}$ being mapped to $Λ_{r_1}$, so the indices $l_1 = |Γ_{r_1}:Λ_{r_1}|$ and $l_2 = |Γ_{r_2}:Λ_{r_2}|$ must be equal. Hence all cusp-widths are equal, and they must coincide with their least common multiple, which is l.

(ii) The order of $ZΛ$ in $Γ/Λ$ is the greatest common divisor of all k such that $Z^k \in Λ$; now $Z^k \in Λ$ if and only if $T^{-1}Z^k T \in Λ$ for all $T \in Γ$ (since Λ is normal), and this is equivalent to the condition that $P^k \in Λ$ for all parabolics P, since they are the conjugates of powers of Z, so $ZΛ$ has order l by Lemma 6.10.7.

(iii) Since $Λ \trianglelefteq Γ$, Γ maps Λ-orbits on $\hat{\mathbb{Q}}$ to Λ-orbits, and permutes them transitively since it permutes $\hat{\mathbb{Q}}$ transitively. An element $T \in Γ$ maps the Λ-orbit $[\infty]$ to itself if and only if $T(\infty) = S(\infty)$ for some $S \in Λ$; this is equivalent to $TS^{-1} \in Γ_\infty$, that is, $T \in Γ_\infty Λ$, so the stabiliser of $[\infty]$ is the subgroup $Γ_\infty Λ$ of Γ. Since there are s Λ-orbits, permuted transitively by

Fig. 6.19

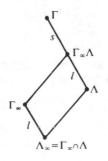

$\Gamma, |\Gamma:\Gamma_\infty\Lambda| = s$. Now Λ is normal in $\Gamma_\infty\Lambda$ (since $\Lambda \lhd \Gamma$), so $\Gamma_\infty\Lambda/\Lambda \cong \Gamma_\infty/(\Gamma_\infty \cap \Lambda) = \Gamma_\infty/\Lambda_\infty$, giving $|\Gamma_\infty\Lambda:\Lambda| = |\Gamma_\infty:\Lambda_\infty| = l$, since ∞ has cusp-width l, by (i). Thus $N = |\Gamma:\Lambda| = |\Gamma:\Gamma_\infty\Lambda| \cdot |\Gamma_\infty\Lambda:\Lambda| = sl$. (These subgroup inclusions are depicted in Fig. 6.19.) $\square$

For example, if Λ is the principal congruence subgroup $\Gamma(n)$ of level n (as defined in §6.9), then in the quotient-group $\Gamma/\Gamma(n) \cong PSL(2, \mathbb{Z}_n)$, Z is mapped to the transformation $\tau \mapsto \tau + 1$ of order n, so Λ has level $l = n$ by Corollary 6.10.8 (ii); thus our two definitions of level agree for principal congruence subgroups, and indeed a theorem of Wohlfahrt (Wohlfahrt [1964], or Newman [1972]) shows that they agree for *all* congruence subgroups of Γ. We can also use Corollary 6.10.8 (iii) to find the parabolic class number s of $\Gamma(n)$; for example, when n is a prime p, the index N is given by (6.9.5), and $N = ls = ns$ implies that

$$s = \begin{cases} \dfrac{p^2 - 1}{2} & \text{if } p \text{ is odd,} \\ 3 & \text{if } p = 2. \end{cases}$$

Theorem 6.10.9. *If Λ is a normal subgroup of level l and of finite index $N > 3$ in Γ, then the Euler characteristic of $\overline{\mathcal{U}}/\Lambda$ is*

$$N\left(\frac{1}{2} + \frac{1}{3} + \frac{1}{l} - 1\right),$$

that is, the genus of $\overline{\mathcal{U}}/\Lambda$ (and hence of Λ) is

$$g = 1 + \frac{N}{2}\left(\frac{1}{6} - \frac{1}{l}\right).$$

Proof. The map $f:\overline{\mathcal{U}}/\Lambda \to \overline{\mathcal{U}}/\Gamma$, given by $[\tau]_\Lambda \mapsto [\tau]_\Gamma (\tau \in \overline{\mathcal{U}})$, is easily seen to be a branched covering, with branch-points over the points $[i]_\Gamma$, $[\rho]_\Gamma$ and $[\infty]_\Gamma$ of $\overline{\mathcal{U}}/\Gamma$, unbranched elsewhere.

As $\overline{\mathcal{U}}/\Gamma$ is a sphere (since J induces a conformal equivalence $\overline{\mathcal{U}}/\Gamma \to \Sigma$), we can calculate the genus of $\overline{\mathcal{U}}/\Lambda$ by determining the number of sheets and the total order of branching of f, and then applying the Riemann–Hurwitz formula (Theorem 4.16.3).

If τ is not fixed by any non-identity element of Γ, and if $T_1, \ldots, T_N$ are coset representatives for Λ in Γ, then

$$f^{-1}([z]_\Gamma) = \{[T_1(z)]_\Lambda, \ldots, [T_N(z)]_\Lambda\},$$

so that away from the branch-points every point in $\overline{\mathcal{U}}/\Gamma$ is covered by N points in $\overline{\mathcal{U}}/\Lambda$; in other words, f is an N-sheeted covering.

To determine the order of branching, we first show that Λ contains neither of the generators X, Y of Γ (defined in §6.8). Let $\theta:\Gamma \to \Gamma/\Lambda$ be the natural homomorphism, mapping X and Y to generators x, y of Γ/Λ as in Lemma 6.9.1. If $X \in \Lambda$ then $x = 1$ and hence Γ/Λ, being generated by an element y satisfying $y^3 = 1$, has order $N \leqslant 3$, against our assumption; thus $X \notin \Lambda$, and similarly $Y \notin \Lambda$ since $x^2 = 1$, so Λ contains neither X nor Y, nor (being normal) any of their conjugates. It follows that for each coset ΛT_j of Λ in Γ, $\Lambda T_j X$ is a *different* coset, say $\Lambda T_{j'}$, with $\Lambda T_{j'} X = \Lambda T_j X^2 = \Lambda T_j$. Thus the cosets are arranged in $\frac{1}{2}N$ pairs ΛT_j, $\Lambda T_{j'}$, and since X generates the stabiliser Γ_i of i in Γ, we see that $f^{-1}([i]_\Gamma)$ consists of the $\frac{1}{2}N$ distinct Λ-orbits $[T_j(i)]_\Lambda = [T_{j'}(i)]_\Lambda$, that is, $[i]_\Gamma$ is covered by $\frac{1}{2}N$ points in $\bar{\mathcal{U}}/\Lambda$. Similarly, $[\rho]_\Gamma$ is covered by $\frac{1}{3}N$ points, while Corollary 6.10.5 implies that the number of points covering $[\infty]_\Gamma = \hat{\mathbb{Q}}$ is the parabolic class number s of Λ, and this is equal to N/l by Corollary 6.10.8 (iii). Thus the total order of branching, which is the number of 'missing points', is

$$\left(N - \frac{N}{2}\right) + \left(N - \frac{N}{3}\right) + \left(N - \frac{N}{l}\right) = N\left(3 - \frac{1}{2} - \frac{1}{3} - \frac{1}{l}\right),$$

so the Riemann–Hurwitz formula gives

$$g = 1 - N + \frac{N}{2}\left(3 - \frac{1}{2} - \frac{1}{3} - \frac{1}{l}\right)$$

$$= 1 + \frac{N}{2}\left(\frac{1}{6} - \frac{1}{l}\right),$$

from which we can calculate the Euler characteristic $2 - 2g$. □

Corollary 6.10.10. *For each prime p, the genus of $\bar{\mathcal{U}}/\Gamma(p)$ is*

$$g = \begin{cases} \dfrac{(p + 2)(p - 3)(p - 5)}{24} & \text{if } p \text{ is odd,} \\[2mm] 0 & \text{if } p = 2. \end{cases}$$

Proof. This follows from the previous Theorem, 6.9.5 (giving N), and the fact that $\Gamma(p)$ has level $l = p$. □

Notice that in Corollary 6.10.10, $\bar{\mathcal{U}}/\Gamma(p)$ has genus 0 if and only if $p = 2$, 3 or 5. By calculating the genus of $\bar{\mathcal{U}}/\Gamma(n)$ for composite n (see Exercise 6M), we see that the only other case giving $g = 0$ is $n = 4$. We will conclude our investigation of Γ by considering the automorphisms and tessellations of the quotient-surfaces corresponding to principal congruence subgroups of small level.

If $\Lambda \trianglelefteq \Gamma$ then, as shown in §5.9, each element ΛT of Γ/Λ acts as an automorphism of $\mathcal{U}/\Lambda$ by mapping each Λ-orbit $[z]_\Lambda$ to $[T(z)]_\Lambda$. We have seen that if Λ has finite index in Γ, then $\bar{\mathcal{U}}/\Lambda$ is a compact Riemann surface formed from $\mathcal{U}/\Lambda$ by filling in finitely many punctures $[r]_\Lambda (r \in \hat{\mathbb{Q}})$; now ΛT permutes these punctures, mapping $[r]_\Lambda$ to $[T(r)]_\Lambda$, and from our description of the charts at the punctures it is easily seen that ΛT induces an automorphism of $\bar{\mathcal{U}}/\Lambda$. If we choose $\tau \in \mathcal{U}$ to be fixed by no non-identity elements of Γ (that is, $\tau \in \mathcal{U} \setminus ([i]_\Gamma \cup [\rho]_\Gamma)$), then ΛT fixes $[\tau]_\Lambda$ if and only if $T \in \Lambda$; hence Γ/Λ acts faithfully on $\bar{\mathcal{U}}/\Lambda$, that is, we have an embedding of Γ/Λ in the automorphism group $\mathrm{Aut}(\bar{\mathcal{U}}/\Lambda)$, and, in particular, taking $\Lambda = \Gamma(n)$ we have

$$PSL(2, \mathbb{Z}_n) \cong \Gamma/\Gamma(n) \leqslant \mathrm{Aut}(\bar{\mathcal{U}}/\Gamma(n)).$$

For example, if $n = 7$ then $\bar{\mathcal{U}}/\Gamma(7)$ has genus $g = 3$ by Corollary 6.10.10; now (6.9.5) gives $|PSL(2, \mathbb{Z}_7)| = 168 = 84(g - 1)$, which is the upper bound for the number of automorphisms of any surface of genus $g = 3$, by Theorem 5.11.1. Thus $PSL(2, \mathbb{Z}_7) \cong \mathrm{Aut}(\bar{\mathcal{U}}/\Gamma(7))$, and moreover we have another proof (confirming Theorem 5.11.5 and §6.9) that $PSL(2, \mathbb{Z}_7)$ is a Hurwitz group; this proof is a little more satisfactory in that it actually specifies a surface of genus 3 on which $PSL(2, \mathbb{Z}_7)$ acts.

If $n = 2, 3, 4$ or 5 then $\bar{\mathcal{U}}/\Gamma(n)$ has genus 0, and hence by Theorem 4.17.2 it is conformally equivalent to the Riemann sphere Σ. By Theorem 4.17.3 (i) we therefore have $\mathrm{Aut}(\bar{\mathcal{U}}/\Gamma(n)) \cong PSL(2, \mathbb{C})$ in these cases, so Corollary 2.13.2 implies that $PSL(2, \mathbb{Z}_n)$, being a finite group of Möbius transformations, must be isomorphic to a finite group of rotations of Σ; it then follows from Theorem 2.13.5 that $PSL(2, \mathbb{Z}_n)$ is cyclic, dihedral, or isomorphic to the rotation group of a regular tetrahedron, octahedron or icosahedron. For example, we showed in (6.9.2) and Theorem 6.9.3 that $PSL(2, \mathbb{Z}_2)$ is isomorphic to the symmetric group S_3, and hence to the dihedral rotation group D_3; we can see the action of $PSL(2, \mathbb{Z}_2)$ on $\bar{\mathcal{U}}/\Gamma(2)$ by considering Fig. 6.18, where there are exactly six rotations of the sphere permuting the three punctures $[0]$, $[1]$ and $[\infty]$. In the cases $n = 3, 4$ and 5 we have $|PSL(2, \mathbb{Z}_n)| = 12, 24$ and 60 by (6.9.5) and Exercise 6L; it is not hard to show that $PSL(2, \mathbb{Z}_n)$ is neither cyclic nor dihedral for $n > 2$ (see Exercise 6N), so in these three cases the corresponding rotation groups must be those of a tetrahedron, octahedron and icosahedron respectively.

We shall illustrate the isomorphism between $PSL(2, \mathbb{Z}_5)$ and the icosahedral rotation group A_5 (see §2.13); the cases $n = 3, 4$ are similar (see

Exercise 6P). Let us put $\Lambda = \Gamma(5)$; having index 60 and level 5, Λ has parabolic class number $\frac{60}{5} = 12$ by Corollary 6.10.8(iii). Thus $\mathcal{U}/\Lambda$ is a sphere with 12 punctures, and we shall find an icosahedron $\mathcal{I}$ inscribed in the sphere $\bar{\mathcal{U}}/\Lambda$, with the 12 punctures $[r]_\Lambda (r \in \hat{\mathbb{Q}})$ as its vertices, such that $\mathcal{I}$ is invariant under the action of $PSL(2, \mathbb{Z}_5) = \Gamma/\Lambda$. We shall do this by choosing a fundamental region Φ for Γ on $\mathcal{U}$ such that three adjacent images of Φ under Γ form a hyperbolic triangle Δ with vertices in $\hat{\mathbb{Q}}$; then Δ projects onto a triangle in $\bar{\mathcal{U}}/\Lambda$ with punctures as its vertices. Since a fundamental region for Λ consists of $|\Gamma:\Lambda| = 60$ copies of Φ, when we form $\mathcal{U}/\Lambda$ by identifying edges of this fundamental region we find that there are $\frac{60}{3} = 20$ copies of Δ, and we shall show that they form the faces of an icosahedron invariant under $PSL(2, \mathbb{Z}_5)$.

Unfortunately, we cannot take Φ to be the Dirichlet region F which we have used so far: a little experiment will soon convince the reader that one cannot form a suitable triangle Δ from three copies of F. Instead, we take

$$\Phi = \{\tau \in \mathcal{U} \mid -1 \leqslant \operatorname{Re}(\tau) \leqslant 0, |\tau| \geqslant 1, |\tau + 1| \geqslant 1\},$$

a fundamental region formed from F by removing the subset $F_+ = \{\tau \in F \mid \operatorname{Re}(\tau) > 0\}$ and replacing it by the congruent set $Z^{-1}(F_+)$ illustrated in Fig. 6.20 (see §3.4). We then let

$$\Delta = \Phi \cup Y(\Phi) \cup Y^2(\Phi),$$

Fig. 6.20

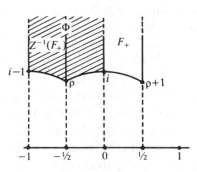

Fig. 6.21

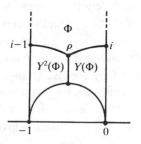

a hyperbolic triangle with vertices at $0, -1$ and ∞, shown in Fig. 6.21.

By an argument used in the proof of Theorem 6.10.9, we may choose coset representatives for Λ in Γ to have the form

$$T_1, T_1 Y, T_1 Y^2, \ldots, T_{20}, T_{20} Y, T_{20} Y^2$$

for suitable elements $T_1, \ldots, T_{20} \in \Gamma$, and hence by Theorem 5.10.9 a fundamental region for Λ is

$$T_1(\Phi) \cup T_1 Y(\Phi) \cup T_1 Y^2(\Phi) \cup \ldots \cup T_{20}(\Phi) \cup T_{20} Y(\Phi) \cup T_{20} Y^2(\Phi)$$
$$= T_1(\Delta) \cup \ldots \cup T_{20}(\Delta).$$

Thus the projections Δ_j of $T_j(\Delta)$ in $\mathcal{U}/\Lambda$ ($j = 1, \ldots, 20$) form a triangular tessellation of the sphere $\bar{\mathcal{U}}/\Lambda$. For example, we can take $T_1 = I \in \Gamma$, and then $T_1(\Delta) = \Delta$ projects onto the triangle Δ_1 with vertices $[0]_\Lambda$, $[-1]_\Lambda$ and $[\infty]_\Lambda$, shown in Fig. 6.22.

Fig. 6.22

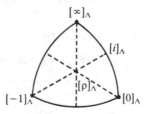

Consider the action of an element $T \in \Gamma$ on a triangle $T_j(\Delta)$: we have $TT_j = ST_k Y^\varepsilon$ for some $S \in \Lambda$, $k = 1, \ldots, 20$, and $\varepsilon = 0, 1, 2$; then

$$TT_j(\Delta) = ST_k Y^\varepsilon(\Delta)$$
$$= ST_k(\Delta)$$

(since Y maps Δ to itself), and this projects to the image Δ_k of $T_k(\Delta)$ in $\bar{\mathcal{U}}/\Lambda$. Thus in the action of Γ/Λ as a group of automorphisms of $\bar{\mathcal{U}}/\Lambda$, the triangles Δ_j are permuted amongst themselves; in fact, Γ/Λ permutes them transitively since Γ permutes the images in $\mathcal{U}$ of Φ transitively. Since Δ_1 has all three of its vertices at punctures, so does each Δ_j; since Γ/Λ permutes the punctures transitively (by Lemma 6.10.1), every puncture is a vertex, so the tessellation of $\bar{\mathcal{U}}/\Lambda$ has exactly twelve vertices. By Corollary 6.10.8(i) all cusp-widths are equal to 5, so exactly five triangles Δ_j meet at each vertex: for example, taking $T_j = Z^{j-1}$ for $j = 1, \ldots, 5$ we see that the vertex $[\infty]_\Lambda$ is surrounded by the five triangles $\Delta_1, \ldots, \Delta_5$.

Now let $f : \bar{\mathcal{U}}/\Lambda \to \Sigma$ be a conformal equivalence; then $f \circ (\Gamma/\Lambda) \circ f^{-1}$ is a group of automorphisms of Σ, and being finite it is conjugate to a subgroup $T \circ f \circ (\Gamma/\Lambda) \circ f^{-1} \circ T^{-1}$ of the rotation group $PSU(2, \mathbb{C})$ for some

$T \in PSL(2, \mathbb{C})$, by Corollary 2.13.2. To put it another way, if we use $T \circ f$ to identify $\overline{\mathcal{U}}/\Lambda$ with Σ, then Γ/Λ is a group of rotations of $\overline{\mathcal{U}}/\Lambda$. Since $Y\Lambda$ permutes the three sides of Δ_1 cyclically, Δ_1 is an equilateral triangle, and hence so are all its images Δ_j. It is easily seen that a tessellation of a sphere by 20 equilateral triangles, five at each vertex, must correspond to an inscribed icosahedron $\mathcal{I}$; thus $PSL(2, \mathbb{Z}_5) = \Gamma/\Lambda$ is contained in the rotation group of $\mathcal{I}$, and comparing orders we see that these two groups are equal.

Had we wished to use the Dirichlet region F rather than Φ, we could have shown that

$$\Pi = \bigcup_{i=0}^{4} Z^i(F)$$

projects onto a pentagon in $\overline{\mathcal{U}}/\Lambda$, and that the 12 images of this pentagon form a dodecahedron (dual to $\mathcal{I}$) rotated by $PSL(2, \mathbb{Z}_5)$. In the cases $\Lambda = \Gamma(3)$ and $\Gamma(4)$, similar methods produce a tetrahedron and an octahedron (or its dual, a cube) with $PSL(2, \mathbb{Z}_3)$ and $PSL(2, \mathbb{Z}_4)$ as their rotation groups.

Thus, having reached the modular group by a long and circuitous route, involving elliptic functions and the classifications of lattices and of tori, we see that this group, so rich in geometric, number-theoretic and group-theoretic structure, embodies some of the earliest topics considered in this book, namely the Riemann sphere and its rotations.

EXERCISES

6A. Verify that if $T(\tau) = (a\bar{\tau} + b)/(c\bar{\tau} + d)$, $a, b, c, d \in \mathbb{Z}$, $ad - bc = -1$, then

$$g_2(T(\tau)) = (c\bar{\tau} + d)^{-4}\overline{g_2(\tau)},$$
$$g_3(T(\tau)) = (c\bar{\tau} + d)^{-6}\overline{g_3(\tau)},$$
$$\Delta(T(\tau)) = (c\bar{\tau} + d)^{-12}\overline{\Delta(\tau)},$$
$$J(T(\tau)) = \overline{J(\tau)}.$$

6B. (i) If $q = e^{2\pi i \tau}$ show that

$$(g_2(\tau))^3 = \frac{2^6}{3^3}\pi^{12}\left(1 + 2^4 \times 3 \times 5 \sum_{k=1}^{\infty} \sigma_3(k)q^k\right)^3,$$

and that

$$27(g_3(\tau))^2 = \frac{2^6}{3^3}\pi^{12}\left(1 - 2^3 \times 3^2 \times 7 \sum_{k=1}^{\infty} \sigma_5(k)q^k\right)^2.$$

Deduce that

$$(2\pi)^{-12}\Delta(\tau) = \frac{1}{12}\sum_{k=1}^{\infty}(5\sigma_3(k) + 7\sigma_5(k))q^k$$

$$+ \text{ power series in } q \text{ with coefficients in } \mathbb{Z}.$$

(ii) Show that if $d \geqslant 1$ is an integer then $d^3(1 - d^2)$ is divisible by 12 and deduce that

$$\Delta(\tau) = (2\pi)^{12}\sum_{k=1}^{\infty}a_k q^k$$

where $a_k \in \mathbb{Z}$.

6C. Let Λ be a subgroup of Γ of index n. Suppose that

$$\Gamma = \Lambda T_1 \cup \Lambda T_2 \cup \ldots \cup \Lambda T_n$$

is the coset decomposition of Γ into right Λ-cosets. Let $Z : \tau \mapsto \tau + 1$ act on these cosets by right multiplication, that is

$$Z : \Lambda T_i \mapsto \Lambda T_i Z.$$

This gives a permutation of the n cosets. Show that if we write this permutation as a product of disjoint cycles then the lengths of these cycles are the cusp-widths of Λ. Deduce that the order of the permutation is the level of Λ and the number of cycles is the parabolic class number of Λ.

6D. For each prime number p let $\Gamma_0(p)$ be the set of those transformations $z \mapsto (az + b)/(cz + d)$ of Γ for which $c \equiv 0 \bmod p$, and let $\Gamma^0(p)$ be the set of transformations with $b \equiv 0 \bmod p$. Prove that $\Gamma_0(p)$ and $\Gamma^0(p)$ are subgroups of Γ which are conjugate in Γ.

Prove that $\Gamma^0(p)$ has index $p + 1$ in Γ by showing that $I, Z, Z^2, \ldots, Z^{p-1}$ and $X : \tau \mapsto -1/\tau$ are right coset representatives for $\Gamma^0(p)$ in Γ.

Using Exercise 6C or otherwise show that $\Gamma^0(p)$ has parabolic class number equal to 2. What is its level?

The following series of problems will enable us to calculate the index of $\Gamma(n)$ in Γ. We would like to thank John Thornton for showing us Exercise 6F.

6E. Show that if X, Y, Z are as in §6.8 then Z and $W = XY^2$ generate the modular group.

6F. Let m_1 and m_2 be a pair of co-prime positive integers. Then there exist $a, b \in \mathbb{Z}$ such that $am_1 + bm_2 = 1$. By calculating $(Z^{m_1})^a(Z^{m_2})^b$ and $(W^{m_1})^a(W^{m_2})^b$ and using Exercise 6E, show that every element of Γ is of the form AB where $A \in \Gamma(m_1)$ and $B \in \Gamma(m_2)$.

The following exercise is purely group-theoretic.

6G. Let H be a group and N_1, N_2 normal subgroups of H with $N_1 N_2 = H$ (where $N_1 N_2 = \{n_1 n_2 \mid n_1 \in N_1, n_2 \in N_2\}$). Prove that

$$H/(N_1 \cap N_2) \cong (H/N_1) \times (H/N_2).$$

6H. Let $G(m)$ denote the subgroup of $SL(2, \mathbb{Z})$ consisting of the matrices

$A \in SL(2, \mathbb{Z})$ such that $A \equiv I \bmod m$. Prove that if $(m_1, m_2) = 1$ then

(i) $G(m_i) \cap G(m_2) = G(m_1 m_2)$,

(ii) $-I \in G(m_1) G(m_2)$.

6I. Show that $G(m)$ is the kernel of the natural group homomorphism from $SL(2, \mathbb{Z})$ to $SL(2, \mathbb{Z}_m)$ (described in §6.9) and hence show that if $m = q_1 \ldots q_s$ is the prime-power decomposition of m ($q_1, \ldots, q_s$ being powers of distinct primes), then $SL(2, \mathbb{Z}_m) \cong SL(2, \mathbb{Z}_{q_1}) \times \ldots \times SL(2, \mathbb{Z}_{q_s})$.

6J. Show by example that if $(m_1, m_2) = 1$ then it is not necessarily true that $\Gamma(m_1) \cap \Gamma(m_2) = \Gamma(m_1 m_2)$. (For this reason $PSL(2, \mathbb{Z}_m)$ does not have the direct product structure of $SL(2, \mathbb{Z}_m)$ described in Exercise 6I.) What can in general be said about the relationship between the subgroups $\Gamma(m_1) \cap \Gamma(m_2)$ and $\Gamma(m_1 m_2)$?

6K. Show that the order of the group $SL(2, \mathbb{Z}_{p^r})$ (p prime) is $p^{3r-2}(p^2 - 1)$. (Hint: first show that a pair $([a], [b])$ of elements of $\mathbb{Z}_{p^r}$, where $[x]$ is the residue class of x, can be the first row of a matrix in $SL(2, \mathbb{Z}_{p^r})$ if and only if a, b and p have no common divisors other than ± 1.)

6L. Show that the index of $\Gamma(n)$ in Γ is given by

$$|\Gamma : \Gamma(n)| = \begin{cases} 6 & \text{if } n = 2 \\ \dfrac{n^3}{2} \displaystyle\prod_{p|n} \left(1 - \dfrac{1}{p^2}\right) & \text{if } n > 2. \end{cases}$$

where p ranges over the distinct prime factors of n.

6M. Find the genus of $\mathcal{U}/\Gamma(n)$.

6N. Show that the only cyclic or dihedral groups which are homomorphic images of the modular group are the cyclic groups of orders $1, 2, 3$ or 6 and the dihedral group of order 6. Deduce that $PSL(2, \mathbb{Z}_n)$ is not cyclic or dihedral if $n > 2$.

6P. Give illustrations of the isomorphisms $PSL(2, \mathbb{Z}_3) \cong A_4$ and $PSL(2, \mathbb{Z}_4) \cong S_4$ analogous to that given for the isomorphism $PSL(2, \mathbb{Z}_5) \cong A_5$ in §6.10.

6Q. Let a_1, a_2, a_3, a_4 be 4 distinct complex numbers and consider the Riemann surface S of the equation $w^2 = (z - a_1)(z - a_2)(z - a_3)(z - a_4)$. Define $Z = 1/(z - a_1)$, $W = w/(a_1 - z)^2$. Show that $(z, w) \mapsto (Z, W)$ induces a bijection from S to a Riemann surface S' of an equation of the form $W^2 = A(Z - b_1)(Z - b_2)(Z - b_3)$, $(b_1, b_2, b_3$ distinct, $A \neq 0)$, and using Theorem 6.5.11 show that S is conformally equivalent to $\mathbb{C}/\Omega$ for some lattice Ω.

APPENDIX 1

A review of complex variable theory

In this appendix we review the theorems of complex variable theory which have been important in the text. We give proofs of some of the results which have been used particularly often.

Theorem A.1. (Cauchy–Riemann equations). *Let $f(z)$ be analytic in a region R and let $u(x, y)$, $v(x, y)$ be the real and imaginary parts of $f(z)$ respectively. Then*

$$\frac{\partial u}{\partial x} = \frac{\partial v}{\partial y}, \quad \frac{\partial u}{\partial y} = -\frac{\partial v}{\partial x}$$

at all points of R. These are called the Cauchy–Riemann equations. Conversely, suppose that u_x, u_y, v_x, v_y exist and are continuous in R and satisfy the Cauchy–Riemann equations. Then f is analytic throughout R. □

The basic theorem of complex integration is *Cauchy's theorem.*

Theorem A.2. *Let $f(z)$ be analytic in the simply connected region A and let γ be a piecewise differentiable closed path lying in A. Then*

$$\int_\gamma f(z)\,dz = 0.$$ □

We give some of the important consequences of Cauchy's theorem.

Theorem A.3. (Cauchy integral formula). *Let $f(z)$ be analytic in a simply connected region A, let γ be a piecewise differentiable simple closed path in A and let b be a point of the region enclosed by γ. Then*

$$f(b) = \frac{1}{2\pi i}\int_\gamma \frac{f(z)}{(z - b)}\,dz.$$

Moreover, we have the following formula for the derivatives of f, obtained by differentiating under the integral sign:

$$f^{(n)}(b) = \frac{n!}{2\pi i} \int_\gamma \frac{f(z)}{(z-b)^{n+1}} \, dz.$$ □

Theorem A.4. (Liouville's theorem). *If f is analytic throughout $\mathbb{C}$ and if there is a real number M such that $|f(z)| \leqslant M$ for all $z \in \mathbb{C}$, then f is constant.* □

Cauchy's theorem has the following converse:

Theorem A.5. (Morera's theorem). *Let f be continuous in a region A and suppose that $\int_\gamma f = 0$ for every closed curve γ in A. Then f is analytic in A.* □

Theorem A.6. *Let γ be a piecewise differentiable simple closed path lying in a region R and enclosing a region A. Let f be analytic in R except possibly for a finite number of singularities in A. Then $\int_\gamma f(z)\, dz = 2\pi i$ {sum of the residues at the singularities of f in A}.* □

The following result is a simple consequence of the previous Theorem. The proof follows that of Theorem 3.6.4.

Theorem A.7. *Let γ be a piecewise differentiable simple closed path lying in a region R and enclosing a region A. Let f be analytic in R except possibly for a finite number of poles in A and suppose that $f(z) \neq 0$ along γ. Then*

$$\int_\gamma \frac{f'(z)}{f(z)} \, dz = 2\pi i (N - P)$$

where N is the sum of the orders of the zeros of f in A and P is the sum of the orders of the poles of f in A, each counted with the correct multiplicity.

□

One of the main applications of complex integration is the existence of power series for analytic functions. If a function g is analytic at a point $z_0 \in \mathbb{C}$ then it can be represented by a power series

$$g(z) = \sum_{n=0}^{\infty} b_n (z - z_0)^n,$$

which converges in some disc with centre at z_0. This leads to a proof of the basic result underlying analytic continuation.

Theorem A.8.   *Let f be analytic in a region R with zeros at a sequence of points z_i which tend to a limit $z^* \in R$. Then f is identically zero in R.*

Proof.   As f is continuous, $f(z^*) = \lim_{i \to \infty} f(z_i) = 0$. If f is not identically zero then we can expand f as a power series about z^*,

$$f(z) = \sum_{n=1}^{\infty} a_n (z - z^*)^n,$$

in which not all the coefficients a_n are zero. If a_m is the first non-zero coefficient then

$$f(z) = \sum_{n=m}^{\infty} a_n (z - z^*)^n = (z - z^*)^m g(z),$$

where g is analytic in R and $g(z^*) = a_m \neq 0$.

   Suppose that $|g(z^*)| = 2\varepsilon$. Then as g is continuous there exists $\delta > 0$ such that if $|z - z^*| < \delta$ then $|g(z) - g(z^*)| < \varepsilon$. Hence in the disc $|z - z^*| < \delta$,

$$\big|\,|g(z)| - 2\varepsilon\,\big| = \big|\,|g(z)| - |g(z^*)|\,\big| \leqslant |g(z) - g(z^*)| < \varepsilon,$$

and therefore $|g(z)| > \varepsilon$ in this disc. Thus $f(z) \neq 0$ in a disc with centre z^* which contradicts the hypothesis that z^* is a limit point of zeros of f. Therefore f is identically zero in R. $\square$

This theorem is used to prove the following result due to Weierstrass.

Theorem A.9.   *An analytic function comes arbitrarily close to any complex value in every neighbourhood of an essential singularity.* $\square$

Our next result is that non-constant analytic functions define open mappings, that is if $A \subseteq \mathbb{C}$ is open and f is analytic in A then $f(A)$ is open. We first point out that we only need to prove this result locally, that is, every $a \in A$ has an open neighbourhood $V_a \subseteq A$ such that $f(V_a)$ is open; for then we can write $A = \bigcup_{a \in A} V_a$ and

$$f(A) = f\left(\bigcup_{a \in A} V_a \right) = \bigcup_{a \in A} f(V_a)$$

which is open.

   The strategy of the proof will be to show that, apart from an additive constant, every non-constant analytic function can be written locally as a composition $p_m \circ \phi$ where $p_m(z) = z^m$, and ϕ' does not vanish. We will then show that ϕ and p_m define open mappings and the result will follow from the observation that the composition of two open mappings is open. The method of proof will show also that f is locally m-to-one.

Theorem A.10. (i) *Let* $R \subseteq \mathbb{C}$ *be a region and let* f *be a non-constant analytic function defined on* R. *Then* $f : R \to \mathbb{C}$ *defines an open mapping.*

(ii) *If* $z_0 \in R$ *and if* $f(z_0) = w_0$ *with multiplicity* m *then there exists a neighbourhood* N *of* z_0 *such that for each* $w \in f(N) \setminus \{w_0\}$, *the set* $f^{-1}(w)$ *contains* m *points in* N.

The proof is divided into three Lemmas.

Lemma 1. *There is a disc* $D \subseteq R$, *with centre* z_0, *such that for all* $z \in D$, f *can be written as*

$$f(z) = w_0 + (\phi(z))^m,$$

where ϕ *is an analytic function whose derivative does not vanish in* D.

Proof. By Theorem A.8 we can find a disc $D \subseteq R$ with centre at z_0 such that $f(z) - w_0$ is non-zero for all $z \in D \setminus \{z_0\}$. Hence in D we can write

$$f(z) - w_0 = (z - z_0)^m (a_0 + a_1(z - z_0) + a_2(z - z_0)^2 + \ldots)$$
$$= a_0(z - z_0)^m g(z),$$

where $m > 0$, $a_0 \neq 0$, g is analytic in D, $g(z_0) = 1$, and $g(z) \neq 0$, for all $z \in D$.

Let $F(z) = g'(z)/g(z)$ and for each $z \in D$ define

$$h(z) = \int_{z_0}^{z} F(z)\, dz,$$

where the integral is taken over a path from z_0 to z lying in D. (By Cauchy's theorem this integral is independent of the path.) Now h is analytic in D and $h'(z) = F(z) = g'(z)/g(z)$. Also

$$\frac{d}{dz}(g(z)e^{-h(z)}) = 0,$$

and so $g(z) = ce^{h(z)}$, where c is a constant. As $g(z_0) = 1$ and $h(z_0) = 0$, $c = 1$ and so

$$g(z) = e^{h(z)}.$$

Choose a value of $a_0^{1/m}$ and define

$$\phi(z) = a_0^{1/m}(z - z_0)e^{h(z)/m}.$$

Then

$$f(z) = w_0 + (\phi(z))^m,$$

as required. □

Lemma 2. *There is an open neighbourhood* V *of* z_0 *such that* $\phi(V)$ *is open and* ϕ *is one-to-one on* V.

Proof.   We know that $\phi(z) \neq 0$ for all $z \in D \setminus \{z_0\}$. Let D_1 be a disc with centre at z_0 which is properly contained in D. Then $\phi(z)$ does not vanish on ∂D_1, the boundary of D_1, and as ∂D_1 is compact the minimum μ of $|\phi(z)|$ is attained on ∂D_1 and $\mu \neq 0$. Let Δ be an open disc with centre at $0(= \phi(z_0))$ and radius μ. We will show that if $w_1 \in \Delta$ then there is just one value of $z \in D_1$ such that $\phi(z) = w_1$. We put $V = \phi^{-1}(\Delta) \cap D_1$ and then ϕ is one-to-one on V, $z_0 \in V$ and $\phi(V) = \Delta$ is open.

To achieve this aim we show that $\phi(z) - w_1$ has precisely one zero in D_1. As $\phi(z) - w_1$ is analytic in D_1, Theorem A.7 implies that the number of zeros of $\phi(z) - w_1$ in D_1 is $N(w_1)$, where

$$N(w) = \frac{1}{2\pi i} \int_{\partial D_1} \frac{(\phi(z) - w)'}{\phi(z) - w} \, dz = \frac{1}{2\pi i} \int_{\partial D_1} \frac{\phi'(z)}{\phi(z) - w} \, dz,$$

where the dash denotes differentiation with respect to z. Note that if $w \in \Delta$ then $\phi(z) - w$ cannot vanish on ∂D_1 for otherwise

$$\min_{z \in \partial D_1} |\phi(z)| \leqslant |w| < \mu,$$

which contradicts the definition of μ.

As ϕ only vanishes at z_0 inside D_1 and as $\phi'(z_0) \neq 0$, ϕ has a simple zero at z_0 and $N(0) = 1$. We now show that $N(w)$ is a continuous function of w, for $w \in \Delta$, and as $N(w)$ is an integer this implies that $N(w) = 1$ for all $w \in \Delta$. Now if $s, t \in \Delta$ then

$$|N(s) - N(t)| = \left| \frac{1}{2\pi i} \int_{\partial D_1} \left(\frac{\phi'(z)}{\phi(z) - s} - \frac{\phi'(z)}{\phi(z) - t} \right) dz \right|$$

$$\leqslant \frac{1}{2\pi} \int_{\partial D_1} \left| \frac{\phi'(z)|t - s| dz}{(\phi(z) - s)(\phi(z) - t)} \right|$$

$$\leqslant \frac{2\pi r M |t - s|}{2\pi \delta^2},$$

where r is the radius of D_1, M is the maximum of $|\phi'(z)|$ on ∂D_1 and $\delta = \min\{|\phi(z) - s|, |\phi(z) - t| \, | z \in \partial D_1\}$. As we saw above $\phi(z) - w$ cannot vanish on ∂D_1 for any $w \in \Delta$ and hence by the compactness of $\partial D_1, \delta$ is non-zero. Therefore $N(w)$ is a continuous function of w. $\square$

Lemma 3.   *For each positive integer m, $p_m(z) = z^m$ defines an open mapping.*

Proof.   Let A be an open set in $\mathbb{C}$. We just need to show that every $z_0 \in A$ has an open neighbourhood W such that $p(W)$ is open. If $z_0 \neq 0$ then $p'_m(z_0) \neq 0$ and the proof follows directly from Lemma 2 by putting $\phi = p_m$.

If $z_0 = 0$ then we let W be an open disc, centre 0, radius $r < 1$, which is contained in A. Then $p_m(W)$ is an open disc, radius $r^m < r$, which proves the result. $\square$

Part (i) of Theorem A.10 now follows as p_m and ϕ are both open. Part (ii) is clear as ϕ is locally one-to-one and p_m is locally m-to-one. More precisely, let P be a disc, with centre w_0, contained in $f(V)$. Then $N = f^{-1}(P)$ has the required properties. $\square$

Theorem A.10 gives an easy proof of the following important result.

Theorem A.11 (maximum-modulus principle). *If $f(z)$ is a non-constant analytic function in a region R then $|f(z)|$ has no maximum in R.*

Proof. If $z_0 \in R$ then as $f(R)$ is open, $f(z_0)$ is an interior point of $f(R)$. Hence there exists a disc with centre at $f(z_0)$ which is contained in $f(R)$. This disc contains points $f(z)$, $(z \in R)$ whose modulus is greater than $f(z_0)$. Hence $|f(z)|$ does not achieve its maximum at the point z_0. As z_0 is an arbitrary point of R, $|f(z)|$ has no maximum in R. $\square$

By Lemma 2, analytic functions f with non-zero derivatives in a region define a local homeomorphism in that region. (As with real variables we can then show that the inverse function is also analytic.) Such functions have an important geometric property which we now describe. A sense-preserving mapping $f: R \to f(R)$ is called *directly conformal* if whenever two differentiable paths γ_1, γ_2 intersect at an angle θ then $f(\gamma_1), f(\gamma_2)$ intersect at the same angle θ. (A sense-reversing mapping with this property is called *indirectly conformal*.)

Theorem A.12. *If $R \subseteq \mathbb{C}$ is a region then $f: R \to f(R) \subseteq \mathbb{C}$ is directly conformal if and only if f is analytic in R and $f'(z) \neq 0$ for all $z \in R$.*

Proof. For a direct proof see Ahlfors [1966]. We outline an alternative proof which uses the induced mapping on the differentials.

First assume that f is analytic and $f'(z) \neq 0$ for all $z \in R$. Let $w = f(z) = u + iv$, where $z = x + iy$. Then by the chain rule,

$$\begin{pmatrix} du \\ dv \end{pmatrix} = \begin{pmatrix} u_x & u_y \\ v_x & v_y \end{pmatrix} \begin{pmatrix} dx \\ dy \end{pmatrix},$$

where we assume that the partial derivatives u_x, u_y, v_x, v_y are evaluated at some given point $z \in R$.

Using the Cauchy–Riemann equations

$$|f'(z)|^2 = u_x^2 + v_x^2 \neq 0$$

and writing J for the above 2×2 matrix we find, again by the Cauchy–Riemann equations, that

$$J = |f'(z)| A,$$

where A is orthogonal (that is, $AA^t = I$) and has determinant equal to $+1$. Thus on the tangent space spanned by the differentials dx, dy, f is a rotation about z followed by a magnification, which implies that f is directly conformal.

For the converse we assume that f is differentiable with respect to x and y (for otherwise our definition of conformality has no meaning), so we assume that the first partial derivatives exist and are continuous. Then direct conformality implies that $JJ^t = (\det J)I$ and $\det J > 0$ as f preserves orientation – see §4.15. Then $J^t = (\det J)J^{-1}$ which gives the Cauchy–Riemann equations. Thus f is analytic and $|f'(z)|^2 = \det J \neq 0$. □

If $f'(z_0) = f''(z_0) = \cdots = f^{(m-1)}(z_0) = 0$ and $f^{(m)}(z_0) \neq 0$ then by Lemma 1, apart from an additive constant, $f = p_m \circ \phi$ where $\phi'(z_0) \neq 0$. As ϕ defines a local homeomorphism, by Lemma 2, we see that the topological character of f in a neighbourhood of z_0 is the same as that of the map $z \mapsto z^m$; that is, f is like a branched covering map. Geometrically, f expands angles by a factor m.

APPENDIX 2

Presentations of groups

Suppose that a group Γ is generated by elements $X_i (i \in I)$ satisfying relations $R_j(X_i) = 1$ $(j \in J)$, where I and J are indexing sets and each $R_j(X_i)$ is a *word* in the generators X_i, that is, a product of finitely many powers (positive or negative) of generators X_i. We say that the relations $R_j(X_i) = 1$ are *defining relations* for Γ if every relation in Γ (in effect, the multiplication table for Γ) can be deduced from them, using only the group axioms; more precisely, this condition states that a word $W(X_i)$ represents the identity element of Γ if and only if it can be transformed, by using the group axioms, to a product of conjugates of words $R_j(X_i)$. We then say that Γ has a *presentation*

$$\Gamma = \langle X_i(i \in I) \mid R_j(X_i) = 1(j \in J) \rangle.$$

For example, finite cyclic and dihedral groups have presentations

$$C_n = \langle X \mid X^n = 1 \rangle,$$
$$D_n = \langle X, Y \mid X^n = Y^2 = (XY)^2 = 1 \rangle,$$

and (as shown in §6.8) the modular group has presentation

$$PSL(2, \mathbb{Z}) = \langle X, Y \mid X^2 = Y^3 = 1 \rangle.$$

The following result shows that homomorphic images of Γ are obtained by adding relations to a presentation for Γ.

Theorem A.13. *If $\Gamma = \langle X_i(i \in I) \mid R_j(X_i) = 1(j \in J) \rangle$ and G is any group, then the following are equivalent:*

(i) *there is an epimorphism $\theta : \Gamma \to G$;*
(ii) *there is a normal subgroup $\Lambda \trianglelefteq \Gamma$ with $\Gamma / \Lambda \cong G$;*
(iii) *G is generated by elements $x_i(i \in I)$ satisfying relations $R_j(x_i) = 1$ for all $j \in J$ (and possibly other relations not implied by these).*

Proof (i) $\Rightarrow$ (ii). This is the first isomorphism theorem, with $\Lambda = \ker(\theta)$.

(ii) $\Rightarrow$ (iii). Since Γ is generated by the elements X_i, satisfying $R_j(X_i) = 1$, Γ / Λ is generated by the cosets $X_i \Lambda$, satisfying $R_j(X_i \Lambda) = 1$. (For example, in the modular group Γ we have $X^2 = 1$, so that $(X\Lambda)^2 = X\Lambda \cdot X\Lambda = X^2\Lambda = \Lambda$, which is the identity in Γ/Λ.) If x_i is the element of G corresponding to $X_i\Lambda$ under the isomorphism $\Gamma/\Lambda \cong G$, then $\{x_i | i \in I\}$ generates G and satisfies $R_j(x_i) = 1$ for all $j \in J$, by the elementary properties of isomorphisms.

(iii) $\Rightarrow$ (i). We define $\theta : \Gamma \to G$ as follows. Each $g \in \Gamma$ can be expressed as a word $W(X_i)$ in the generators X_i, so we define $\theta(g)$ to be the corresponding element $W(x_i)$ of G. To show that this is well defined, suppose that $g = W_1(X_i) = W_2(X_i)$ in Γ (for example, $XY = YX^{-1}$ in D_n). Then $W_1(X_i)W_2(X_i)^{-1} = 1$ in Γ, and this relation must be a consequence of the defining relations $R_j(X_i) = 1$ of Γ; the same argument which derives $W_1(X_i) W_2(X_i)^{-1} = 1$ in Γ from these relations can also be used to derive $W_1(x_i)W_2(x_i)^{-1} = 1$ in G from the relations $R_j(x_i) = 1$ (we simply replace X_i by x_i throughout), so $W_1(x_i) = W_2(x_i)$ and hence $\theta(g)$ is independent of the word chosen to represent g. By construction, $\theta(gh) = \theta(g)\theta(h)$ for all $g, h \in \Gamma$, so θ is a homomorphism. Since the image $\theta(\Gamma)$ contains a set of generators x_i for G, θ must be an epimorphism. $\square$

A simple example of this is the way in which the finite cyclic group $G = C_2 = \langle x | x^2 = 1 \rangle$ can be formed from a larger cyclic group $\Gamma = C_6 = \langle X | X^6 = 1 \rangle$ by adding an extra relation, say $X^2 = 1$ (or even $X^4 = 1$, which, together with $X^6 = 1$, implies $X^2 = 1$). For less trivial examples, see §5.11 and §6.9, and for a more detailed treatment of presentations, see Johnson [1980], Lyndon & Schupp [1977] or Magnus, Karass & Solitar [1966].

APPENDIX 3

Resultants

Let F be any field, and let

$$a(x) = a_m x^m + \ldots + a_1 x + a_0,$$
$$b(x) = b_n x^n + \ldots + b_1 x + b_0$$

be polynomials in $F[x]$, with $a_m, b_n \neq 0$. We need necessary and sufficient conditions, in terms of the coefficients a_i and b_j, for $a(x)$ and $b(x)$ to have a non-constant common factor in $F[x]$.

Suppose that

$$a(x)c(x) = b(x)d(x), \tag{1}$$

where $c(x)$ and $d(x)$ are non-zero polynomials in $F[x]$, with

$$\left. \begin{array}{c} \deg(c) < n, \\ \deg(d) < m. \end{array} \right\} \tag{2}$$

Since $F[x]$ is a unique factorisation domain, both sides of (1) have the same factorisation in $F[x]$, so the irreducible factors of $a(x)$ all divide $b(x)d(x)$, and hence each divides $b(x)$ or $d(x)$. Since $d(x)$ has degree less than that of $a(x)$, they cannot all divide $d(x)$, so at least one divides $b(x)$. Thus $a(x)$ and $b(x)$ have a common factor.

Conversely, if $p(x)$ is a common factor of $a(x)$ and $b(x)$, then

$$a(x) = p(x)d(x)$$

and

$$b(x) = p(x)c(x)$$

for some non-zero polynomials $c(x)$ and $d(x)$ satisfying (2); clearly, (1) also holds, so (1) and (2) are necessary and sufficient conditions for $a(x)$ and $b(x)$ to have a common factor.

To see whether $c(x)$ and $d(x)$ exist, we write

$$c(x) = c_{n-1} x^{n-1} + \ldots + c_1 x + c_0$$

and

$$d(x) = d_{m-1} x^{m-1} + \ldots + d_1 x + d_0,$$

with $c_i, d_j \in F$ and possibly $c_{n-1} = 0$ or $d_{m-1} = 0$. Then substituting in (1) and equating coefficients, we see that

$$
\begin{aligned}
a_m c_{n-1} &= b_n d_{m-1} \\
a_{m-1} c_{n-1} + a_m c_{n-2} &= b_{n-1} d_{m-1} + b_n d_{m-2} \\
a_{m-2} c_{n-1} + a_{m-1} c_{n-2} + a_m c_{n-3} &= b_{n-2} d_{m-1} + b_{n-1} d_{m-2} + b_n d_{m-3} \\
&\vdots \\
a_0 c_1 + a_1 c_0 &= b_0 d_1 + b_1 d_0 \\
a_0 c_0 &= b_0 d_0.
\end{aligned}
$$

We can take all terms across to the left, and regard this as a system of $m + n$ simultaneous linear equations in the $m + n$ variables $c_{n-1}, \ldots, c_0$, $-d_{m-1}, \ldots, -d_0$, with coefficients a_i, b_j. Then we have seen that $a(x)$ and $b(x)$ have a common factor if and only if this system has a non-trivial solution in F (that is, with not all c_i or $d_j = 0$); this is equivalent to the vanishing of $\det M$, where M is the $(m + n) \times (m + n)$ matrix

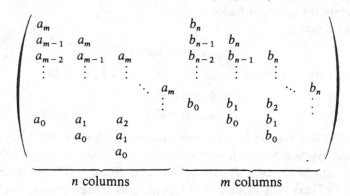

$$
\underbrace{}_{n \text{ columns}} \underbrace{}_{m \text{ columns}}
$$

given by the coefficients of the system of equations. We define $\det M$ to be the *resultant*

$$
R = R(a, b)
$$

of $a(x)$ and $b(x)$; it vanishes if and only if $a(x)$ and $b(x)$ have a common factor, or equivalently a common root x (possibly in some extension field of F), so we can eliminate x from the pair of equations

$$
a(x) = 0, \quad b(x) = 0
$$

to obtain a single equation

$$
R = 0,
$$

as in §3.17. Notice that R is a polynomial in the variables a_i and b_j, with integer coefficients.

Suppose that we multiply M on the left by the square matrix

$$J = \begin{pmatrix} 1 & & & & \\ & 1 & \cdot & & \\ & & \cdot & \cdot & \\ & & & 1 & \\ x^{m+n-1} & x^{m+n-2} & \cdots & x & 1 \end{pmatrix}.$$

Since $\det J = 1$, we have $R = \det M = \det J . \det M = \det(JM)$. Now the entries in the last row of JM are

$$x^{n-1}a(x), x^{n-2}a(x), \ldots, a(x), x^{m-1}b(x), x^{m-2}b(x), \ldots, b(x),$$

while the other rows are identical to those of M. Hence, if we expand $R = \det(JM)$ using the entries of its last row and their cofactors, we see that

$$R = u(x)a(x) + v(x)b(x), \tag{3}$$

where $u(x)$ and $v(x)$ are polynomials in x, whose coefficients are polynomials in a_i and b_j.

We now give two applications of resultants.

(i) Let $p(z)$ be a polynomial of degree k in $\mathbb{C}[z]$. Then $p(z)$ has a repeated root if and only if it shares a factor with its derivative $p'(z)$, or equivalently, if and only if $p'(z)$ and $b(z) = kp(z) - zp'(z)$ have a common factor (this second condition is simpler to use, since $b(z)$ has degree less than the degree k of $p(z)$). Thus $p(z)$ has a repeated root if and only if the resultant $R = R(p', b)$ vanishes; this resultant, or rather a suitable multiple of it, is called the *discriminant* of $p(z)$.

For instance, let $k = 3$ and let

$$p(z) = 4z^3 - c_2 z - c_3$$

as in §6.2. Then

$$p'(z) = 12z^2 - c_2$$

and

$$b(z) = -2c_2 z - 3c_3,$$

of degrees $m = 2$ and $n = 1$, so R is the 3×3 determinant

$$R = \begin{vmatrix} 12 & -2c_2 & \\ & -3c_3 & -2c_2 \\ -c_2 & & -3c_3 \end{vmatrix}$$

$$= -4(c_2^3 - 27c_3^2)$$

$$= -4\Delta_p,$$

where Δ_p is the discriminant of $p(z)$ as defined in §6.2. This gives an alternative proof of Corollary 6.2.6, that $p(z)$ has distinct roots if and only if $\Delta_p \neq 0$.

(ii) Let $a(z, w)$ and $b(z, w)$ be elements of $\mathbb{C}[z, w]$, that is, polynomials in z and w with coefficients in $\mathbb{C}$. We can also regard them as polynomials in w, whose coefficients $a_i = a_i(z)$ and $b_j = b_j(z)$ are in the polynomial ring $\mathbb{C}[z]$, and hence in the field $F = \mathbb{C}(z)$ of rational functions of z.

Suppose that a and b are co-prime in $\mathbb{C}[z, w]$. Then it follows easily from Gauss's lemma that they are also co-prime in $F[w]$, so their resultant R is a *non-zero* element of F. Moreover, our construction of R as a determinant shows that R is a polynomial $R(z)$ in z, with coefficients in $\mathbb{C}$.

Now suppose also that $a(z, w) = 0 = b(z, w)$ for some particular choice of $z, w \in \mathbb{C}$. Since (3) gives

$$R(z) = u(z, w)a(z, w) + v(z, w)b(z, w)$$

for suitable polynomials $u, v \in \mathbb{C}[z, w]$, we see that $R(z) = 0$ for all such z. However, $R(z)$ is a non-trivial polynomial in z, so there are only finitely many such elements z in $\mathbb{C}$. Thus we have proved:

Theorem A.14.   *Let $a(z, w)$ and $b(z, w)$ be co-prime elements of $\mathbb{C}[z, w]$. Then there are at most finitely many $z \in \mathbb{C}$ for which the equations*

$$a(z, w) = 0, \quad b(z, w) = 0$$

have a common root $w \in \mathbb{C}$.

(This result is used in §4.14.)

APPENDIX 4

Modern developments

Finally, we will discuss some further developments of the theories described in this book. For recent references and more details the reader is referred to Bers [1972], Thurston [1982], Beardon [1983].

In Chapter 2 we investigated the group $PSL(2, \mathbb{R})$ in detail and in Chapter 5 we considered discrete subgroups of $PSL(2, \mathbb{R})$ with particular reference to plane hyperbolic geometry. As we have already mentioned, the connection between hyperbolic geometry and $PSL(2, \mathbb{R})$ was formulated by Poincaré and published in 1882. In a paper published a year later Poincaré studied discrete subgroups of $PSL(2, \mathbb{C})$ using 3-dimensional hyperbolic geometry (Poincaré [1883]). As the topology of 3-dimensional manifolds was so little understood at the end of the nineteenth century this interesting connection between 3-dimensional manifolds and discrete subgroups of $PSL(2, \mathbb{C})$ was rather neglected. In recent years 3-dimensional topology has advanced considerably and this connection is proving to be of great importance.

We consider $\mathbb{R}^3 \cup \{\infty\}$ as the one-point compactification of $\mathbb{R}^3$, (see §1.2). We identify $x + iy \in \mathbb{C}$ with the point $(x, y, 0) \in \mathbb{R}^3$ so that $\mathbb{C} \cup \{\infty\}$ is a subset of $\mathbb{R}^3 \cup \{\infty\}$; we also denote *upper half 3-space* $\{(x, y, u) \in \mathbb{R}^3 | u > 0\}$ by $\mathcal{U}_3$. We now describe how elements of $PSL(2, \mathbb{C})$ act as directly conformal (angle- and orientation-preserving) transformations of $\mathcal{U}_3$.

If $T \in PSL(2, \mathbb{C})$ then by exercises $2E - G$, T is a product of an even number of inversions in circles (where this includes reflections in lines, §2.7). Every circle in $\mathbb{C}$ is the equatorial circle of a unique sphere in $\mathbb{R}^3$ and every line in $\mathbb{C}$ belongs to a unique plane in $\mathbb{R}^3$ perpendicular to $\mathbb{C}$. Each inversion in a circle extends to an inversion in the corresponding sphere or reflection in the corresponding plane. Each such inversion or reflection gives an indirectly conformal transformation of $\mathbb{R}^3$ so that T extends to a directly conformal transformation of $\mathcal{U}_3$ to itself.

We can make $\mathcal{U}_3$ into a model of hyperbolic 3-space by defining the length $h(\beta)$ of a piecewise continuously differentiable path $\beta(t) = (x(t), y(t),$

$u(t)$), $(0 \leqslant t \leqslant 1)$ by

$$h(\beta) = \int_0^1 \frac{\sqrt{((dx/dt)^2 + (dy/dt)^2 + (du/dt)^2)}}{u} \, dt.$$

We can define hyperbolic surface-area and volume similarly. $PSL(2,\mathbb{C})$ is then the group of orientation-preserving isometries with respect to the hyperbolic metric.

Let Γ be a discrete subgroup of $PSL(2,\mathbb{C})$. As we saw in §5.6, Γ need not act discontinuously on $\mathbb{C} \cup \{\infty\}$. However it does act discontinuously on $\mathscr{U}_3$ and conversely every subgroup of $PSL(2,\mathbb{C})$ acting discontinuously on $\mathscr{U}_3$ is discrete in $PSL(2,\mathbb{C})$; (this is analogous to Theorem 5.63(i)). Now let $\alpha \in \mathbb{R}^3 \cup \{\infty\}$ and let (T_n) be a sequence of distinct elements of Γ. If $(T_n(\alpha))$ has a limit point $z_0 \in \mathbb{R}^3 \cup \{\infty\}$ then $z_0 \in \mathbb{C} \cup \{\infty\}$. The set of all such limit points is called the *limit set* $L(\Gamma)$; it is a closed subset of $\mathbb{C} \cup \{\infty\}$ and its complement in $\mathbb{C} \cup \{\infty\}$ is called the *ordinary set* (or *regular set*) $O(\Gamma)$. If $O(\Gamma) \neq \varnothing$ then Γ is called a *Kleinian group* and then Γ acts properly discontinuously on the plane set $O(\Gamma)$. For example if $\Gamma = \Omega$ is a lattice then $L(\Omega) = \{\infty\}$ and $O(\Omega) = \mathbb{C}$. If Γ is a Fuchsian group then $L(\Gamma) \subseteq \mathbb{R} \cup \{\infty\}$; if $L(\Gamma) = \mathbb{R} \cup \{\infty\}$ then $O(\Gamma)$ has two components, the upper and lower half-planes, whereas if $L(\Gamma)$ is a proper subset of $\mathbb{R} \cup \{\infty\}$ then $O(\Gamma)$ has just one component. It is quite possible for $O(\Gamma)$ to have an infinite number of components when Γ is not Fuchsian.

A conjugate of a Fuchsian group in $PSL(2,\mathbb{C})$ has its limit set lying on a circle or line and will clearly have the same algebraic and geometric characteristics as a Fuchsian group. For an example of a Kleinian group which is not conjugate in $PSL(2,\mathbb{C})$ to a Fuchsian group (and is more complicated than a lattice) consider $2g > 2$ circles $C_1, C_2, \ldots, C_{2g}$ such that the domain exterior to C_j contains every C_k, $(k \neq j)$. Let $T_1, \ldots, T_g$ be Möbius transformations such that T_j maps the exterior of C_j to the interior of C_{g+j}. The group generated by $T_1, \ldots, T_g$ is a Kleinian group which, in general, is not conjugate to a Fuchsian group in $PSL(2,\mathbb{C})$. Such a group is called a *classical Schottky group*; a non-classical Schottky group can be obtained by replacing the C_i by Jordan curves.

We can also consider discrete subgroups of $PSL(2,\mathbb{C})$ for which $O(\Gamma) = \varnothing$. An example is the Picard modular group $PSL(2,\mathbb{Z}[i])$ considered in §5.6. (Some authors use the term Kleinian group for arbitrary discrete subgroups of $PSL(2,\mathbb{C})$, calling those for which $O(\Gamma) = \varnothing$ Kleinian groups of the first kind and those for which $O(\Gamma) \neq \varnothing$ Kleinian groups of the second kind.)

If Γ is an arbitrary discrete subgroup of $PSL(2,\mathbb{C})$ then $\mathscr{U}_3/\Gamma$ is a 3-

manifold. Recent work of W. Thurston shows that most 3-manifolds can be obtained in this way (Thurston [1982]). If Γ is a Kleinian group then $(\mathscr{U}_3 \cup O(\Gamma))/\Gamma$ is a 3-manifold with boundary $O(\Gamma)/\Gamma$. This boundary is a union of connected surfaces and by an extension of Theorem 5.9.1 we can show that these are Riemann surfaces and that the projection from $O(\Gamma)$ to $O(\Gamma)/\Gamma$ is holomorphic.

We now restrict attention to the case where Γ is finitely generated. If Γ is a Fuchsian group with a Dirichlet region having a finite number of sides then by Theorem 5.8.7, Γ is finitely generated. The proof of the theorem is easily adapted to show that a discrete subgroup of $PSL(2, \mathbb{C})$ which has a Dirichlet region – this will be a hyperbolic polyhedron in $\mathscr{U}_3$ – with a finite number of sides is finitely generated. For Fuchsian groups the converse is true; every Dirichlet region for a finitely generated Fuchsian group has a finite number of sides. However, for discrete subgroups of $PSL(2, \mathbb{C})$ the converse is false; in 1966, L. Greenberg showed that there exist finitely generated Kleinian groups possessing no fundamental polyhedron with a finite number of sides.

One of the most important results concerning finitely generated Kleinian groups is Ahlfors' finiteness theorem, (1964):

Let Γ be a finitely generated Kleinian group. Then $O(\Gamma)/\Gamma$ is a finite union of connected Riemann surfaces. Each such surface can be obtained from a compact surface by removing a finite number of points and there are at most a finite number of points in $O(\Gamma)/\Gamma$ over which the natural projection from $O(\Gamma)$ is branched.

References

W. Abikoff, 1980. *The Real Analytic Theory of Teichmüller Space. Lecture Notes in Mathematics, Volume 820*, Springer-Verlag, Berlin, Heidelberg, New York.

L.V. Ahlfors, 1964. 'Finitely generated Kleinian groups,' *Amer. J. Math.*, **86**, 413–29.

L.V. Ahlfors, 1966, *Complex Analysis* (2nd Edn). McGraw-Hill, New York.

N. Alling, 1981. *Real Elliptic Curves*. North-Holland Publishing Company, Amsterdam, New York, Oxford.

T.M. Apostol, 1963, *Mathematical Analysis, A Modern Approach to Advanced Calculus*. Addison-Wesley, Reading, Massachussets.

T.M. Apostol, 1976. *Modular Functions and Dirichlet Series in Number Theory*. Springer-Verlag, New York, Heidelberg, Berlin.

M.A. Armstrong, 1979. *Basic Topology*. McGraw-Hill, London.

A.F. Beardon, 1983. *The Geometry of Discrete Groups*. Springer-Verlag, New York, Heidelberg, Berlin.

A.F. Beardon, 1984. *A Primer on Riemann Surfaces. London Mathematical Society Lecture Note Series 78*. Cambridge University Press, Cambridge.

E.T. Bell, 1965. *Men of Mathematics* (2 vols.) Penguin, Harmondsworth, Middlesex.

L. Bers, 1972. 'Uniformization, moduli, and Kleinian groups.' *Bull. London Mathematical Society*, **4**, 257–300.

A. Borel, 1956. 'Groupes linéaires algébriques.' *Ann. of Math.* (2), **64**, 20–82.

H. Cohn, 1967. *Conformal Mapping on Riemann Surfaces*. McGraw-Hill, New York.

J.H. Conway & S.P. Norton, 1979. 'Monstrous moonshine.' *Bull. London Math. Soc.*, **11**, 308–39.

H.S.M. Coxeter, 1969. *Introduction to Geometry* (2nd Edn). Wiley, New York.

L.E. Dickson, 1901. *Linear Groups with an Exposition of the Galois Field Theory*. Teubner (Dover reprint 1958 with an introduction by W. Magnus).

P. du Val, 1964. *Homographies, Quaternions and Rotations*. Oxford University Press, Oxford.

P. du Val, 1973. *Elliptic Functions and Elliptic Curves. London Mathematical Society Lecture Note Series 9*. Cambridge University Press, Cambridge.

H.M. Farkas & I. Kra, 1980. *Riemann Surfaces*. Springer-Verlag, New York, Heidelberg, Berlin.

L.R. Ford, 1951. *Automorphic Functions* (2nd Edn). Chelsea, New York.

A.R. Forsyth, 1918. *Theory of Functions of a Complex Variable.* Cambridge University Press (reprinted by Dover Publications Inc. in 2 volumes 1965).

W.J. Harvey (Ed.), 1977. *Discrete Groups and Automorphic Functions.* Academic Press, London.

N. Jacobson, 1951. *Lectures in Abstract Algebra.* Van Nostrand, Princeton.

D.L. Johnson, 1980. *Topics in the Theory of Group Presentations. London Mathematical Society Lecture Note Series 42.* Cambridge University Press, Cambridge.

F. Klein, 1893. *On Riemann's Theory of Algebraic Functions and their Integrals.* Reprinted by Dover Publications Inc. (1963), New York.

F. Klein, 1913. *Lectures on the Icosahedron* (2nd Edn). Kegan Paul, London. Reprinted by Dover Publications Inc. (1956), New York.

S. Lang, 1978. *Elliptic Curves, Diophantine Analysis.* Springer-Verlag, Berlin, Heidelberg, New York.

J. Lehner, 1964. *Discontinuous Groups and Automorphic Functions.* American Mathematical Society, Providence, Rhode Island.

J. Lehner, 1966. *A Short Course in Automorphic Functions.* Holt, Rinehart and Winston Inc., New York.

R.C. Lyndon & J.L. Ullman, 1967. 'Groups of elliptic linear fractional transformations.' *Proc. Amer. Math. Soc.,* **18**, 1119–24.

R.C. Lyndon & P.E. Schupp, 1977. *Combinatorial Group Theory.* Springer-Verlag, Berlin, Heidelberg, New York.

A.M. Macbeath, 1961. 'On a theorem of Hurwitz.' *Proc. Glasgow Math. Assoc.,* **5**, 90–6.

W. Magnus, 1974. *Noneuclidean Tesselations and Their Groups.* Academic Press, New York.

W. Magnus, A. Karass & D. Solitar, 1966. *Combinatorial Group Theory,* Interscience, New York.

B. Maskit, 1971. 'On Poincaré's theorem for fundamental polygons.' *Advances in Mathematics,* **7**, 219–30.

W.S. Massey, 1967. *Algebraic Topology: An Introduction.* Harcourt, Brace and World, Inc., New York.

L.J. Mordell, 1969. *Diophantine Equations.* Academic Press, London, New York.

M. Newman, 1972. *Integral Matrices,* Academic Press, New York.

H. Poincaré, 1882. 'Théorie des groupes Fuchsiens.' *Acta Math.,* **1**, 1–62.

H. Poincaré, 1883. 'Mémoire sur les groupes Kleinéens.' *Acta Math.,* **3**, 49–92.

R. Rankin, 1977. *Modular Forms and Functions.* Cambridge University Press, Cambridge.

J.S. Rose, 1978. *A Course on Group Theory.* Cambridge University Press, Cambridge.

W. Rudin, 1974. *Real and Complex Analysis* (2nd Edn). Tata McGraw-Hill, New Delhi.

B. Schoeneberg, 1974. *Elliptic Modular Functions.* Springer-Verlag, Berlin, Heidelberg, New York.

G. Springer, 1957. *Introduction to Riemann Surfaces.* Addison-Wesley, Reading, Massachusetts.

W.P. Thurston, 1982. 'Three dimensional manifolds, Kleinian groups and hyperbolic geometry.' *Bull. Amer. Math. Soc.*, **6**, 357–81.

H. Weyl, 1955. *The Concept of a Riemann Surface* (3rd Edn). Addison-Wesley, Reading, Massachusetts.

K. Wohlfahrt, 1964. 'An extension of F. Klein's level concept.' *Illinois J. Math.*, **8**, 529–35.

Index of symbols

$\mathbb{C}$ 1

Σ 1

$\pi: S^2 \to \Sigma$ 2

$J: \Sigma \to \Sigma$ 4

$\mathbb{C}(z)$ 8

$\deg(f)$ 8

$v_a(f)$ 10

$\mathrm{Aut}(\Sigma)$ 17

$GL(2, \mathbb{C})$ 18

$SL(2, \mathbb{C})$ 18

$PGL(2, \mathbb{C})$ 18

$PSL(2, \mathbb{C})$ 19

$GL(n, F)$ 19

$SL(n, F)$ 19

$\overline{\mathrm{Aut}(\Sigma)}$ 19

$PGL(n, F)$ 20

$PSL(n, F)$ 20

$PG(1, \mathbb{C})$ 20

$PG(n-1, F)$ 20

R_θ 20

S_r 20

T_t 21

λ 24, 27

$(z_0, z_1; z_2, z_3)$ 24, 27

$I_{\mathbb{C}}$ 28

$G(X)$ 30

$\mathrm{tr}(A)$ 33

$\mathrm{Rot}(\Sigma)$ 40

$PSU(2, \mathbb{C})$ 40

$SO(3, \mathbb{R})$ 41

$U(n, \mathbb{C})$ 41

$SU(n, \mathbb{C})$ 41

$PSU(n, \mathbb{C})$ 41

Q_8 42

D_n 45

$\mathscr{F}$ 49

$\mathcal{O}$ 49

$\mathscr{I}$ 49

$\mu(R)$ 50

$\mathscr{I}(T)$ 54

Ω_f 56

S^1 60

m_g 60

$\Omega(\omega_1, \omega_2)$ 65

$z_1 \sim z_2$ 66

$D(\Omega)$ 68

X/G 71

$\mathrm{ord}(f)$ 73

$\|f\|$ 81

$\|f\|_E$ 81

$\mathrm{Log}(z)$ 83

$S(z)$ 87, 89

$Z(z)$ 89

$P(z)$ 89

Ω_r 90

$\sum_{\omega \in \Omega}$ 91

$\sum'_{\omega \in \Omega}$ 91

$\prod_{\omega \in \Omega}$ 91

$\prod'_{\omega \in \Omega}$ 91

$F_N(z)$ 92

$\wp(z)$ 92

$\sigma(z)$ 93

$\zeta(z)$ 94

$G_k(\Omega)$ 96

e_j 97

$E(\Omega)$ 98

$E_1(\Omega)$ 98

η_j 101

$V(l_1, b_1; \ldots; l_s, b_s)$ 106

$\hat{f}: \mathbb{C}/\Omega \to \Sigma$ 107

E 109

$E_{\mathbb{R}}$ 111

$\hat{E}_{\mathbb{R}}$ 119

$\hat{E}_{\mathbb{Q}}$ 119

$\mathscr{D}$ 123

$(D_1, f_1) \sim (D_2, f_2)$ 124

$\Gamma(z)$ 127

$\ln(r)$ 127

D_J 128

L_J 128

$\simeq$ 141

$\gamma\delta$ (product of paths)   147

γ^{-1} (inverse of path)   147

$\pi_1(X, a)$ 148

$\pi_1(X)$ 148

$n_a(\delta)$ 149

g (genus)   163

$f \sim_a g$ 177

$[f]_a$ 177

$\mathcal{M}$ 177

$D(m)$ 177

$\psi: \mathcal{M} \to \Sigma$ 177

$\phi: \mathcal{M} \to \Sigma$ 178

$\mathcal{M}(m)$ 179

$\mathcal{M}_A$ 179

$\mathcal{S}$ 183

$\mathcal{S}(m)$ 184

$\mathcal{S}_A$ 184

C_A 185

J_f 191

$\mathcal{A}$ 193

$\chi(S)$ 194

$\cong$ (conformal equivalence)   198

$\mathcal{U}$ 199

Aut S 200

$\tilde{f}$ 203

$\hat{S}$ 210

$C_G(g)$ 220

$e(\beta)$ 222

$h(\gamma)$ 222

$\rho(z, w)$ 224

$\eta(z, w)$ 226

$\tau(z, w)$ 226

$\mu(E)$ (hyperbolic area)   228

$L(\Gamma)$ 235, 332

$\mathbb{Z}[i]$ 235

$PSL(2, \mathbb{Z}[i])$ 235

$N_G(H)$ 240

$D_p(\Gamma)$ 241

$(g; m_1, \ldots, m_r)$ 257

$(g; m_1, \ldots, m_r; s)$ 262

τ (modulus)   272

Γ (modular group)   272

R_g 273

T_g 273

Δ_p 274, 329

$\Delta(\Omega)$ 276

$J(\Omega)$ 276

$g_2(\tau)$ 276

$g_3(\tau)$ 276

$\Delta(\tau)$ 276

$J(\tau)$ 277

$\sigma_3(k)$ 282

$\sigma_5(k)$ 282

F 284

ρ 284

X, Y, Z 289

$\lambda(\tau)$ 293

λ_α 294

$\psi: SL(2, \mathbb{Z}) \to S_3$ 294

$\Phi(\tau)$ 294

$\Gamma(2)$ 294

$G * H$ 299

G_n 300

θ_n 300

$SL(2, \mathbb{Z}_n)$ 300

ϕ_n 300

$\Gamma(n)$ 300

$PSL(2, \mathbb{Z}_n)$ 300

$\mathcal{U}_r$ 303

$\overline{\mathcal{U}}_r$ 303

$\overline{\mathcal{U}}$ 303

$\hat{\mathbb{Q}}$ 303

s 303

N 305

l_i 306

l 308

$\Gamma_0(p)$ 316

$\Gamma^0(p)$ 316

$G(m)$ 316

$R(a, b)$ 328

$\mathcal{U}_3$ 331

$O(\Gamma)$ 332

Index of names and definitions

Abikoff, W., 274
absolute convergence (of an infinite product), 84
abstract Riemann surface, 169
addition theorem, 115
addition theorem for $\wp$, 118
Abel, N.H, 72
Ahlfors, L., 333
algebraic function, 184
algebraic group, 119
amplitude (of a cusp), 306
analytic at ∞, 5
analytic atlas, 168
analytic continuation, 125
analytic continuation along a path, 138
analytic function, 1
analytic function on a Riemann surface, 172
analytic function element, 123
analytic germ, 177
angle (between curves), 37
angle (between H-lines), 229
anti-automorphism (of Σ), 19
antipodal map, 40
antipodal point, 40
atlas, 168
automorphic function, 231, 266
automorphism (of a lattice), 120
automorphism (of a Riemann surface), 200
automorphism (of the Riemann sphere), 17

basis (for a lattice), 59, 65
Beardon, A.F., 225, 331
Bers, L., 274
Bolyai, J., 221
branch, 140
branch-point, 13, 107, 207
branch-point of order $k - 1$, 18, 183, 207
branched covering, 13, 207
branched Riemann surface (of a germ or equation), 184

C^∞ atlas, 191
Cauchy integral formula, 318

Cauchy–Riemann equations, 318
Cauchy's theorem, 318
centraliser, 220
characteristic subgroup, 269
chart, 168
circle (in S^2 or Σ), 21
classical Schottky group, 332
closed path, 138
commutator subgroup, 270
compact, 3
compatible (atlases), 168, 191, 192
complete global continuation, 184
complex plane, 1
complex projective line, 20
complex structure, 169
concyclic, 28
conformal equivalence, 198
conformal homeomorphism, 198
conformal map, 36
conformally equivalent, 198
congruence (modulo a lattice), 66
congruence (modulo $\mathbb{Z}$), 62
congruence subgroup, 301
congruent sides, 247
congruent vertices, 245
conjugacy class, 32
conjugate elements (of a group), 32
conjugate point, 28
convergence (of an infinite product), 83, 85
Conway, J.H., 283
coordinate transition function, 168
co-prime (polynomials), 8
covering map, 13, 206
covering space, 13
covering surface, 206
critical point, 185
cross-ratio, 24, 27
crystallographic restriction, 120
cusp, 306
cycle (of vertices), 245
cyclic quadrilateral, 28

D-neighbourhood, 177
defining relations, 325

degree (of an algebraic function), 184
degree (of a rational function), 8
differential equation for $\wp(z)$, 97
dihedral group, 45, 46
direct analytic continuation, 124
direct meromorphic continuation, 124
directly conformal, 37, 323
Dirichlet, G.P.L., 68
Dirichlet polygon, 68
Dirichlet region (for a Fuchsian group),
  241
Dirichlet region (for a lattice), 68
Dirichlet tessellation, 241
disc (in Σ), 30
disc centred at a, 177
discontinuous group action, 208
discrete subgroup, 61
discrete subset, 57
discriminant, 274, 329
doubly periodic, 59
duplication theorem, 118

edge (of a polygonal subdivision), 193
Eisenstein series, 96
elementary neighbourhood, 206
elliptic curve, 109
elliptic cycle, 245
elliptic cyclic group, 232
elliptic function, 72
elliptic integral, 72
elliptic transformation, 35, 219
elliptic vertex, 245
entire function, 292
Euclidean length, 222
Euler, L., 116
Euler characteristic, 194
Euler–Poincaré formula, 196
extended complex plane, 1, 2

face (of a polygonal sundivision), 193
Fagnano, C.G., 116
fibre, 206
field, 6, 98
Fischer–Griess monster group, 283
fixed-point, 32
Fourier, J.B.J., 64
free product, 299
Fuchs, L., 231
Fuchsian group, 231
function element, 123
fundamental group, 148
fundamental parallelogram, 66
fundamental polygon, 66
fundamental region (for a Fuchsian group),
  240
fundamental region (for a lattice), 66

Galois group, 52, 190
gamma-function, 127
Gauss, C.F., 72, 271
Gauss–Bonnet formula, 229
Gaussian integers, 235
general linear group, 18, 19
genus (of a Fuchsian group), 307
genus (of a surface), 163, 193
germ, 177
great circle, 50
Greenberg, L., 333

H-circle, 227
H-invariant, 225
H-line, 224
H-line segment, 223
H-reflection, 237
Hadamard's gap theorem, 134
Hamilton, W.R., 42
Heine–Borel theorem, 3
holomorphic function, 1, 173
homotopic, 141
homotopy, 141
homotopy class, 141
Hurwitz, A., 264
Hurwitz formula, 196
Hurwitz group, 264, 300
hyperbolic area, 228
hyperbolic circle, 227
hyperbolic cyclic group, 232
hyperbolic distance, 224
hyperbolic geometry, 221
hyperbolic length, 222
hyperbolic line, 224
hyperbolic line segment, 223
hyperbolic metric, 224
hyperbolic plane, 224
hyperbolic polygon, 228
hyperbolic surface-area, 332
hyperbolic transformation, 35, 219
hyperbolic volume, 332
hyperbolic 3-space, 331
hyperbolically convex, 230
hyperbolically starlike , 230

indirectly conformal, 37, 323
infinite product, 83, 85
inverse (of a path), 147
inversion (in a circle), 28
isolated vertex, 269
isometric circle, 54
isometry, 37

Jacobi, C.G.J., 72
Jacobian, 191

k-transitive, 23

Klein, F., 167
Kleinian group, 332

lattice, 59,65
Legendre's relation, 102
Lehner, J., 247
length (of an orbit), 44
level, 301, 308
lift, 203
limit set, 235, 332
linear fractional transformation, 17
Liouville's theorem, 319
Lobatchewsky, N.I., 221
local coordinate, 168
local uniformising parameter, 183
locally finite fundamental region, 244
logarithmic derivative, 88
loxodromic transformation, 35, 220
lune, 50
Lyndon, R.C., 42

Macbeath, A.M., 269
Maskit, B., 259
maximum-modulus principle, 323
meromorphic at ∞, 5
meromorphic continuation, 125
meromorphic continuation along a path,
    138
meromorphic continuation of a germ, 179
meromorphic function (on a Riemann
    surface), 174
meromorphic function (on Σ), 6
Möbius, A.F., 17
Möbius band, 191
Möbius transformation, 17
modular function, 276
modular group, 232, 272
moduli space, 273
modulus (of a lattice), 272
monodromy group, 166
monodromy theorem, 146
monster simple group, 283
Morera's theorem, 319
multiple point, 7
multiplicity, 6
multiplier, 34

n-manifold, 168
natural boundary, 133
non-Euclidean geometry, 221
norm, 81
normal convergence (of an infinite
    product), 86
normal convergence (of an infinite series),
    81
normaliser, 240

Norton, S., 283
null-homotopic, 143
number of sheets, 206

omits a value, 292
one-point compactification, 4
open mapping, 12, 320
orbit, 44
orbit-space, 71
order (of a branch-point), 13, 207
order (of a meromorphic function at a
    point), 10, 215
order (of a rational function), 8
order (of a transformation), 35
order (of an elliptic function), 73
ordinary set (of a Kleinian group), 332
orientable, 192
orientation, 193
orientation-preserving, 192

parabolic class number, 303
parabolic cyclic group, 232
parabolic subgroup, 303
parabolic transformation, 35, 219
parabolic vertex, 255
partial fractions, 16
path, 137
period (of a Fuchsian group), 246
period (of a function), 56
period (of a transformation), 35
periodic function, 56
Picard, C.E., 289
Picard's theorem, 292
Poincaré, H., 221, 259, 331
point at ∞, 2
polygonal subdivision, 193
presentation, 238, 325
principal congruence subgroup, 301
principal part, 7
principal value of log z, 83
principle of permanence of identical
    relations, 180
product (of paths), 147
projective general linear group, 18, 20
projective special linear group, 19, 20
projective special unitary group, 41
properly discontinuous group action, 232
Puiseux, V., 183
Puiseux series, 183
Pythagoras' theorem, 68

quaternion algebra, 42
quaternion group, 42
quotient-space, 71, 248

Radó, T., 194
Ramanujan's tau-function, 283

ramified covering map, 207
rational elliptic curve, 119
rational function, 8
real elliptic curve, 111
real lattice, 111
real meromorphic function, 111
real projective plane, 71
real rectangular lattice, 113
real rhombic lattice, 113
reduced word, 297
region, 1, 5
regular covering space, 209
regular function, 1
regular icosahedron, 49
regular octahedron, 49
regular point (for a function element), 133
regular point (for an algebraic function), 185
regular set (of a Kleinian group), 332
regular tetrahedron, 49
resultant, 116, 328
Riemann, B., 3, 167
Riemann mapping theorem, 199
Riemann space, 273, 285
Riemann sphere, 3, 169
Riemann surface, 169
Riemann surface of $\log z$, 150
Riemann surface of $z^{1/q}$, 154
Riemann surface of $\sqrt{p(z)}$, 157
Riemann zeta-function, 91, 280
Riemann–Hurwitz formula, 196
Riemann–Roch theorem, 106
right translation, 60
rotation (of Σ), 40

Schottky group, 332
Schwarz, H.A., 201, 264
Schwarz's lemma, 201
sheaf (of germs), 177
side (of a fundamental region), 245
signature, 257, 262
similar lattices, 202
simple path, 138
simple point, 7
simply connected, 143
simply periodic, 59
singular point, 133
smooth atlas, 191
smooth structure, 191
smooth surface, 191
space of moduli, 273
special linear group, 18, 19
special orthogonal group, 41

special unitary group, 41
spherical triangle, 50
stabiliser, 30
stereographic projection, 1
sufficient for continuation within a region, 180
surface, 71, 168
symmetric open set, 61

Teichmüller, O., 273
Teichmüller space, 273
tessellation, 67
Thompson, J.G., 283
3-manifold, 331
Thurston, W., 333
topological group, 60
torus, 70
total order of branching, 197
trace, 33
transitive, 22
triangle group, 238, 300

Ullman, J.L., 42
unbranched Riemann surface (of a germ or an equation), 179
uniform convergence, 80, 81
uniform convergence on compact sets, 80, 82
uniformisation, 267
uniformisation theorem, 199–200
unimodular matrix, 20
unitary group, 41
unitary matrix, 41
universal covering surface, 210
upper half 3-space, 331

vertex (of a Dirichlet region), 244, 245
vertex (of a hyperbolic polygon), 229
vertex (of a polygonal subdivision), 193

Weierstrass' M-test, 81
Weierstrass normal form, 274
Weierstrass pe-function, 92
Weierstrass sigma-function, 93
Weierstress' theorem, 320
Weierstrass zeta-function, 94
Weyl, H., 167
width (of a cusp), 306
winding number, 142, 149
Wohlfahrt, K., 310
word, 325

Γ-automorphic, 266